Psychopathology of Human Adaptation

Kittay Scientific Foundation Symposia Published by Plenum Press

Nutrition and Mental Functions • 1975
Edited by George Serban

Animal Models in Human Psychobiology • 1976
Edited by George Serban and Arthur Kling

Psychopathology of Human Adaptation • 1976
Edited by George Serban

Psychopathology of Human Adaptation

Edited by
George Serban
New York University Medical Center

PLENUM PRESS · NEW YORK AND LONDON

Library of Congress Cataloging in Publication Data

Main entry under title:

Psychopathology of human adaptation.

Proceedings of the 3d of a series of meetings; proceedings of the 2d are entered under title: Animal models in human psychobiology.
Includes bibliographies and indexes.
1. Psychology, Pathological–Congresses. 2. Adjustment (Psychology)–Congresses. 3. Psychology, Physiological–Congresses. 4. Behavior therapy–Congresses. I. Serban, George, 1926- II. Kittay Scientific Foundation. [DNLM: 1. Psychopathology–Congresses. 2. Stress, Psychological–Congresses. 3. Adaptation, Psychological–Congresses. WM100 P9894 1975]
RC454.4.P8 616.8'9'07 76-40455
ISBN 0-306-30917-3

Proceedings of the Third International Symposium of the Kittay Scientific Foundation held April 6-8, 1975 in New York, New York

The editor gratefully acknowledges the contribution of Hoffmann-La Roche toward this symposium

Manuscripts prepared by Arlyne Zimmermann, Director of Communications, Kittay Scientific Foundation

A Division of Plenum Publishing Corporation
227 West 17th Street, New York, N.Y. 10011

Printed in the United States of America

Dedication

This book is dedicated to researchers in the field of neurophysiology and psychopathology of the brain (as related to stress), who, through their perseverance, ingenuity, and creativity have advanced our knowledge of human behavior.

Special mention should be made of the late Dr. James Olds, an outstanding researcher and scientist, whose contribution was so significant in this area. His premature death is an irreplaceable loss to science.

Foreword

Undoubtedly this symposium will prove to be an important landmark in the development of our understanding of the psychopathology of human adaptation in general, as well as of the general adaptation syndrome and stress in particular. It was organized to give an opportunity to an international group of experts on adaptation and stress research to present summaries of their research that could then later be exhaustively analyzed.

The carefully structured program brings out three major aspects of adaptation to stress in experimental animals and man.

The first section deals with the neurophysiology of stress responses, placing major emphasis upon the neuroanatomical and neurochemical aspects involved.

The second section is devoted to the psychology and psychopathology of adaptive learning, motivation, anxiety, and stress.

The third section examines the role played by stress in the pathogenesis of mental diseases.

Many of the relevant subjects receive particularly detailed attention. Among these, the following are especially noteworthy:

The existence of reward and drive neurons.

Constitutional differences in physiological adaptations to stress and distress.

Motivation, mood, and mental events in relation to adaptive processes.

Peripheral catecholamines and adaptation to underload and overload.

Selective corticoid and catecholamine responses to various natural stimuli.

The differentiation between eustress and distress.

Resistance and overmotivation in achievement-oriented activity.

The dynamics of conscience and contract psychology.

Sources of stress in the drive for power.

Advances in the therapy of psychiatric illness.

The application of experimental studies on learning to the treatment of neuroses.

The recurrent dilemmas of behavioral therapy.

The affective significance of uncertainty.

Perhaps the most striking general statement that can be made about the symposium is that its arrangement leads naturally from neuroanatomical, neurochemical, and neurophysiological attributes of stress to the major clinical problems of psychopathology.

It would be redundant for me to go into the details of each paper. Similarly, it would seem arbitrary to pick out certain subjects or authors for special comment. Hence, I will review certain basic questions that have repeatedly been dealt with by different authors in different terms and in different connections, and which thus may still appear to be unresolved or even paradoxical.

It is well worth noting that, as the president of the Foundation, Sol Kittay, aptly points out in his inaugural comments, "Throughout human history, man was faced with continuous environmental changes to which he was able to successfully adapt. Yet, apparently never before did he have to deal with the accelerated pace and the magnitude of these changes" (in other words "future shock," as Alvin Toffler would say). This rate of change has now become so impressive as to cast doubt upon man's ability to cope with it. We have to face not only environmental pollution, overpopulation, and constant socioeconomic crises, but also the superindustrialization of many parts of the Western world. While improving the standard of living, these developments have also made obsolete older techniques and skills—indeed, to a large extent, the need for work itself.

The improved technology makes it increasingly less difficult to comply with the constant pressure of labor for "less work and more pay." However, at the same time, it deprives man more and more of his most stabilizing guidelines of conduct, which through one behavioral code or another give him motivation and a feeling of security through the knowledge that he is doing his duty and accomplishing a useful task.

As Montaigne said, "No wind blows in favor of the ship that has no port of destination," and being deprived of work means just that for the vast majority of people. Up to now this problem was of great importance, especially in connection with the psychologic disturbances and somatic diseases that tend to occur after compulsory retirement at age 65. However, it may be said that, by the very nature of our "social progress," semiretirement will soon begin when a young man finishes his education and starts out on a career. Hence his training must prepare him not only for "work" but also for "leisure." We have to know what to do with our leisure time, for those who do not use their pent-up energy turn to behavior destructive to themselves and/or to society, such as alcohol and drug abuse or violence.

Several authors point out that stress is an indispensable part of life and too little of it can be just as harmful as too much. In technical language, sensory deprivation, insufficient nervous input, monotony, and aimlessness are as likely to cause mental and physical derangements as overwork.

At the Symposium on "Society, Stress, and Disease" held in Stockholm under the auspices of the WHO, we tried to define stress as "the nonspecific response of the body to any demand." If we make no demands on either body or mind, life is not only deprived of purpose and satisfaction, but becomes physically impossible. Even during sleep and various states of profound relaxation (Transcendental Medication, Zen, Yoga, autohypnosis), demands on the cardiovascular, respiratory, and nervous systems are by no means at a standstill, although they drop far below the levels necessary for maximum activity while coping with extremely stressful situations.

These considerations raised by the definition of stress have focused attention upon two basic questions:

1. What shall we understand by nonspecificity?
2. How much stress is desirable?

The discussions have, I believe, largely confirmed earlier views according to which the concept of nonspecificity is a relative one.

It has been pointed out, with much justification, that in practice stress can never be a completely stereotyped and hence a totally nonspecific reaction form, since every stressor has some specific effects of its own; in addition to the nonspecific manifestations of stress, cold causes shivering, most of the acute infections induce fever, and overexertion of the musculature in the right arm produces selective fatigue in its muscles. However, it should not be forgotten that stress must always be regarded as an abstraction; that which is common to all agents and remains if we disregard their specific effects. This consideration seemingly weakens the value of the stress concept. But let us not forget that life itself is an abstraction; it never occurs unless attached to some living being which exhibits its own specific characteristics, those of a man, woman, cat, dog, or microbe.

This Symposium clearly brings out the fact that in man, with his highly developed nervous system, neurogenic and particularly psychogenic stressors are especially common. The phenomenon of "nervous arousal" is associated with almost all types of stress conditions, even physical trauma or loss of blood. When nervous arousal is artificially blocked as far as possible (by various pharmacologic or other means) many stressors virtually lose their ability to produce the typical somatic manifestations of stress, such as increased ACTH, corticoid, or catecholamine production.

ACTH release has, however, been shown to occur under the influence of stressors in animals with a completely deafferented hypothalamus which cannot be influenced by nervous arousal. Here, we must assume that the "first mediator" of stress, the message of a demand for adaptive work, must reach the critical region of the median eminence and the pituitary gland through the only remaining connections to the rest of the body, namely, the humoral substances carried by blood vessels. Similarly, proving conclusively that stress is not merely a "nervous" response, stress occurs under deep anesthesia, and even in lower animals and plants which have no nervous system.

On the whole, the concept of stress as we now understand it is most readily grasped if we compare it to that of energy consumption. Energy must be utilized for any type of demand made upon a living or even inanimate machine, but the optimal rate of consumption and the results of it will largely depend upon the structure of the machine and the manner in which it is set to utilize the energy made available to it.

These considerations help us to understand the relevance to stress and human adaptation of many among the papers in this volume which deal with "specific stressors," yet I think the latter could better be designated as specific agents whose stressor effect is greatly modified by or dependent upon their associated specific actions, be they psychogenic or somatic. Evidently, the nonspecific or stressor effect of anxiety (e.g., catecholamine or corticoid secretion) can be efficiently combated by tranquilizers or psychotherapeutic measures, whereas that of acute, severe infections is often most easily blocked by antibiotics. Neither of these therapeutic agents are "antistressors" in the strict sense of the word; they merely block certain specific effects of agents which would otherwise provoke a nonspecific stress response.

Adaptation is largely a problem of acquiring resistance to the stressor effect of various situations and agents we are likely to meet in coping with the demands of daily life. This Symposium furnishes an abundance of data concerning the mechanism, prophylaxis, and therapy of various life situations that require adaptation, especially with regard to psychological and psychiatric conditions. But I think this is the point where the Introduction to this Symposium should stop, leaving the reader to turn to the individual papers which reflect, more exactly than could be done by any one person, the opinions of those many eminent experts who have described their views in their own words.

Hans Selye, C.C.

Université de Montréal
Institut de médecine et de chirurgie expérimentales
Montréal, Canada

Participants

John W. Atkinson
Professor of Psychology, University of Michigan

D.E. Berlyne
Professor of Psychology, University of Toronto

Wagner H. Bridger
Professor of Psychiatry and Neuroscience, Albert Einstein College of Medicine

Eugene B. Brody
Professor & Chairman, Department of Psychiatry, University of Maryland

William E. Bunney, Jr.
Chief, Adult Psychiatry Branch, National Institute of Mental Health

Leonard Cook
Assistant Director of Pharmacology, Hoffman-LaRoche

Samuel A. Corson
Professor of Psychiatry (Physiology) and Biophysics, The Ohio State University

Borje Cronholm
Director, Department of Psychiatry, Karolinska Institutet, Stockholm, Sweden

Richard de Charms
Professor of Education, Graduate Institute of Education, Washington University

Bruce P. Dohrenwend
Professor of Social Science, Department of Psychiatry, Columbia University

Jarl Dyrud
Professor of Psychiatry, University of Chicago

Marianne Frankenhaeuser
Professor & Head of Experimental Psychology Research Unit, Swedish Medical Research Council

Alfred M. Freedman
Professor & Chairman, Department of Psychiatry, New York Medical College

Sebastian P. Grossman
Professor of Biopsychology, University of Chicago

John Hakes
Medical Director, Pfizer, Inc.

Hans Hippius
Director Nervenklinik, University of Munich, Munich, Germany

Howard F. Hunt
Chief of Psychiatric Research (Psychiatry), New York State Psychiatric Institute

Martin Katz
Chief, Clinical Research Branch, National Institute of Mental Health

Seymour S. Kety
Professor of Psychiatry, Director, Psychiatry Research Laboratories, Massachusetts General Hospital

Sol Kittay
President, The Kittay Scientific Foundation

Arthur Kling
Professor of Psychiatry, Rutgers Medical School

Eric Klinger
Professor of Psychology, University of Minnesota, Morris

Lawrence C. Kolb
Commissioner, New York State Mental Hygiene, and Professor of Psychiatry, Columbia University

Richard S. Lazarus
Professor of Psychology, University of California at Berkeley

Theodore Lidz
Professor of Psychiatry, Yale University School of Medicine

Juan Lopez-Ibor
Chairman, Department of Psychiatry and Medical Psychology, Complutense University, Madrid, Spain

Sidney Malitz
Professor & Acting Chairman, New York State Psychiatric Institute, and Department of Psychiatry, Columbia University

Isaac M. Marks
Reader in Experimental Psychopathology, Institute of Psychiatry, University of London, and Consultant Psychiatrist, Bethlem-Maudsley Hospital, London, England

John W. Mason
Scientific Advisor, Division of Neuropsychiatry, Walter Reed Army Institute of Research, Walter Reed Army Medical Center

David C. McClelland
Professor of Psychology, Harvard University

Neal E. Miller
Head of a Laboratory of Physiological Psychology, Rockefeller University

O. Hobart Mowrer
Professor of Psychology, University of Illinois at Champagne-Urbana

James Olds†
Professor of Behavioral Biology, California Institute of Technology

Pierre Pichot
Professor & Chairman, Department of Psychiatry, University of Paris, Paris, France

Chester M. Pierce
Professor of Education and Psychiatry, Harvard University

Melvin Sabshin
Medical Director, American Psychiatric Association

Edward Sachar
Director of Psychiatry, Bronx Municipal Hospital Center

Hans Selye
Professor & Director, Institute of Experimental Medicine & Surgery, University of Montreal

George Serban
Medical Director, The Kittay Scientific Foundation, and Associate Clinical Professor of Psychiatry, New York University Medical Center

Eliot Stellar
Provost and Professor of Physiological Psychology, University of Pennsylvania

Elmer L. Struening
Director, Epidemiology of Mental Disorders Unit, New York State Department of Mental Hygiene

Elliot S. Valenstein
Professor of Neuroscience and Psychology, University of Michigan

Louis Jolyon West
Professor & Chairman, Department of Psychiatry, University of California at Los Angeles

Stewart G. Wolf
Director, Marine Biomedical Institute, and University of Texas, Medical Branch

Joseph Wolpe
Professor of Psychiatry, Temple University School of Medicine, Pennsylvania, and Eastern Pennsylvania Psychiatric Institute

Contents

II. PSYCHOPATHOLOGY OF ADAPTIVE LEARNING: MOTIVATION, ANXIETY, AND STRESS

III. CLINICAL MODIFICATION OF BEHAVIOR

Psychopathology of Human Adaptation

Psychological and Physiological Mechanisms in Human Adaptation and Maladaptation

SOL KITTAY, GEORGE SERBAN, LAWRENCE C. KOLB, and MELVIN SABSHIN

I do not think that there could be a more important or timely choice for an international scientific meeting than the subject of stress and its effect on the human condition.

Throughout human history man was faced with continuous environmental changes to which he was able to successfully adapt. Yet, apparently never before did he have to deal with the accelerated pace and the magnitude of these changes, changes which have now become so impressive as to question the entire human ability to adapt, not to mention the present international socioeconomic situation. Related to it are a few general societal factors which emerge clearly as ones deeply affecting the daily human emotional equilibrium of living.

Let's take some aspects of the continuous process of superindustrialization of our country. The pride of our intellectual and creative capacity which

SOL KITTAY • President, Kittay Scientific Foundation; GEORGE SERBAN • Medical Director, Kittay Scientific Foundation; LAWRENCE C. KOLB • Commissioner of New York State Mental Hygiene and Professor of Psychiatry, Columbia University; MELVIN SABSHIN • Medical Director, American Psychiatric Association.

brought about a definite mass improvement in our standard of living has also made obsolete older techniques with their skills and jobs, thereby exacerbating the conflict between corporations and unions. Furthermore, it has led to a constant shift of economic and social forces resulting in a serious impact on man's sense of security.

Parallel with this instability, man, rationally or irrationally, confident in his successful control of his environment felt free to overpollute, overcrowd, and overpopulate the world, straining, thereby, his own capability to adapt.

An overoptimistic mankind hoped to strike a happy balance between overpopulation and productivity, only to discover of late that the natural resources are now gradually dwindling, or at least no longer available at the same rate and price as before.

As if this were not enough in taxing the human capability for adaptation to stress, these economic and social convolutions are supposed to take place serenely in the shadow of possible atomic confrontation between superpowers—a reality which only reinforces man's anxiety concerning his future security.

What appears to make the whole picture even more dramatic is the fact that everyone of us is acutely aware of his delicately balanced stability in a volatile world. Everyday mass media keeps us abreast of our changing world, and more than just a few news articles which have appeared recently are most disquieting.

If a day in the life of Ivan Denisovich was a day of sheer physical survival, with various unbearable physical and psychological stressors, a day in the life of an executive is a psychological survival dealing with various subtle and sophisticated stressors. Without any intention of equating them, I think that both represent levels of stress expressed in different terms. It is a well-known fact that highly pressured businessmen, executives, and others occupying positions of great responsibility suffer from heart attacks and ulcers, which, if I understand correctly, are due to stress. But even for the average citizen, the conditions of life, with its crowding, noise, and daily competition for wages and a better standard of living—in a world of changing ethics and values—creates stress. Nobody appears to be immune from it in the particular international crisis. Apparently, some of us develop physical illnesses, others withdraw from the social scene in order to avoid these insurmountable stresses, and still others adapt by becoming maladapted.

To me the question is why this diversity of reaction, when the biological response appears to be the same? Secondly, what coping mechanisms should be developed in order to control stress? These are two of the questions which a distinguished international body of scientists hope to raise and investigate in this volume.

Sol Kittay

The study of stress as related to diseases of adaptation has received considerable attention lately in the biological and psychological fields.

Extrapolated from physiology to psychology, the concept of stress has undergone significant changes leading to conflicting definitions and contradictory results.

Stress was equated psychologically with frustration, anxiety, and distress. Funkenstein identified it in terms of an unidimensional response to frustration, while Rosenzweig considered it a multidimensional response, including in its structure some pertinent psychological variables. This approach attempted to take into account the personality variables in the response to threat, suggesting their role in mediating stress. Other researchers went further to clarify the stress response, as resulting from a demand perceived as a threat by the individual.

Psychological stress then could be translated in terms of a state of awareness of an unpleasant demand which might cause degradation of performance, failure, or disease. In other words, a failure of adaptation expressed mainly by disorganized behavior. Successful coping with environmental demands will be considered adaptive behavior.

Some researchers question the whole concept of stress as applied to psychology, to the extent which it merely replaces the older concept of psychic conflict, without elucidating any additional mechanisms.

Regardless of the merit of this approach, the intrapsychic mechanism differentiating the adaptive from the maladaptive responses still remains to be identified.

If the environmental demands are understood as social-interpersonal factors acting as stressors, the response would depend on these variables: (1) past experiences (level of conditioning) (2) self-appraisal (evaluation of capacity to respond to demands), and (3) the need to overcome the situation (motivational drive).

In this sense stress is directly linked to motivation. Yet the concept of motivation is as ill-defined and elusive as stress itself. For instance, should motivation be restricted to need achievement or should it be constructed in a broader sense? (i.e., as a mental construct of an act which gives direction and selectivity to the striving inner force which actualizes it.)

These different conceptualizations have some definite implications for the profile of the psychological response. Need achievement as related to task performance attempts to explain the frustration-anxiety aspect resulting from the interference with the realization of the goal. It still does not explain the intrapsychic mechanism deciding the adaptive versus nonadaptive behavior. Apparently we need a different set of psychological constructs to correlate stress with a particular disease of psychological nonadaptation, though stress appears to be a necessary condition of disease but not the determinant one.

Why are some people coping with anxiety-frustration, striving to overcome the obstacle (goal-directed) while others withdraw from the stressful situation in order to reduce the tension? It is obvious that certain individuals have different personality patterns which might be considered ("built-in responses to stress") at least to particular types of stress.

In the same vein, why do some people during wartime develop combat neurosis, while others attempt to become heroes. This holds in spite of the fact that either group might have developed at one time or another "normal combat neurosis." Where is the baseline for the psychological processes which determine the groups?

From the pure physiological reaction to physical stressors, stress as a concept was applied for the understanding of various psychological situations like crowding, lack of privacy, change of social interaction, and sexual role due to changes in social values, or reaction to a novel situation, boredom, etc. It demonstrates the existence of a mediating biological mechanism, a condition not sufficient for the explanation of the psychological response.

Let us assume that in all these conditions we have a physiological response (increase of blood pressure, pulse rate, adrenalin). Are these sufficient reasons for maladaptive reaction? Certainly the element of time–chronicity of exposure to stress could be considered a factor determining the direction toward producing a disease. Another one should be the intensity of the stressor. Yet these factors are not always assessed carefully. It appears that the concept of stress made a better impact in psychosomatic medicine where the relationship between the stressor and the response is better understood.

Conversely, the attempt to classify neurosis as a disease of general adaptation does not appear satisfactory since it reduces the mental conflict to behavior. The correction of behavior does not necessarily correct the process of thinking. Though this classification works well in desensitizing individuals from fears, anxiety, or inadequate sexual patterns, it does not solve problems of personality, conflict of human interaction, etc. (guilt).

These problems presented by the conceptualization of stress in psychological terms are discussed in this volume. Certainly there are no easy answers for each one of these variables of stress in psychology. It's very easy to make theoretical concepts; it's very hard to do experiments which can be applied to the framework of human life.

George Serban

My interest in the subject of stress goes back to World War II. While serving as a psychiatrist I saw enormous numbers of young people in the Navy who had been in combat and were returned home because of a very, very stereotyped reaction, that in my previous experience was unnoticeable on the civilian scene. All these people had a series of symptoms so consistent in their expression that the history really was predictable. They were suffering from a general restlessness and apprehensiveness. They regularly reported traumatic nightmares that repeated the life-threatening perception. Finally, they were extraordinarily sensitive to any stimuli in the periphery that seemed to record or relate to the sounds or other perceptions they had experienced in battle, and they were also extraordinarily sensitive to the auditory and visual productions of strife or battle shown in movies. This clinical picture was stereotypic.

I had the opportunity of seeing every survivor of a sunken destroyer and aside from about 5%, every man had the same set of symptoms. I saw those people within five days of the ship sinking. Three months later they were all reviewed once more. I presume through the processes of "extinction" which occurred during their leaves at home, all but seven of the 90 recovered. But in addition, there appeared many, many people with these symptoms who had the whole gamut of psychopathology which we discover as we examine patients in civilian life. This was psychopathology which had either been recognized before in the individual or had been relighted by the acutely life-threatening event, the stressful event.

It amazed me when the wise men of the worldwide psychiatric profession decided a few years ago—literally with the stroke of a pen—to remove the diagnosis of acute stress reactions. Fortunately I can tell you that with a bit of fighting over the last several years this reaction type will reappear in the nomenclature of psychiatric disturbance now being prepared. I'm not sure whether any of those making the decision had ever served in a period when catastrophies were occurring regularly, either in wartime or civilian life.

Our program for today is very complex. This book begins with an examination of the meaning of stress. When one talks about stress in various settings, its meaning is extraordinarily difficult. For instance, much of the work on stress in schizophrenia is confused. Many of those who have engaged in stress studies in relation to this syndrome have failed to define what is stressful for the idiosyncratic thinker with whom he is relating. Certainly we in the clinical field who have treated such patients intensively have had to spend many months in order to discover the significant stressful events, the stressful perceptions and cognitions of the person with this extraordinarily complicated thinking disorder.

We move then from studying the physiological responses to stress to the motivational and psychological processes connected with stress. Finally, we shall concern ourselves with the matters of the varieties of interventions to relieve the pathological aftereffects of stress customarily used in the field, ranging from those deriving from learning theory to those derived from psychoanalytic theory and their interrelationships.

In regard to these matters, some of the best thinkers and theoreticians are contributors to this volume. There has been an enormous enthusiasm in this country in recent years regarding behavior therapy.

Having made two visits to Russia in the last ten years and also having gone alone, without colleagues, and spending a long time with small groups in that country (who eventually led me out in the snow-covered fields where their interpreter was not available), I discovered that some Russian colleagues who for 50 years had had to do their therapeutic work under the guise of relating to a particular theoretical construct, were distressed with the inadequacy of their therapeutic results taken in the Pavlovian framework. They wanted to know what might be done to change their processes in order to be more effective. I would think that in the future we shall see a wedding of the therapeutic concepts presently existing in the psychoanalytic and behavioral fields. From such a wedding we shall proceed to produce more effective means to intervene and to relieve people of stress.

Lawrence C. Kolb

On behalf of the American Psychiatric Association I would like to congratulate the Kittay Scientific Foundation, its International Board, and the Steering Committee for conceptualizing, planning, and implementing a conference on the subject of adaptation. I had been a participant at the earlier Stanford conference on coping and adaptation and I am pleased that other participants of that particular conference are also represented in this book, as well as the Stanford volume, which has just been published, and which serves as a fine complement to the issues raised in the present work.

I have been deeply interested in research on healthy adaptation and coping and I am delighted to see that the subject has been examined closely by such an eminent group of investigators.

I know that it's gratuitous to remind the reader about the precarious status of current and future behavioral science research in the United States. I think most of us are aware of the depth of the antiintellectual, antiscientific, and antirational forces that have been so strong in our country during the past decade. The danger of weakening the relatively thin reed of behavioral science research has been quite high up to the present time. The American Psychiatric Association, among others, is engaged in trying to reverse this cycle and I think that there are some positive signs that the nadir of irrationality has passed and that we are beginning to return to a more rational research policy. I do hope that this change will involve new support for both basic and applied research but with perhaps more of a clear-cut priority setting than existed in the 1950s and 60s. In my judgment, the areas covered in this particular volume do indeed have a high priority at both the basic and applied levels. Our lack of empirical data and our conceptual dilemmas in the realm of psychological adaptation do have profound impact upon almost all areas of clinical psychiatry.

One of the major problems in assessing therapeutic outcome or assessing the impact of any primary preventive program in psychiatry is the paucity of empirical data that permits us to make useful comparisons of individuals after they have had therapeutic intervention whether primary or secondary in nature. More significantly, we are still weak in fundamental concepts that would permit us to develop empirical analyses useful in the facilitation of better methods to assess the quality of psychiatric care, including outcome criteria. This cannot be emphasized too strongly and my concern is that this lack of fundamental concepts simply has not been recognized clearly by many of our colleagues.

I believe that improved understanding of adaptation would be a strong stimulus for our efforts to assess the results of all of our interventions in the psychiatric field. In my opinion, the clarification of adaptation and its relationship to psychopathology will emerge from new theories integrating biological, psychological, and social variables as they affect the adaptational processes. I hope that this volume will make a contribution toward such integration in addition to clarifying our understanding of the various subsystems included in adaptation to stress.

Melvin Sabshin

Neurophysiological Mechanisms of Adaptive Behavior

Some Experimental Observations on the Neuroanatomical Substrates of Learned Adaptive Behaviors

SEBASTIAN P. GROSSMAN

INTRODUCTION

Adaptation is a concept which has been defined quite differently in a number of scientific disciplines. The neurophysiologist uses the term to describe a response decrement that occurs, purely as a consequence of repeated stimulation of a sensory receptor or nerve fiber. Such a response decrement is, of course, also "adaptive" in the sense that most behavioral scientists use the term, since it prevents excessive use of a system and thus may prevent harmful consequences. However, this is excess meaning so far as the neurophysiologist is concerned. He, in fact, goes to considerable trouble to distinguish adaptation from the response decrement which can be seen in the central nervous system after it has been established that a stimulus does not have "significance" (i.e., does not portend food, shelter, a mate, an enemy, etc.). This process of *habituation* is, of course, quite closely related to adaptation as the behaviorist uses the term. Continued responses to insignificant stimuli or, conversely, a loss of responses to stimuli that are of potential significance for the survival of the individual or the

SEBASTIAN P. GROSSMAN · Committee on Biopsychology, The University of Chicago, Chicago, Illinois.

perpetuation of the species is a generally acceptable if not terribly informative definition of maladaptive behavior.

However, adaptive behavior is not merely the automatic result of properly habituating sensory pathways. Significant sensory input has to be recognized, categorized, and acted upon, and the response to stimuli which appear to be insignificant or to have lost their significance must be inhibited. In primitive organisms this is accomplished to a significant degree by genetically programmed mechanisms. The resulting "instinctive" behaviors are characterized by a degree of stability which minimizes the possibility of maladaptive responses to significant aspects of the environment. However, the very stability of the programming can itself become maladaptive because it does not provide for adjustments to unusual stimulus situations that were not incorporated into the genetic programs. This lack of flexibility increasingly interferes with adaptive behavior as organisms become more complex and venture into more diverse environments.

Survival comes to depend more and more on a behavioral repertoire which is an end result of the voluntary selection of responses that are found, by trial and error, to aid survival and the elimination of responses that may threaten it. This selection process involves every aspect of the nervous system, including the reception and processing of information, the organization of responses, feedback concerning the consequences of specific behaviors, and storage of relevant information for future reference. Or, in psychological terminology: sensation, perception, learning, memory, motivation, reward, punishment, etc.

To review the "neurophysiological basis of adaptive behavior" would involve a review of neurophysiology and related fields that would take several years and volumes. A version of such a review that would be sufficiently condensed to fit into this symposium would undoubtedly not be helpful. I have therefore decided to focus on the enormously complex interpretational problems that invariably arise in the context of research in this field, and to do so by considering in some detail the behavior of an animal which appears to have lost its capacity for adaptive behavior following limited surgical intervention in its central nervous system. The research I intend to discuss has been done in collaboration with E. Kent (Kent and Grossman, 1973; Kent, Rezak, and Grossman, 1973) and G. Alheid (Alheid and Grossman, 1974, and unpublished observations).

AN EXAMPLE OF AN ORGANISM THAT FAILS TO ADAPT

The basic preparation is a rat which has undergone stereotaxic surgery that results in the transection of a significant portion of the lateral connections of the diencephalon. The surgery is performed with the aid of a retractable tungsten wire knife 125 μ in diameter (about the diameter of a human hair) which severs fibers of passage without producing significant direct damage to cellular components of the area of the cut (Fig. 1). This encephalotome is used to make a 2–4-mm cut in the parasagittal plane along the lateral border of the diencephalon. The cut extends from the base of the brain to the ventral surface of the medial lemniscus (Fig. 2). Depending on the precise location and extent of the transection, an interesting behavioral syndrome results that may include aphagia and adipsia, motor dysfunctions including akinesia and catatonia, and sharply

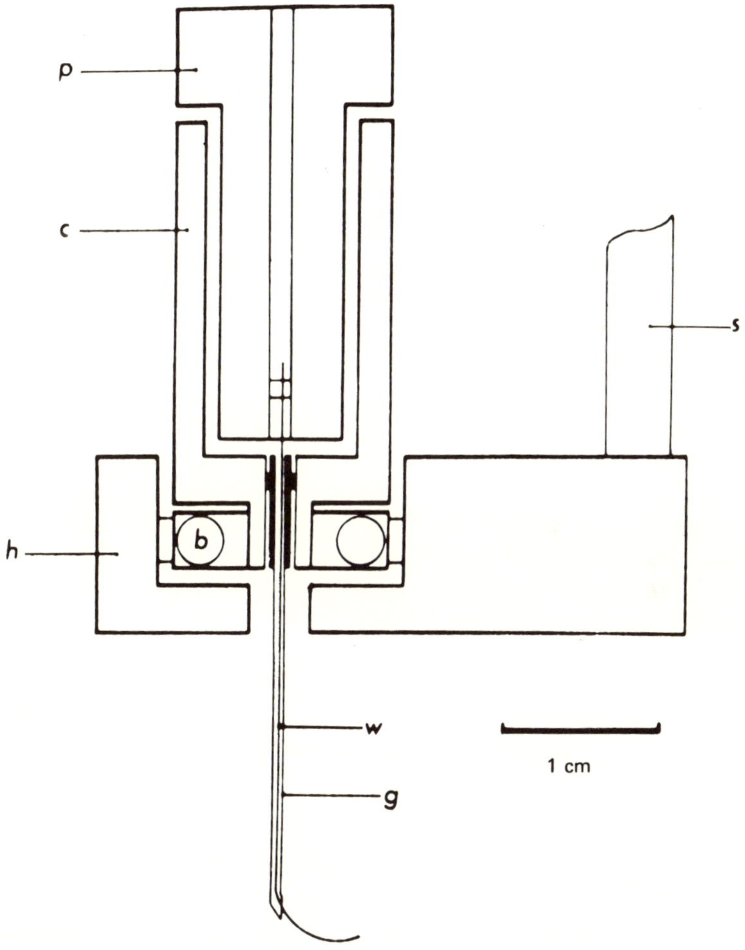

Fig. 1. Schematic representation of the encephalotome that was used to transect the lateral connections of the hypothalamus. s, stereotaxic holder; p, rotating piston of encephalotome; c, stationary cylinder; h, horizontal guide; g, guide; w, wire knife.

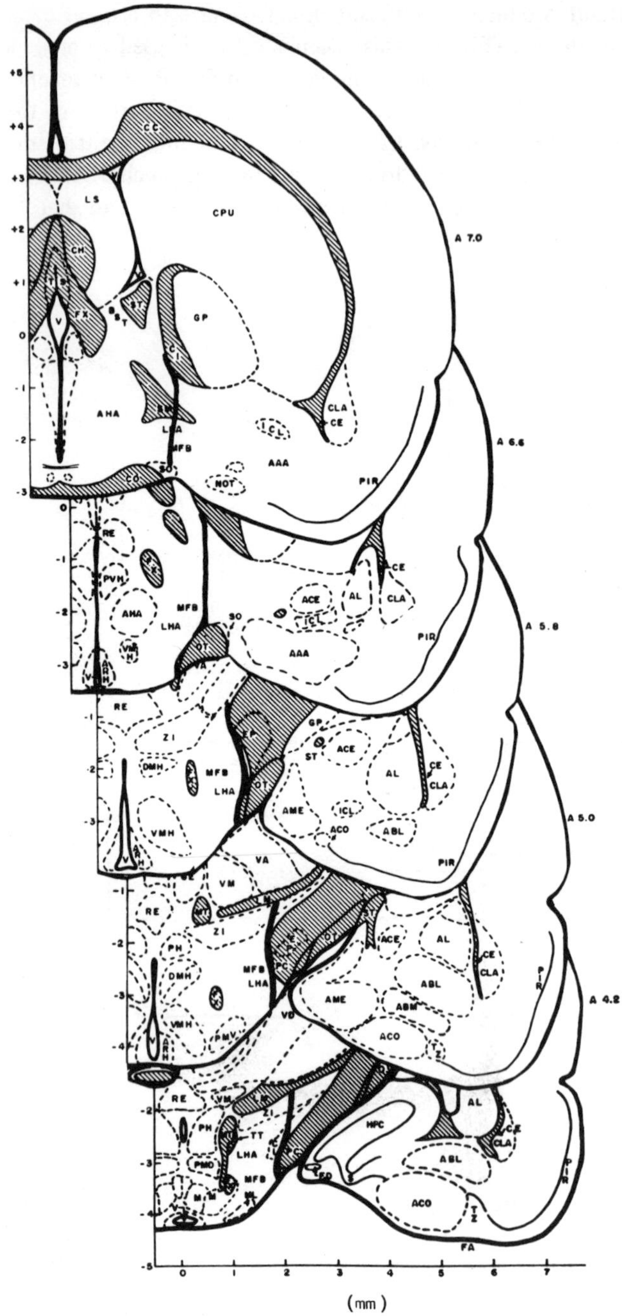

Fig. 2. Schematic representation of the location of the transection superimposed on sections from the deGroot (1959) atlas of the rat brain.

reduced general arousal. In a significant number of cases, these sensory, motor, and arousal deficits are absent or disappear after a few days of postsurgical care but complex adaptive responses to the environment are abolished. It is the behavior of these animals which I would like to discuss.

When tested several weeks after surgery, such animals eat and drink normally, groom themselves well, are normally active or even somewhat hyperactive, and show no detectable signs of gross sensory or motor dysfunctions. They orient well to stimuli of various modalities, do not assume or maintain abnormal postures, are not akinetic, do not display abnormal resistance to displacement of limbs and show no discernible arousal deficits. When placed in an observation cage together with sham-operated controls that bear comparable external scars but have no neurological damage, these animals cannot be identified even by a trained observer.

Yet such animals no longer perform and appear incapable of relearning or learning de novo complex adaptive behaviors. We observed this initially in the context of experiments (Kent and Grossman, 1973) which were designed to demonstrate whether rats that were aphagic and adipsic after the transection of lateral projections from the hypothalamus would lever-press to receive "rewarding" electrical stimulation of the intact lateral hypothalamus. Although the medial forebrain bundle is largely intact in these animals and hypothalamic norepinephrine levels are within the normal range, we observed that "self-stimulation" behavior was effectively eliminated in all of our animals. Continued testing revealed that the animals would not perform or reacquire the behavior even after voluntary food and water intake had reappeared.

We then asked whether the impairment might be peculiar to brain-stimulation rewarded behavior or to the specific response of lever-pressing and found a much more general impairment than we had expected. Not only did our rats fail to lever-press for brain-stimulation rewards, but they also no longer emitted this response when rewarded by food or water or by escape from painful footshock. They also showed comparable impairments when a much simpler behavior—running down a straight alley toward a goal box—was used instead of the lever-press. Even the animals' ability to inhibit spontaneous locomotor activity appeared to be lost or severely impaired in a simple "passive" avoidance paradigm which required only that the rat remain on a "safe" platform surrounded by an electrified grid floor.

We were intrigued by our observations and have invested considerable time and effort to elucidate the nature of the behavioral deficit and the anatomy of the pathways responsible for it. This research has led to some interesting observations although I still find it difficult to provide a precise description of the functional impairment or an exact definition of its anatomical basis.

On the behavioral end, we have found that aphagia and adipsia, although often present, are not a necessary part of the more general deficit in complex learned behavior. Some animals eat and drink quite normally within a day or two after surgery but show a persisting loss of complex behavior. That this is not merely a reflection of the fact that eating and drinking may be quite simple behaviors and may be extensively overlearned is indicated by the fact that some animals are aphagic and adipsic for many weeks and months without showing major deficits in complex brain-stimulation or shock-escape rewarded behaviors. Indeed, I (Grossman and Grossman, 1970) have observed that a group of rats which were trained to avoid shock in a shuttlebox six weeks after surgery learned the response somewhat better than normal controls even though all experimental animals had been aphagic and adipsic and required gastric feeding for several days or weeks after the operation. Intraspecific aggressive behaviors (fighting in a food-competition situation) also appeared normal in these animals.

The loss of complex, learned behaviors following our transection is not an all-or-none effect. Some animals do, indeed, show a complete and apparently permanent loss of all complex, learned behaviors. Others do not perform or learn certain behaviors but are merely impaired in learning some simple tasks. Some of our smaller cuts produce impairments which appear to be peculiar to acquisition. Behaviors which are not entirely lost often recover to some extent with repeated testing and may show some improvement as a function of the surgery-test interval even when the animal is not given opportunities to "practice."

There is no indication that the severity or persistance of the impairment might be related to the nature of the motivational state or reinforcer used (except, of course, in the case of aphagic and adipsic animals which do not emit any food-or-water rewarded responses until voluntary ingestive behaviors have recovered but may show little or no impairment in shock-escape or shock-avoidance situations). The nature of the sensory input or the topography of the response also does not appear to be a determining factor although this issue is complicated by the fact that more complex behaviors tend to be more severely affected. It seems more probable, at this time, that gradations in the severity of the impairments seen after anatomically similar cuts may be related to the specificity of the stimulus-response association required. A severe impairment or loss of adaptive behavior is most likely when specific responses are required within a limited time after the presentation of a stimulus. Some degree of adaptive behavior often remains when the situation does not require the elaboration of overt behaviors (e.g., "passive" avoidance which requires merely inaction). Some sparing is also often seen when reinforcement is contingent on the achievement of a generally specified goal regardless of the specific behaviors involved (e.g., locating and swimming toward a "safe" platform in a pool of cold water or escaping from painful footshock by entering a "safe" compartment), particularly when a good deal of time is permitted to complete the task.

On the anatomical side of things, we have attempted to identify the pathways that are responsible for the observed loss or impairment of adaptive behavior by investigating the effects of smaller cuts along the lateral border of the diencephalon. This work (Alheid and Grossman, 1974; and unpublished observations) has significantly contributed to the demonstration of graded behavioral effects described above and has provided some tantalizing hints concerning the pathway responsible for them but a number of unresolved questions remain as we shall see in a moment. We do know that the loss of complex behavior is related to the transection of fibers which enter or leave the brainstem laterally at the level of the ventromedial nucleus of the hypothalamus or slightly posterior to it. To produce the most severe effects the cuts must involve a fairly large area, indicating that the fiber system which is responsible for the behavioral impairment may be diffuse.

A comparison of our anatomical data with recent mappings of the catecholaminergic pathways that ascend in the hypothalamus (Jacobowitz and Palkovits, 1974; Palkovits and Jacobowitz, 1974) suggest that our cuts should transect a significant portion of dopaminergic nigrostriatal projections (as well as dopaminergic projections to the amygdaloid complex). Biochemical assays (Alheid *et al.*, unpublished) have supported this interpretation in showing a severe depletion of striatal dopamine after our behaviorally effective cuts. Some of the more anterior cuts also interfere to some extent with the pallidofugal projections to the substantia nigra and related nuclei, thus further isolating the striatum from its normal interconnections with the lower brainstem.

We (Kent *et al.*, 1973) have found that small cuts in the coronal plane just rostral to the pars compacta of the substantia nigra abolished lever-pressing for shock escape much like our larger parasagittal cuts do (and also produced aphagia and adipsia). Similar cuts rostral to the pars reticulata of the substantia nigra did not have such effects. These observations further implicate the dopaminergic projections which originate in the lateral portions of the substantia nigra and turn medially into area A-10 before ascending in the dorsolateral hypothalamus. However, we also found that injections of 6-hydroxydopamine (6-OHDA) into the substantia nigra (which should selectively destroy catecholaminergic components of the area) reproduced the aphagia and adipsia syndrome but had little effect on the performance or acquisition of the escape response. More extensive behavioral observations on the behavior of these animals are needed before we can reach any firm conclusions but the presently available data suggest that the loss of escape behavior seen after our coronal cuts may be due to the interruption of fibers which approximate the course of the dopaminergic nigrostriatal projections but may not, themselves, be aminergic. The cholinergic pallidonigral projections fit this description well but we have no independent evidence as yet to suggest that they might specifically be responsible.

That an interruption of the afferent and/or efferent components of the feedback loop which interconnects the striatum with the lower brainstem might be responsible for the loss of complex behavior is an appealing hypothesis in view of the rapidly increasing evidence that this structure may serve much more complex functions than traditional discussions of the extrapyramidal motor system assign to it.

The effects of striatal lesions on complex behavior are as yet not well understood. We (Neill and Grossman, 1970; Neill *et al.,* 1974*a*) as well as others (Green *et al.,* 1967; Kirkby and Kimble, 1968) have found that some striatal lesions severely impair or abolish the acquisition of simple avoidance behaviors. Striatal lesions also have been reported to interfere with the acquisition of appetitively reinforced behaviors (e.g., Thompson and Mettler, 1963; Potegal, 1969) especially when the problems are complicated by alternation, delay, or reversal requirements (Chorover and Gross, 1963; Schwartzbaum and Donovick, 1968, Mikulas and Isaacson, 1965). There are reports of negative results in both avoidance (Albert and Bignami, 1968; Winocur and Mills, 1969) and appetitive test paradigms (Thompson *et al.,* Rich, 1962) and various hypotheses have been advanced to account for these discrepancies (e.g., Potegal, 1969). It appears likely, however, that differences in lesion size and location, interacting, perhaps with variations in task complexity may account for them.

The results of electrical stimulation experiments have supported the hypothesis that the striatum may contribute importantly to learning-related processes. A single brief pulse of electrical stimulation applied bilaterally to the striatum of rats or mice shortly after a training trial results in a significant impairment in performance on retention trials given one or several days later. Some investigators have demonstrated this phenomenon in "passive avoidance" experiments (e.g., Wyers *et al.,* 1968; Wyers and Deadwyler, 1971; Zornetzer and Chronister, 1973) where only a single learning trial and single pulse of stimulation are given. Others (e.g., Herz and Peeke, 1971; Peeke and Herz, 1971; Grinberg-Zylberbaum *et al.,* 1974) have used complex appetitive learning paradigms and have given electrical stimulation during or shortly after each trial.

Dopaminergic components of the striatum have been specifically implicated by the observation that intraventricular (Cooper *et al.,* 1972; Cooper *et al.,* 1973), intrastriatal (Neill *et al.,* 1974*b*), or intranigral (Fibiger *et al.,* 1974) injections of 6-OHDA prevent the acquisition of avoidance responses. Simple food-reinforced instrumental behaviors appear to be similarly affected by intranigral 6-OHDA injections (Fibiger *et al.,* 1974). Previously learned avoidance behavior may be little affected by such treatments (Fibiger *et al.,* 1974) suggesting that the animal is capable of performing the instrumental response required in these experiments. This interpretation has been supported by experiments (Zis *et al.,* 1974) which show that rats with 6-OHDA induced damage to the nigrostriatal projections can learn an avoidance response after treatment with L-dopa in combination with a peripheral decarboxylase inhibitor. This compound is converted to dopamine in the brain and may act on receptor sites in the striatum after the normal innervation is reduced by 6-OHDA treatments.

A significant involvement of the projections to the striatum is also suggested by Routtenberg and Holzman's (1973) report of amnesia following electrical stimulation of that part of the substantia nigra which gives rise to the nigrostriatal pathway and not after stimulation of adjacent areas. Reports of learning deficits following lesions in the substantia nigra (e.g., Mitcham and Thomas, 1972; Thompson, 1969) are also compatible with such an interpretation.

That the cholinergic components of the striatum may also play an important role in learning-related processes is suggested by several observations. We (Neill and Grossman, 1970) have found that the inhibitory effects of some striatal lesions are mimicked by intrastriatal injections of the anticholinergic compound scopolamine. Deadwyler *et al.*, (1972) reported that posttrial injections of the cholinomimetic agent carbachol into the striatum produced amnesic effects similar to those seen after single-pulse electrical stimulation. Lesions in the parafascicular-center median portion of the thalamus which may be a major source of cholinergic inputs to the striatum (Mehler, 1966; Olivier *et al.*, 1970) produce memory disturbances in man (e.g., Spiegel *et al.*, 1955) as well as in rats (Cardo, 1960) and in cats (Pechtel *et al.*, 1955). Microinjections of cholinergic or anticholinergic compounds into this portion of the thalamus modify the acquisition of appetitively as well as aversively reinforced behaviors without affecting related unconditioned behaviors (Grossman *et al.*, 1965; Grossman and Peters, 1966).

It is apparent from this brief and necessarily selective review of recent research on striatal functions that the severe loss of complex, learned behavior which is seen after our transection surgery may well be the result of an interference with striatal mechanisms which are, in some important fashion, related to the process of learning, memory formation, or retrieval. However, we cannot, at this time, reject a number of alternative interpretations regarding the nature of the dysfunction or its anatomical substrate.

According to the more recent mappings of the course of the catecholaminergic pathways through the diencephalon (Palkovits and Jacobowitz, 1974; Jacobowitz and Palkovits, 1974) our parasagittal cuts may interrupt some noradrenergic projections to the telencephalon which travel in the dorsal noradrenergic bundle, particularly components of it which project across the lateral border of the hypothalamus. Damage to the ventral noradrenergic bundle which courses more medially through the lateral hypothalamus, on the other hand, is quite unlikely. Our biochemical assays support this conclusion by showing that our cuts do not significantly reduce hypothalamic norepinephrine. Their effects on telencephalic levels of this amine have not yet been established.

Medial forebrain bundle lesions which result in massive forebrain depletions of norepinephrine (e.g., Heller *et al.,* 1966) reduce food and water intake (Morgane, 1961*a*) and reactivity to sensory input (Marshall, *et al.,* 1971) and result in some impairments in aversively (Coscina and Balagura, 1970) as well as appetitively reinforced behavior (Olds and Hogberg, 1964; Morgane, 1961*b*; Chase and Moore, 1968). However, the impairments do not appear to be in any way comparable to the severe deficits or even total loss of complex behavior which is seen after some of our parasagittal transections. We (Ross *et al.,* 1975) have recently cut the rostrally projecting components of the medial forebrain bundle by a circular cut beneath the septal area and found that this did not result in impaired acquisition or performance of a variety of appetitively or aversively motivated behaviors (in fact, facilitatory effects were seen in some paradigms). Kraly and Blass (1974) have similarly reported that massive lesions in the anterior lateral hypothalamus which interrupted the MFB anterior to the feeding "center" did not interfere with the performance of food-rewarded lever-pressing. In view of the important role which noradrenergic pathways are assumed to play in reward- (e.g., Stein, 1968) or learning- (e.g., Kety, 1972) related processes, it will nonetheless be important to further investigate the effects of our cuts on telencephalic norepinephrine, and to hold in abeyance any judgments concerning the role of noradrenergic pathways in the behavioral pathology that results from our transection surgery.

Cuts along the lateral border of the anterior hypothalamus and preoptic region do not produce a loss or severe impairment of learned behavior, suggesting that the ventral amygdalofugal (and -petal) pathway and ansa peduncularis of Maynert are probably not responsible for the effect. This observation is of significance since we (Grossman *et al.,* 1975) have observed severe impairments in the acquisition of various active as well as passive avoidance behaviors following lesions restricted to the periamygdaloid piriform cortex (which, in the rat, is the site of origin and termination of the ventral amygdalofugal pathway). Since the precise course of this diffuse interconnection between the temporal lobe and hypothalamus is as yet not described in detail in the rat (see Lammers, 1972; Hall, 1972 for review), the possibility of an involvement of posterior portions of the ventral amygdalofugal projections in our behaviorally effective cuts cannot be ruled out entirely.

CONCLUSIONS

Where then do we stand?

On the behavioral side, we have an animal that is severely impaired or incapable of performing complex learned behaviors and cannot relearn them or learn them de novo. In a significant number of cases, this syndrome is not accompanied by obvious impairments in arousal, or sensory-motor functions and we have reason to believe that the animals are, in fact, capable of performing the lost behaviors. Why then do they not do so reliably? Recent experimental observations suggest that the striatum and its connections with the brainstem may play an important role in associative processes. A large clinical literature which indicates that dementia is a significant symptom in many cases of advanced Parkinsonism and Huntington's Chorea supports such an interpretation. It is therefore tempting to suggest that our animals may find it difficult or impossible to perceive the relationships between stimuli and appropriate responses to them, to appreciate the consequences of particular behaviors, or to select behaviors that are appropriate to the possibly proper perceptions of such relationships. I should emphasize that our behavioral observations do not entirely rule out alternative interpretations which might be discussed more profitably in terms of higher-order sensory or motor dysfunctions.

On the anatomical side, we similarly have some promising leads but no certainty. We clearly are interfering with afferent and efferent connections of the striatum and there is experimental and clinical evidence that this may produce the types of behavioral deficits which we observe in our animals. We cannot yet, however, rule out possible contributions by noradrenergic pathways that have been widely discussed as a possible substrate of "reward"-related processes and cannot yet entirely discount other fiber systems which interconnect the brainstem with the temporal lobe and/or other telencephalic structures.

REFERENCES

Albert, M., and Bignami, G. (1968). Effects of frontal median cortical and caudate lesions on two-way avoidance learning by rats. *Physiol. Behav.* **3**, 141–147.

Alheid, G., and Grossman, S. P. (1974). Aphagia and adipsia produced by knife cuts ventral to the globus pallidus. *Proc. Soc. Neurosci. 4,* 115 (abstract).

Cardo, B. (1960). Action de lésions thalamiques et hypothalamiques sur le conditionnement de fuite et la différenciation tonale chez le rat. *J. Physiol. (Paris) 52,* 537–553.

Chase, P., and Moore, R. Y. (1968). Medial forebrain bundle and dorsomedial tegmentum lesions–effect on operant behavior and activity in the rat. *Commun. Behav. Biol., Part A 1,* 133–141.

Chorover, S., and Gross, C. (1963). Caudate nucleus lesions: Behavioral effects in the rat. *Science 141,* 826–827.

Cooper, B. R., Breese, G. R., Howard, J. L., and Grant, L. D. (1972). Effect of central catecholamine alterations by 6-hydroxy-dopamine on shuttle box avoidance acquisition. *Physiol. Behav. 9,* 727–731.

Cooper, R. R., Breese, G. R., Grant, L. D., and Howard, J. L. (1973). Effects of 6-hydroxydopamine treatments on active avoidance responding: Evidence for involvement of brain dopamine. *J. Pharmacol. Exp. Ther. 185,* 358–370.

Coscina, D. V., and Balagura, S. (1970). Avoidance and escape behavior of rats with aphagia produced by basal diencephalic lesions. *Physiol. Behav. 5,* 651–658.

Deadwyler, S. A., Montgomery, D., and Wyers, E. J. (1972). Passive avoidance and carbachol excitation of the caudate nucleus. *Physiol. Behav. 8,* 631–635.

Fibiger, H. C., Phillips, A. G., and Zis, A. P. (1974). Deficits in instrumental responding after 6-hydroxydopamine lesions of the nigroneostriatal dopaminergic projection. *Pharmacol. Biochem. Behav. 2,* 87–96.

Green, R. H., Beatty, W. W., and Schwartzbaum, J. S. (1967). Comparative effects of septohippocampal and caudate lesions on avoidance behavior in rats. *J. Comp. Physiol. Psychol. 64,* 444–452.

Grinberg-Zylberbaum, Y., Carranza, M. B., Cepeda, G. V., Vale, T. C., and Steinberg, N. N. (1974). Caudate nucleus stimulation impairs the processes of perceptual integration. *Physiol. Behav. 12,* 913–918.

Grossman, S. P., and Grossman, L. (1970). Surgical interruption of the anterior or posterior connections of the hypothalamus: Effects on aggressive and avoidance behavior. *Physiol. Behav. 5,* 1313–1317.

Grossman, S. P., and Peters, R. (1966). Acquisition of appetitive and avoidance habits following atropine-induced blocking of the thalamic reticular formation. *J. Comp. Physiol. Psychol. 61,* 325–332.

Grossman, S. P., Grossman, L., and Walsh, L. L. (1975). Functional organization of the rat amygdala with respect to avoidance behavior. *J. Comp. Physiol. Psychol. 88,* 829–850.

Grossman, S. P., Freedman, P., Peters, R., and Willer, H. (1965). Behavioral effects of cholinergic stimulation of the thalamic reticular formation. *J. Comp. Physiol. Psychol. 59,* 57–65.

Hall, E. (1972). Some aspects of the structural organization of the amygdala. In B. E. Eleftheriou (Ed.), *The Neurobiology of the Amygdala.* New York: Plenum Press.

Heller, A., Seiden, L. S., and Moore, R. Y. (1966). Regional effects of lateral hypothalamic lesions on brain norepinephrine in the cat. *Int. J. Neuropharmacol. 5,* 91–101.

Herz, M. J., and Peeke, H. V. S. (1971). Impairment of extinction with caudate nucleus stimulation. *Brain Res. 33,* 519–522.

Jacobowitz, D. M., and Palkovits, M. (1974). Topographic atlas of catecholamine and acetylcholinesterase-containing neurons in the rat brain. I. Forebrain (telencephalon, diencephalon). *J. Comp. Neurol. 157,* 13–28.

Kent, E. W., and Grossman, S. P. (1973). Elimination of learned behaviors after transection of fibers crossing the lateral border of the hypothalamus. *Physiol. Behav. 10,* 953–963.

Kent, E. W., Rezak, M., and Grossman, S. P. (1973). Transection and chemical lesion of nigrostriatal pathways: A comparison of effects on learned behavior. *Proc. Soc. Neurosci. 3,* 410 (abstract).

Kety, S. S. (1972). The possible role of the adrenergic systems of the cortex in learning. In I. J. Kopin (Ed.), *Research Publication of the Association for Nervous and Mental Diseases.* Baltimore: Williams and Wilkins, pp. 376–389.

Kirkby, R. J., and Kimble, D. P. (1968). Avoidance and escape behavior following striatal lesions in the rat. *Exp. Neurol. 20,* 215–227.

Kraly, F. S., and Blass, E. M. (1974). Motivated feeding in the absence of glucoprivic control of feeding in rats. *J. Comp. Physiol. Psychol. 87,* 801–807.

Lammers, H. J. (1972). The neural connections of the amygdaloid complex in mammals. In B. E. Eleftheriou (Ed.), *The Neurobiology of the Amygdala.* New York: Plenum Press, pp. 123–144.

Marshall, J. F., Turner, B. H., and Teitelbaum, P. (1971). Sensory neglect produced by lateral hypothalamic damage. *Science 174,* 523–525.

Mehler, W. R. (1966). Further notes on the centre median nucleus of Luys. In D. P. Purpura and M. D. Yahr (Eds.), *The Thalamus.* New York: Columbia University Press, pp. 109–127.

Mikulas, W. L., and Isaacson, R. L. (1965). Impairment and perseveration in delayed tasks due to bilateral lesions of the caudate nucleus in rats. *Psychonom. Science 3,* 485–486.

Mitcham, J. C., and Thomas, Jr., R. K. (1972). Effects of substantia nigra and caudate nucleus lesions on avoidance learning in rats. *J. Comp. Physiol. Psychol. 81,* 101–107.

Morgane, P. J. (1961*a*). Medial forebrain bundle and "feeding centers" of the hypothalamus. *J. Comp. Neurol. 117,* 1–25.

Morgane, P. J. (1961*b*). Electrophysiological studies of feeding and satiety centers in the rat. *Am. J. Physiol. 201,* 838–844.

Neill, D., and Grossman, S. P. (1970). Behavioral effects of lesions or cholinergic blockade in the dorsal or ventral caudate. *J. Comp. Physiol. Psychol. 71,* 311–317.

Neill, D. B., Boggan, W. O., and Grossman, S. P. (1974*a*). Behavioral effects of amphetamine in rats with lesions in the corpus striatum. *J. Comp. Physiol. Psychol. 86,* 1019–1030.

Neill, D. B., Boggan, W. O., and Grossman, S. P. (1974*b*). Impairment of avoidance performance by intrastriatal administration of 6-hydroxy-dopamine. *Pharmacol. Biochem. Behav. 2,* 97–103.

Olds, M. E., and Hogberg, D. (1964). Subcortical lesions and mass retention in the rat. *Exp. Neurol. 10,* 296–304.

Olivier, A., Parent, A., and Poirier, L. J. (1970). Identification of the thalamic nuclei on the basis of their cholinesterase content in the monkey. *J. Anat. 106,* 37–50.

Palkovits, M., and Jacobowitz, D. M. (1974). Topographic atlas of catecholamine and acetylcholinesterase-containing neurons in the rat brain. II. Hindbrain (mesencephalon, rhombencephalon). *J. Comp. Neurol. 157,* 29–42.

Pechtel, C., Masserman, J. H., Schreiner, L., and Levitt, M. (1955). Differential effects of lesions of mediodorsal nuclei of thalamus on normal and neurotic behavior in cats. *J. Nerv. Ment. Dis. 121,* 26–33.

Peeke, H. V. S., and Herz, M. J. (1971). Caudate nucleus stimulation retroactively impairs complex maze learning in the rat. *Science 173,* 80–82.

Potegal, M. (1969). Role of the caudate nucleus in spatial orientation of rats. *J. Comp. Physiol. Psychol. 69,* 756–764.

Ross, J. F., Grossman, L., and Grossman, S. P. (1975). Some behavioral effects of transection of ventral or dorsal fiber connections of the septum. *J. Comp. Physiol. Psychol. 89,* 5–19.

Routtenberg, A., and Holzman, N. (1973). Memory disruption by electrical stimulation of substantia nigra, pars compacta. *Science 181,* 83–85.

Schwartzbaum, J. S., and Donovick, P. J. (1968). Discrimination reversal and spatial alternation associated with septal and caudate dysfunction in rats. *J. Comp. Physiol. Psychol. 65,* 83–92.

Spiegel, A., Wycis, H. T., Orchinik, C. W., and Freed, H. (1955). The thalamic and temporal orientation. *Science 121,* 771–772.

Stein, L. (1968). Chemistry of reward and punishment. In D. H. Efron (Ed.), *Psychopharmacology: A Review of Progress, 1957–1967.* Washington: U.S. Government Printing Office, 105–123.

Thompson, R. (1969). Socialization of the "visual memory system" in the white rat. *J. Comp. Physiol. Psychol.* (Monograph) *69* (4, Part 2).

Thompson, R. L., and Mettler, F. A. (1963). Permanent learning deficit associated with lesions in the caudate nuclei. *Am. J. Ment. Defic. 67,* 526–535.

Thompson, R., Baumeister, A. A., and Rich, I. (1962). Subcortical mechanisms in a successive brightness discrimination habit in the rat. *J. Comp. Physiol. Psychol. 55,* 478–481.

Winocur, G., and Mills, J. A. (1969). Effects of caudate lesions on avoidance behavior in rats. *J. Comp. Physiol. Psychol. 68,* 552–557.

Wyers, E. J., and Deadwyler, S. A. (1971). Duration and nature of retrograde amnesia produced by stimulation of caudate nucleus. *Physiol. Behav. 6,* 97–103.

Wyers, E. J., Peeke, H. V. S., Williston, J. S., and Herz, M. J. (1968). Retroactive impairment of passive avoidance by stimulation of the caudate nucleus. *Exp. Neurol. 22,* 350–366.

Zis, A. P., Fibiger, H. C., and Phillips, A. G. (1974). Reversal by L-dopa of impaired learning due to destruction of the dopaminergic nigro-neostriatal projection. *Science 185,* 960–962.

Zornetzer, S. F., and Chronister, R. B. (1973). Neuroanatomical localization of memory disruption: relationship between brain structure and learning task. *Physiol. Behav. 10,* 747–750.

The Role of Learning in Physiological Response to Stress

NEAL E. MILLER

One of the functions of any society is to protect its members from extreme forms of physical stress, such as cold, tissue damage, pain, and infectious disease. These are the types of stresses that have been studied most intensively in the laboratory, especially by the brilliant work of Selye (1956). However, as our modern technological societies have become better at protecting all but the most disadvantaged of their members from such physical stresses, other more psychological types of stress that involve learning become relatively more important. In the first part of this paper I shall concentrate on one of the best understood of these, namely, fear and the physiological responses to it. In the second part I shall deal with attempts to apply learning more directly to the modification of physiological responses.

EFFECTS OF LEARNING ON FEAR-INDUCED PHYSIOLOGICAL CHANGES

Fear as a Learnable Stress

Fear, or anxiety as it is called when its source is vague, is agreed to be significant for psychopathology by clinicians whose views are as divergent as

---

NEAL E. MILLER • The Rockefeller University, New York, New York.

those of Freud (1936) and Wolpe (1958). An important quality of fear as a stress and a source of psychopathology is that it readily can be learned as a new response to a previously neutral stimulus situation. But the importance of learning does not denigrate that of innate factors; it may be much easier to learn to fear some stimuli than others (Miller, 1951).

Most experimental studies have inforced the learning of fear by painful electric shocks. In these studies many of the physiological responses to the unconditioned stress of pain are conditioned to the stimulus situation as part of the learned stress of fear. But other reinforcements are also effective in eliciting fears. For example, when Fuller (1967) reared dogs in isolation in an extremely barren environment, he found that suddenly introducing them into the complex normal environment was a trauma that taught them a long-lasting fearfulness which inhibited many forms of normal behavior. We need more experimental research on the stressful effects of sudden drastic changes such as those encountered by Fuller's dogs. In the light of the accelerated rate of change in the modern world, we need additional clinical and epidemiological studies of the effects of social changes and informational overload.

Clinical observations describe many other sources of fear, such as loss of love, loss of prestige, loss of money, threats of physical harm or death, and even examinations in school. These less well understood sources need to be studied in much more detail. We need to learn more about the ethology of human fear.

Physiological Consequences of Fear

When a strong fear is learned, it involves a variety of significant physiological changes that were described by Cannon (1929) as preparations for flight or fight, and have been further delineated by subsequent investigators, including Mason who will describe some of his work later. Some of these effects are the release of adrenalin into the bloodstream by the adrenal medulla and noradrenalin by the terminals of sympathetic nerves. These substances, in turn, speed up the heart and increase its stroke volume, increase blood pressure, increase the coagulability of the blood, and help to mobilize energy by release of fatty acids from the fat pads and glycogen from the liver. They also produce vasoconstriction in the gut and vasodilatation in the large skeletal muscles. The powerful hormones, ACTH and the corticosteroids, are released, facilitating energy metabolism and exerting anti-inflammatory and immuno-suppressant effects. There are yet other complex effects such as those on the kidney via ADH and the thyroid, or on the let-down of milk from the mammary glands. As Professor Frankenhaeuser may tell us, some of these impressive physiological reactions may not be specific to the aversive state of fear but may occur also in states of joyful arousal.

Pathological Consequences of Extreme Fear

Studies of fear in combat, like those referred to by Dr. Kolb, show that it can produce virtually all of the major symptoms of neurosis and even psychosis. These include a pounding heart and rapid pulse, dryness of the throat and mouth, a strong feeling of muscular tension, trembling, exaggerated startle, a sinking feeling of the stomach, perspiration, a frequent need to urinate, irritability, aggression, an overpowering urge to cry, run, or hide, confusion, feelings of unreality, feeling faint, nausea, fatigue, depression, slowing down of movements and thoughts, restlessness, loss of appetite, insomnia, nightmares, interference with speech, the use of meaningless gestures and the maintenance of peculiar postures, and sometimes stuttering, mutism, and amnesia (Miller, 1951). Presumably most of these are innate responses to fear or to the conflict induced by it.

Clinical observations backed up by experimental evidence show that the stress of fear can produce a variety of psychosomatic symptoms and also other medical problems that have not hitherto usually been considered to be psychosomatic. Stomach lesions, experimental evidence for which will be presented later, are one example. Sudden death through fibrillation of a partially damaged heart is another. Recently, Lown *et al.* (1973) have shown that either stimulation of the stellate ganglion of the sympathetic nervous system or placing a dog in a chamber he has learned to fear will, under suitable circumstances, reduce the threshold for fibrillation down to physiological levels. Furthermore, a number of studies summarized by Schiavi and Stein (1975) have shown that avoidance learning, a fear-inducing procedure, increases the susceptibility of animals to experimental infections and to implants of malignant tumors. Some such effect might be expected through the action of corticosteroids from the adrenal glands on the thymus and leucocytes as described by Selye (1956), but some effects of fear on the immune system show up even in adrenalectomized animals.

Effects of strong, chronic fear on the immune system would be expected to have a wide variety of medically significant consequences. That this is indeed true is suggested by the fact that epidemiological evidence summarized by Rabkin and Struening (1975) shows that sociological stresses, such as those produced by drastic social changes, are associated with a wide range of medical problems. Such stresses certainly involve learned responses and probably involve a considerable amount of anxiety. Although it is hard to eliminate other confounding factors, the congruence between the epidemiological results and the experimental ones is highly suggestive.

Counterconditioning of Fear

Fear and its accompanying physiological and psychosomatic effects can not only be acquired by learning but, within limits, can be eliminated by incompatible learned responses. One example of this is counterconditioning. In a classical experiment of this type, Pavlov (1927) used, as a signal that an extremely hungry dog would be fed, a painful stimulus such as a strong electric shock that ordinarily would elicit symptoms of fear. He reports that as the conditioned response of turning toward the food and salivating was acquired, the symptoms of fear disappeared. To quote him:

> Subjected to the very closest scrutiny, not even the tiniest and most subtle objective phenomenon usually exhibited by animals under the influence of strong injurious stimuli can be observed in these dogs. No appreciable changes in the pulse or in the respiration occur in these animals, whereas such changes are always most prominent when the noxious stimulus has not been converted into an alimentary conditioned stimulus (p. 30, 1960 edition).

Despite the great theoretical use that has been made of counterconditioning (e.g., Wolpe, 1958), there has been relatively little detailed experimental analysis of this phenomenon in this country with modern, sophisticated physiological and biochemical techniques.

Effects on Fear-Induced Pathology of Learning Discriminations and Coping Responses

There are yet other ways in which learning can influence fear. Two of them are illustrated by studies of fear in combat (Miller, 1959, pp. 267–268). One of these is learning what to expect and when to expect it. Learning a discrimination between when it is dangerous and when it is safe allows the combatant to be afraid only when it is dangerous instead of continuously. Another of these is learning what to do to cope with the danger. Such a response not only reduces the danger but greatly reduces the fear compared with the situation of having to stand by helplessly and do nothing but worry.

The first of these factors was investigated in my laboratory by Dr. Arlo Myers, who trained thirsty rats to drink in special restraining cages. Then in trials with the water bottles absent the rats were given strong electric shocks in the restraining cages. Pairs of rats had electrodes on their tails wired in series so that they received exactly the same strength of shock. One member of each pair had a light as a signal preceding the shock so that he could learn the discrimination of when it was dangerous and when it was safe. The other member of each pair did not have any light as a signal so that he could not learn the discrimination. Then the water bottles were restored and the amount that the thirsty rats drank was tested on trials without further shocks. The suppression of drinking during these tests was used as a measure of fear. This is a standard test, commonly referred to as a Conditioned Emotional Response or CER.

As Fig. 1 shows, during tests with the danger signal off, the animals that had been exposed to signaled and hence predictable shocks drank more water, meaning that they showed less fear than those that had been exposed to unsignaled and hence unpredictable shocks. During tests when the danger signal was on, the situation was slightly reversed. Overall, the animals exposed to the predictable shocks drank more water, indicating that they experienced less fear. In these tests the danger signal was on and off for equal times, but in most situations the danger signal is off most of the time so that the overall difference will be more strongly in favor of the signaled, predictable condition.

Since this early study, a number of investigators in various laboratories have secured similar results. Using a tone, which is much more perspicuous for the rat than a light, Weiss, working in my laboratory, has secured the results shown in Fig. 2 (Weiss, 1970). You can see that, although pairs of rats in different soundproof boxes received exactly the same strength of shock because

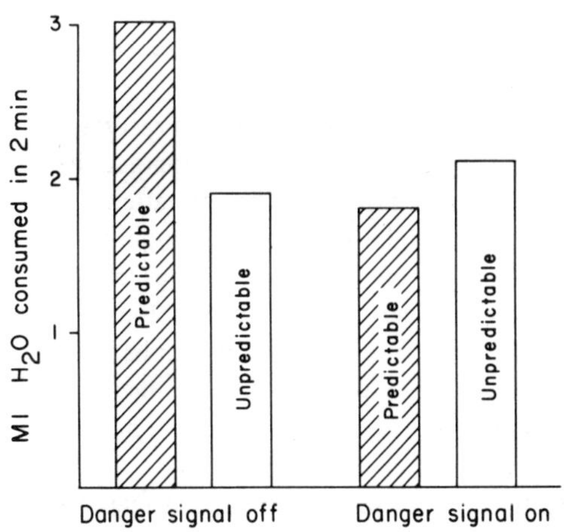

Fig. 1. Rats given electric shocks that are preceded by a signal and hence predictable learn the discrimination of showing fear, indicated by a suppression of drinking, only when the danger signal is on. Rats for whom the signal is unrelated to the shock, so that it is unpredictable, show continuous fear. (Data from Myers, 1956.)

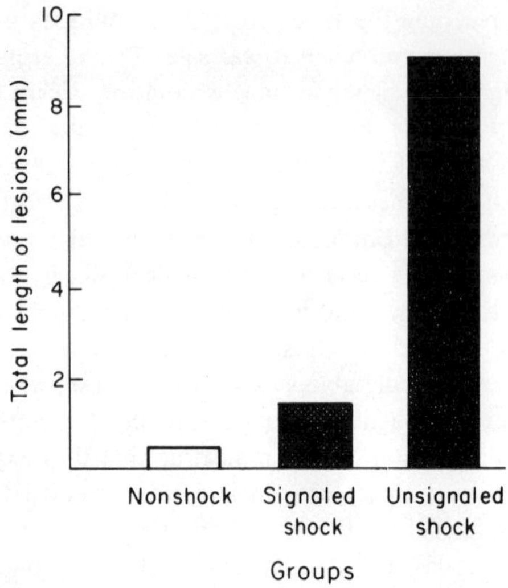

Fig. 2. The amount of stomach lesions produced by signaled (predictable) as compared with unsignaled (unpredictable) electric shocks of equal physical intensity. (Data from Weiss, 1970; figure from Miller, 1972.)

their tails were wired in series, those for whom the shock was not signaled had far more stomach lesions than those for whom it was signaled. In fact, the signaled group did not differ much from the control animals who were not shocked. In addition to showing a striking effect of being able to learn a discrimination on the strength of fear, this study shows that chronic fear can produce stomach lesions.

Similar results were secured with other measures of the stress of fear. Compared with the signaled rats that had the opportunity to learn the discrimination, those that did not showed a higher elevation of plasma corticosterone (Weiss, 1970). Dollard and Miller (1950) have emphasized the value in reducing unrealistic fears of teaching the patient during psychotherapy to discriminate more accurately between dangerous and safe situations.

The other of the two observations in combat, the value of learning a coping response, has also been verified in experiments by Weiss (1968) in my laboratory. He found that, as measured by stomach lesions, plasma corticosterone, and the suppression of eating, the rats that had an opportunity to learn a simple coping response showed much less fear than yoked controls who received exactly the same shocks but had no coping response (Weiss, 1971). Furthermore, compared with control animals who received no shocks, the animals who had no opportunity to learn a coping response, and hence were helpless, showed a depletion of norepinephrine in the brain, while those who had the opportunity to learn a simple and effective coping response showed elevated levels of norepinephrine (Weiss *et al.,* 1970). This result has been replicated in two additional experiments, one of which controlled for activity (Weiss *et al.,* 1975, 1976). These effects on the level of norepinephrine are interesting because the drugs, which if given to the wrong patients will produce a depression, are those that deplete norepinephrine or interfere with its action at the synapse, while the drugs that are useful in treating depression, or that produce elation, are those that increase the level of norepinephrine in the brain and its effectiveness at the synapse (Schildkraut, 1969). Of course, effects on other monoamines also may be involved.

The experiments on coping responses confirm clinical observations on the value of learning such responses in dealing with fear. The results on the increase in norepinephrine levels suggest that there may even be some psychological advantage from meeting and successfully coping with a manageable stress of fear. In the complete absence of stressful challenges, some of the joy of life may be lost.

Role of Brain Amines in Aftereffects of Stress and in Habituation to Stress

If, as suggested by the foregoing results, a stress-induced depletion of norepinephrine (and possibly other monoamines) is involved in situationally produced depressions, stresses that deplete such amines should depress also the rat's ability to mobilize himself for vigorous behavior. Further work by Glazer *et al.* (1975) and Weiss *et al.* (1976) has shown that this is indeed the case. Strong unpredictable and unavoidable electric shocks that deplete the levels of norepinephrine have the aftereffect of causing rats temporarily to be unable to learn a shuttle avoidance task that involves the moderately effortful response of crossing a hurdle even though they are physically able to perform that response. Such rats are, however, able to learn a much less effortful response. Drugs that prevent the depletion of monoamines protect the rat from these aftereffects of stress, and drugs that deplete norepinephrine and other monoamines closely mimic these aftereffects.

From the phenomenon of enzyme induction one might expect that rats which were exposed to strong inescapable shocks on each day for a series of days would increase their capacity to synthesize norepinephrine and, perhaps, other monoamines, and hence would suffer less depletion after a later exposure. Further work from my laboratory has shown that this is the case (Weiss *et al.,* 1975). Rats that were exposed to severe electric shocks each day for 14 days and then tested after similar shocks on the 15th day showed a higher level of activity in the brain of the enzyme tyrosine hydroxylase, which synthesizes norepinephrine, than did rats that received the shocks for the first time on the 15th day. As would be expected from this increased synthesis, the previously shocked rats showed considerably less depletion of norepinephrine in the brain than those not previously shocked. Furthermore, there was an unpredicted effect. Compared to the rats shocked for the first time, the previously exposed ones showed a reduction in the rate of reuptake of norepinephrine into the presynaptic terminals. Both this reduction in reuptake and the lessened depletion would be expected to have the same effect of making norepinephrine more effective at the synapse and therefore, according to our hypothesis, making the behavior of previously exposed rats more normal than that of ones shocked for the first time. This prediction was confirmed. Rats that had received the strong unavoidable shocks for the first time failed to learn when they were taken out of this shock apparatus and tested shortly afterwards in the different shuttle-avoidance one; those that had received strong shocks on the preceding 14 days showed normal avoidance learning.

As an additional pertinent fact, not related to our main argument, previous habituation to shock also considerably reduced the degree to which the strong shocks on the 15th day elevated the level of plasma corticosterone.

Finally, if the foregoing beneficial effects of prior habituation to stress were truly a function of the depletion of norepinephrine, rather than of some other effects of the daily stresses, exposing the rat to prior drug-induced depletions should have similar beneficial effects on the rat's behavior after severe shock. Tetrabenazine, a drug that produces a rapid but transient depletion of norepinephrine and also of other monoamines, was used to test this prediction, which was indeed confirmed. Prior treatment by this drug increased the rat's ability to withstand the aftereffects of the stress of exposure to strong unpredictable and unavoidable electric shocks.

While this paper emphasizes the role of learning in responses to pain and fear, the work that we have just described illustrates the role of unlearned mechanisms. Although we cannot do justice to such factors here, we should point out that genetic and other organic factors produce differences in the strength of fear that is likely to be elicited and learned so that some individuals may be more susceptible to fear than others and a given individual may be more susceptible at some times than at others.

Returning to our main theme of learning, Dollard and Miller (1950, p. 132) have suggested that a child's early experiences may teach it a general habit either of apathy and helplessness or of responding actively to find a way out of a painful situation. Subsequent experimental work has shown that, under the proper circumstances, rats can indeed learn to resist pain and fear (Miller, 1960; Feirstein and Miller, 1963).

Fear Motivates New Learning

Mowrer's (1939) learning theory analysis of Freud's (1936) book on *The Problem of Anxiety* led to the hypothesis that fear can function as a drive and a reduction in the strength of fear can function as a reward to produce learning and maintain performance. Some time ago I secured convincing experimental evidence verifying this hypothesis (Miller, 1948). Page vii of the Foreword to Miller (1971) calls attention to later evidence refuting possible criticisms of this early study. We already have seen that the fact that fear can be learned is one of its significant properties; the fact that it can motivate the learning of new responses is another significant property. Dollard and Miller (1950) and Miller (1975) have used these properties in the explanation of a variety of neurotic phenomena. But fear can motivate the learning of adaptive as well as of nonadaptive responses. Examples are driving carefully and buying insurance. In one study of fear in aerial combat (Wickert, 1947), 49% of the officers reported that mild fear made them perform better, and 34% reported that very strong fear made them perform better. Only 11% reported deleterious effects from mild fear and 25% from very strong fear. In many situations, the crucial point is not whether a person is afraid but what fear has motivated him to learn to do. Such learning to behave to the stress of fear either adaptively or maladaptively may account for some of the individual difference that President Kittay has pointed out in his Introduction to this Symposium.

Brain Cells That Fire to Fear

One of the difficulties in determining the detailed role of fear in the learning of new responses, and in studying some of the interactions between fear and conflict that are relevant for psychopathology, has been the lack of an independent moment-to-monent measure of fear (Miller, 1959). I have spent considerable time unsuccessfully looking for such a measure, but it appears that Vertes and I are on a track that may conceivably lead to one (Vertes and Miller, 1976). We discovered in the freely moving rat large cells in the region of the nucleus reticularis pontis caudalis that will respond to a conditioned stimulus for electric shock but not to the arousal produced by a conditioned stimulus for water. The firing of these cells is not affected by whether the rat responds to the danger signal by increased movements or by the opposite response of "freezing"; these cells differ in other ways from those cells that respond to movements. Figure 3 shows records of the responses of one such cell. At the top you can see the marked and sustained increase in firing to a tone that is the conditioned stimulus for electric shock, and below you can see much less firing to a light that

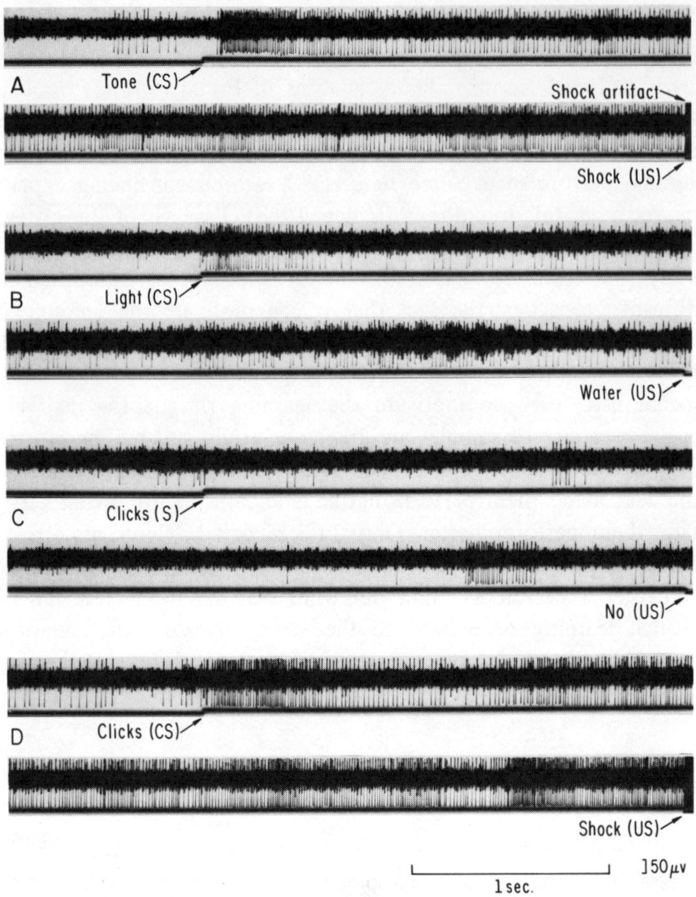

Fig. 3. Differential response of a neuron in the reticular formation of the rat. (A) To tone as a conditioned stimulus (CS) for electric shock, (B) light as a CS for water, (C) clicks as a neutral stimulus, and (D) the same clicks after pairing with shock. Each pair of traces is a continuous strip of an entire test trial. (From Vertes and Miller, 1976.)

is a conditioned stimulus for water under conditions of severe thirst. Next you see much less firing to a clicker, first used as a neutral stimulus. Finally, you see greatly increased firing after the same clicker has been paired with electric shock. Ten such cells were first informally identified and then formally tested. In the latter tests, for each one of the 10 cells the response to the conditioned stimulus for the electric shock (a different stimulus for different animals) was reliably greater ($p < 0.001$ in each case) than the response either to the CS for water or to the neutral CS. We hope to study the firing of such cells in avoidance learning, in conflict behavior, and in a variety of other conditions relevant to psychopathology.

Two Contrasting Patterns of Responses

Fear has an innate tendency to activate two incompatible patterns of behavioral response (Miller, 1975). One of these involves the heightened physical activity of fighting or fleeing, which frequently is accompanied by loud vocalization. The other involves freezing, arresting all motion and remaining mute; in its extreme form, we have death-feigning. According to Corson (this volume, p. 85), the physiological responses, such as increases in heart rate and in the secretion of ADH, that are a preparation for intense activity frequently precede the activation pattern but not the freezing one. Learning can determine which of these two responses is the more likely to occur (Miller and Weiss, 1969) and by doing so presumably will have a considerable effect on which type of physiological responses will occur. Brener (1974) and Brener *et al.* (1974) have shown that training in activity or immobility does indeed affect the subsequent heart rate of animals trained under paralysis by curare. Thus we see one way in which the responses to the stress of fear are not completely nonspecific; their pattern can be changed by learning.

EFFECTS OF SPECIFICALLY REINFORCING PHYSIOLOGICAL RESPONSES

We have seen that there are a number of ways in which learning can affect the strength and duration of fear, and how the strength and duration of fear can have a variety of effects ranging from changes in the immune system, through motivation to learn neurotic symptoms, to the production of stomach lesions. What happens if we try to apply learning directly to the modification of a physiological response? For example, try to use reward or escape from punishment to teach an increase in heart rate.

Different Ways of Producing Effect

If we try to reward changes in heart rate, there are four possibilities:

(1) The subject may be unable to learn anything that changes his heart rate.
(2) The subject may learn a skeletal response, such as muscular exertion, that

has metabolic and physical consequences that increase the demand on the circulatory system and hence indirectly result in an increased heart rate.

(3) The subject may learn some centrally patterned response, like the previously mentioned preparation for fight or flight, that involves simultaneous commands some of which go out over the autonomic nervous system to increase heart rate and others of which go out over the somatic nervous system to the skeletal muscles.

(4) The subject may learn increased heart rate as a more or less specific response.

Rats Paralyzed by Curare

In order to differentiate the second alternative, mediation by the effects of overt muscular contraction, from the third and fourth, my students and I performed a series of experiments on rats whose skeletal muscles were paralyzed by curare, a drug that leaves the autonomic system and the brain relatively unaffected. These experiments seemed to show that animals paralyzed in this way and maintained on artificial respiration could learn increases when increases were rewarded and decreases when decreases were rewarded in a variety of autonomically mediated visceral responses. If this learning had been via the third mechanism, especially any central patterns as massive as those of flight-fight versus freezing, we would have expected it to involve simultaneous changes in a number of different visceral responses; instead, the learning appeared to be able to become remarkably specific (Miller, 1969).

Later we had difficulty in replicating these experiments and, looking back, saw that over several years there had been a progressive decline in the size of the differences learned (Miller, 1972; Miller and Dworkin, 1974). Since then, Dr. Barry Dworkin and I have vastly improved our technique of maintaining rats paralyzed by curare in excellent behavioral condition, as indicated by the ability to learn large, discriminative, classically conditioned responses. We may even be obtaining some instrumental learning of changes in heart rate, but, having seen apparently positive results fade out a number of times in the last few years, we do not want to make any claim yet.

Nonparalyzed Animals

Meanwhile, experiments on animals not paralyzed by curare have been repeated often enough to rule out definitely the first of the alternatives above by showing that changes in some visceral responses, such as heart rate and blood pressure, can indeed be produced by instrumental training procedures. Miller and

Carmona (1967) have used water rewards to train thirsty dogs to either increase or decrease their rates of salivation, and Shapiro and Herendeen (1975) have been able to use food to reward a decrease in salivation—a change opposite to what one would expect by classical conditioning. Benson *et al.* (1969) have used escape from and avoidance of electric shock to train squirrel monkeys first to increase and later to decrease their arterial blood pressure, and Harris *et al.* (1973) have used a combination of shock avoidance and food reward to teach baboons to produce 33-mm Hg increases in blood pressure maintained for the entire period of 12-hour sessions. Engel and Gottlieb (1970) have used escape and avoidance of electric shock to train rhesus monkeys to speed up their heart rate during certain sessions and to slow it down during others.

Human Experiments

Similarly, work on human subjects, which began somewhat before the first publications on animals, has continued to show that in one way or another (i.e., alternatives 2, 3, or 4) rewards for changes in visceral responses can produce such changes (Kimmel, 1967). Recently, work on the specificity of such changes has made explanations in terms of (2) and even (3) less attractive. Crider *et al.* (1969) have summarized four studies showing that training to change the galvanic skin response does not affect other autonomically mediated responses such as heart rate or finger blood volume, and Shapiro *et al.* (1970) have shown that subjects can learn to produce modest changes in blood pressure without changing their heart rate and in heart rate without changing their blood pressure. In patients paralyzed by polio or muscular dystrophy, Dworkin, Pickering, Brucker, and Miller (unpublished) have observed similar learning of modest changes in blood pressure without changes in heart rate.

How Augmented Feedback Helps Human Visceral Learning

A novice learning to shoot foul shots in basketball does not have good voluntary control over where the ball will go. When he sees it going near to the hoop, this partial success serves as a reward so that he is more likely to repeat the movements that he has just made. When he sees it missing widely, that failure serves as a punishment so that he is less likely to repeat that response. When he finally sees the ball swish through the hoop, that signal of success serves as a strong reward so that he is more likely to repeat that response. Thus he gradually learns to improve his voluntary control. But if he were blindfolded so that he did not have any knowledge of results or, in other words, any feedback, he could not learn.

Our perceptions of many visceral responses are notoriously inaccurate; without knowledge of results or, in other words, accurate feedback, we are like the blindfolded foul-shooter and do not learn. In many cases, however, modern instrumentation can give us better feedback and thus, figuratively speaking, remove the blindfold. Such instrumentally augmented feedback about a biological function has been called biofeedback. For a person who wants to achieve voluntary control, feedback indicating success should serve as a reward, just as does seeing a basketball swish through the hoop or a tennis serve hit the corner for an ace. As we have seen in the previous section, there is considerable evidence that the use of such feedback can enable human subjects to acquire in one way or another a certain amount of voluntary control over at least some visceral functions. There is also evidence that the use of instrumentation to improve feedback can help both human and animal subjects to learn to improve their visceral perception (Ádám, 1967). To the extent that people can learn to improve their visceral perception, they can dispense with the need for the feedback provided by elaborate instrumentation.

Therapeutic Applications of Visceral Learning

The possibility of using visceral learning to treat certain psychosomatic symptoms has caught the fancy of the public media who, characteristically, have made the story more interesting by exaggerating the claims and deleting the qualifying phrases so that there is a danger of raising impossible hopes that will produce disillusionment that will interfere with the hard work necessary to discover the best way of producing visceral learning and of disentangling its therapeutic applications from placebo effects (Miller, 1975).

Perhaps the best therapeutic results to date are those reported by Engel and Bleecker (1974) who have trained five patients to control premature ventricular contractions (PVCs) and have found lowered PVC activity on clinical tests outside the laboratory during the considerable number of months that these patients have been followed up. One of them was found to have retained good control for as long as one year and another for as long as five years. That these were not merely nonspecific placebo effects is indicated by the fact that the patients were able voluntarily to turn the PVCs either on or off on request. Pickering and Miller (unpublished) have replicated the learning of similar control by two patients. One of these, who was in bigeminy most of the time, learned by watching his ECG on an oscilloscope which yielded a picture something like the record reproduced in Fig. 4, showing performance during the latter part of training. The top part of the continuous strip of record shows performance during rest. PVCs are indicated by the large spikes alternating with normal sinus beats indicated by the small spikes. Such a rhythm is called bigeminy. In the next portion of the record, the patient is instructed to try to suppress the PVCs;

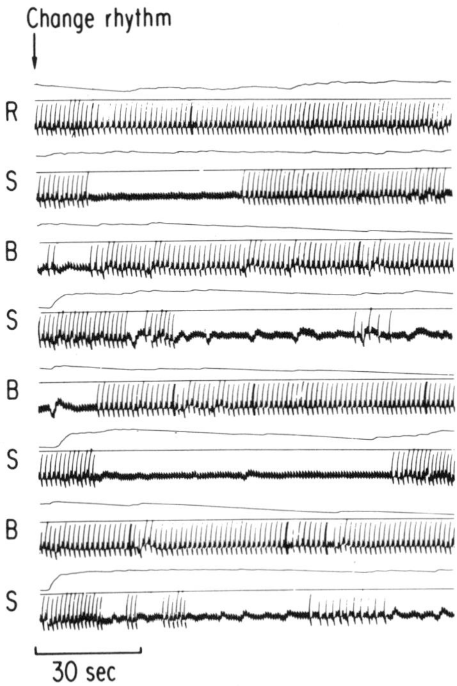

Fig. 4. Learned voluntary control over heart rhythm. Premature ventricular contractions (PVCs) are indicated by large spikes; normal sinus rhythm by their absence. Line above the ECG indicates skin conductance. (R) period of rest, (S) patient instructed to produce PVCs, (B) patient instructed to produce bigeminy or, in other words, alternating PVCs and normal beats. Entire figure is from continuous strip of record. (Recorded by Pickering; figure from Miller, 1975.)

normal sinus rhythm shows up in the middle and at the very end. When instructed to go back into bigeminy at the beginning of the next strip, he promptly does so. Succeeding strips alternating between instructions for sinus and bigeminy show considerable, but not perfect, voluntary control. The line above the ECG indicates skin conductance, which usually increases markedly as soon as the patient is trying to suppress the PVCs and drifts back downward during the rest period. This result suggests that the PVCs are suppressed by an increase in sympathetic activity.

This patient's PVCs were asymptomatic; he could not learn to discriminate without the aid of the oscilloscope when he was in bigeminy and when he was in normal sinus rhythm. Without the aid of the scope his control was much poorer. Therefore, aside from achieving an enormous boost in morale, this patient probably could not make any specific therapeutic use of his training.

Another patient, who suffered from paroxysmal attacks of bigeminy that made him feel weak, learned to control these by learning to speed up his heart.

He showed unusual talent at acquiring this skill, being able to produce an increase of 20 beats per minute without any obvious signs of increased muscular tension or changes in breathing. Since he already could recognize his PVCs and learned to recognize the increase in heart rate, he was able to use that increase to stop attacks outside the laboratory.

Engel *et al.* (1974) also report promising results in training a number of patients suffering from organically based difficulties with fecal incontinence to control their anal sphincters, the internal one of which is presumed to be under autonomic control.

The urethral sphincters are exclusively innervated by the autonomic nervous system, but virtually everyone learns voluntary control of this response under the favorable conditions of high motivation and immediate knowledge of results. For some time there was discussion about whether such control was exercised indirectly via contractions of abdominal muscles that increased the pressure on the bladder, initiating a reflex response. But in an experiment that was overlooked until recently because of disciplinary barriers, Lapides *et al.* (1957) completely paralyzed 16 human volunteer subjects, some by succinlycholine and some by curare, so that they had to be maintained on artificial respiration. Upon command these paralyzed subjects could initiate urination as fast as usual and terminate it in about twice the usual time. Thus this visceral response can be performed in the complete absence of skeletal movement. Incidentally, Mowrer and Mowrer's (1938) invention of a device to treat bedwetting by sensing the first drops of moisture and sounding a buzzer may be considered the first use of electronically augmented feedback in therapeutic visceral training. The results of considerable research on this technique has been summarized by Lovibond (1964). A study by Baker (1969) has shown that its use does not produce symptom substitution, but instead a general reduction in symptoms of maladjustment.

Brucker, one of our collaborators at the Goldwater Memorial Hospital, has discovered that some patients with severed spinal cords have unusual ability to learn to increase their abnormally low blood pressure. One of these patients, whose spinal cord had been severed at T4 (about the middle of the chest) by a gunshot wound, had a strong ambition to use his powerful arms and shoulders to walk with crutches and braces but had been unsuccessful, in spite of extensive efforts by physical therapy, because he tended to faint from low blood pressure whenever he was helped into an upright posture. After relatively few sessions of practice during which Brucker gave him information on his blood pressure, the patient was able to produce an abrupt enough rise so that he could perceive it and practice further on his own. Figure 5 shows a record of a voluntary increase in diastolic blood pressure produced as a prompt response to a request. At first, the heart rate starts to speed up and the blood pressure actually falls, but eventually the diastolic blood pressure increases a total of 24 mm Hg and the heart rate falls, probably as a result of the sino-cardiac reflex. This patient is able

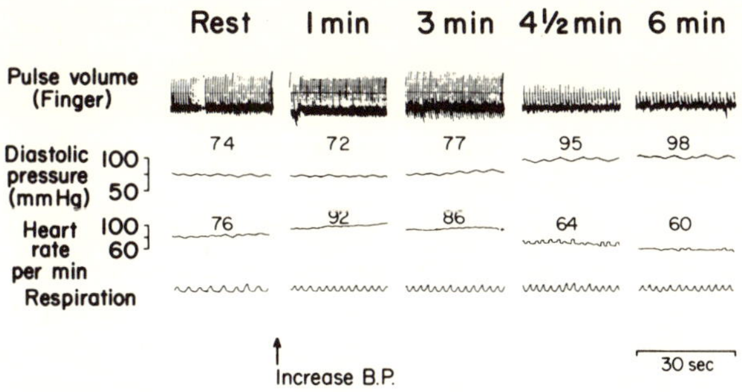

Fig. 5. Learned voluntary control over blood pressure by patient with spinal lesion at T4. The patient was trained by Brucker at Goldwater Memorial Hospital and recorded by Pickering in the author's laboratory.

now to walk with crutches. At first he had to stop walking occasionally when he felt his blood pressure falling, in order to raise it, but now he does not have to do this, either because, as he believes, keeping the pressure up has become automatic like balancing on a bicycle, or perhaps because of some physiological readjustment to the upright posture.

Similar therapeutic results have been secured with a patient suffering a lesion between C4 and C5, whose blood pressure fell so low when her feet were placed in a normal sitting position that she had to go around in a wheelchair with her feet elevated above her head. After learning to raise her systolic blood pressure voluntarily 20 mm Hg or more, she can not sit with her feet down for several hours at a time. Yet other patients who did not show serious symptoms of hypotension have demonstrated the ability to learn to raise their systolic blood pressure on command by approximately 15 mm Hg. In the light of these results, it seems possible that spinal lesions have produced an antihomeostatic effect which, in comparison with the nonlesioned patients, shows up on the one hand as a greater spontaneous variability in blood pressure and on the other as a greater ability to learn large voluntary increases. Are there drugs that can produce an antihomeostatic effect and thus facilitate therapeutic visceral learning?

Need to Control for Placebo Effects

In each of the foregoing examples–PVCs, fecal incontinence, and postural hypotension of spinal patients–the subject's acquisition of the ability to show clear-cut, specific, voluntary control over an objectively measured physiological response makes it seem unlikely that the results were a mere nonspecific placebo effect. Nevertheless, it is hoped that more rigorous controls for placebo effects can and will be introduced into subsequent studies.

Some of the other preliminary claims for therapeutic effects of biofeedback training, for example, Sargent *et al.*'s (1972) report that training in warming the hands aids in the control of migraine headaches, are in urgent need of substantiation by specific controls for placebo effects and by follow-up studies to determine the persistence of the effect.

The potency of placebo effects on headaches and on many other symptoms has been pointed out in an excellent summary by Shapiro (1960). One of the mechanisms of certain placebo effects may be the ability of a coping response, and also of attention by an important authority figure, to reduce the stress of fear (Miller, 1975). Whatever its basis, the placebo effect shows the therapeutic importance of a purely psychological factor. Instead of being veiwed merely as a source of error to be ruled out, the powerful placebo effect needs to be investigated in its own right.

The placebo effect is not limited to subjective reports of symptoms; it can affect physically measured physiological functions. This is clearly illustrated in the control groups that receive only sugar pills in double-blind evaluations of antihypertensive drugs. In one such study, Grenfell *et al.* (1963) found that the average reduction in blood pressure of 48 patients receiving only placebo medication was 25 mm Hg in systolic and 12 mm Hg in diastolic pressure.

In trying to train patients with essential hypertension to lower their blood pressure we have had apparently excellent results with one patient who learned considerable specific voluntary control and concurrently showed a general reduction in clinically measured baseline pressure. But we have failed with 26 other patients to secure either specific voluntary changes large enough to be therapeutically significant or general declines in baseline large enough to be beyond the range of placebo effects. Benson *et al.* (1971) published a study in which the specific changes during training sessions were relatively small and the larger declines in baseline levels were within the range of placebo effects; the senior author of that study is not currently enthusiastic about the ability of presently available techniques to train patients to control essential hypertension (personal communication).

Some people ask, "Why be so concerned about placebo effects?" The answer is that it is a waste of time to use expensive equipment and time-consuming training procedures if the effects are only placebo ones that could be produced much more economically by giving sugar pills. On the other hand, if the effect, although small, is a genuine specific one of learned control, it is possible that further research may improve our understanding of how to teach the patient to produce considerably larger therapeutic changes.

Implications for Etiology

The fact that visceral responses can be modified in one way or another by instrumental learning opens up new theoretical possibilities for the etiology of psychosomatic symptoms. It opens up the possibility that psychosomatic symptoms can be strengthened by rewards, including escape from punishment. Psychiatrists frequently call such rewards secondary gains. To cite an example which has been used before (Miller, 1969), suppose a child is terror-stricken at the thought of going to school in the morning because he is completely unprepared for an important examination. The strong fear elicits a variety of fluctuating symptoms, such as a queasy stomach at one time and pallor and faintness at another, at which point his mother, who is particularly concerned about cardiovascular symptoms, says: "You are sick and must stay at home." The child feels a great relief from fear, which should reinforce the cardiovascular responses producing pallor and faintness. If such experiences are repeated frequently enough, the child theoretically should learn to respond to similar situations with pallor and faintness. Similarly, another child whose mother ignores the vasomotor symptoms but is particularly concerned by signs of gastric distress, should learn stomach symptoms. Let me emphasize that at present these are theoretical possibilities that need verification by clinical research.

The foregoing theoretical possibilities may explain some of the individual differences in response that President Kittay's introductory remarks described as an important problem.

But if psychosomatic symptoms are learned under conditions of strong motivation and reward, and if these conditions currently prevail in the patient's life, the therapist who tries to reverse this learning may not have strong enough motivations and rewards at his disposal to do so unless he teaches his patient some other way to solve his emotional problem. Perhaps it is no accident that most of the examples that I have cited as being promising therapeutic applications of visceral learning appear to involve compensation for an organic deficit.

ACKNOWLEDGMENT

Work from the author's laboratory reported in this paper was supported by USPHS grants MH 13189 and MH 19183.

REFERENCES

Ádám, G. *Interoception and Behaviour.* (1967). Budapest: Akadémiai Kiadó.

Baker, B. (1969). Symptom treatment and symptom substitution in enuresis. *J. Abnorm. Psychol. 74,* 42–49.

Benson, H., Herd, A. J., Morse, W. H., and Kelleher, R. T. (1969). Behavioral induction of arterial hypertension and its reversal. *Am. J. Physiol. 217,* 30–34.

Benson, H., Shapiro, D., Tursky, B., and Schwartz, G. E. (1971). Decreased systolic blood pressure through operant conditioning techniques in patients with essential hypertension. *Science 173,* 740–742.

Brener, J. (1974). A general model of voluntary control applied to the phenomena of learned cardiovascular change. In *Cardiovascular Psychophysiology* (P. A. Obrist *et al.,* Eds.). Chicago: Aldine, pp. 365–391.

Brener, J., Eissenberg, E., and Middaugh, S. (1974). Respiratory and somatomotor factors associated with operant conditioning of cardiovascular responses in curarized rats. In *Cardiovascular Psychophysiology* (P. A. Obrist *et al.,* Eds.). Chicago: Aldine, pp. 251–275.

Cannon, W. B. (1929). *Bodily Changes in Pain, Hunger, Fear and Rage.* New York: Appleton.

Crider, A., Schwartz, G. E., and Shnidman, S. (1969). On the criteria for instrumental autonomic conditioning: A reply to Katkin and Murray. *Psychol. Bull. 71,* 455–461.

Dollard, J., and Miller, N. E. (1950). *Personality and Psychotherapy.* New York: McGraw-Hill.

Engel, B. T., and Bleecker, E. R. (1974). Application of operant conditioning techniques to the control of the cardiac arrhythmias. In *Cardiovascular Psychophysiology* (P. A. Obrist *et al.,* Eds.). Chicago: Aldine, pp. 456–476.

Engel, B. T., and Gottlieb, S. H. (1970). Differential operant conditioning of heart rate in the restrained monkey. *J. Comp. Physiol. Psychol. 73,* 217–225.

Engel, B. T., Nikoomanesh, P., and Schuster, M. M. (1974). Operant conditioning of rectosphincteric responses in the treatment of fecal incontinence. *New Engl. J. Med. 290,* 646–649.

Feirstein, A. R., and Miller, N. E. (1963). Learning to resist pain and fear: Effects of electric shock before versus after reaching goal. *J. Comp. Physiol. Psychol. 56,* 797–800.

Freud, S. (1936). *The Problem of Anxiety.* New York: Norton.

Fuller, J. L. (1967). Experimental deprivation and later behavior. *Science 158,* 1645–1652.

Glazer, H. I., Weiss, J. M., Pohorecky, L. A., and Miller, N. E. (1975). Monoamines as mediators of avoidance-escape behavior. *Psychosom. Med. 37,* 535–543.

Grenfell, R. F., Briggs, A. H., and Holland, W. C. (1963). Antihypertensive drugs evaluated in a controlled double-blind study. *South. Med. J. 56,* 1410–1415.

Harris, A. H., Gilliam, W. J., Findley, J. D., and Brady, J. V. (1973). Instrumental conditioning of large-magnitude, daily, 12-hour blood pressure elevations in the baboon. *Science 182,* 175–177.

Kimmel, H. D. (1967). Instrumental conditioning of autonomically mediated behavior. *Psychol. Bull. 67,* 337–345.

Lapides, J., Sweet, R. B., and Lewis, L. W. (1957). Role of striated muscle in urination. *J. Urol. 77,* 247–250.

Lovibond, S. H. (1964). *Conditioning and Enuresis.* New York: Macmillan.

Lown, B., Verrier, R., and Corbalan, R. (1973). Psychologic stress and threshold for repetitive ventricular response. *Science 182,* 834–836.

Miller, N. E. (1948). Studies of fear as an acquirable drive: I. Fear as motivation and fear reduction as reinforcement in the learning of new responses. *J. Exp. Psychol. 38,* 89–101.

Miller, N. E. (1951). Learnable drives and rewards. In *Handbook of Experimental Psychology* (S. S. Stevens, Ed.). New York: Wiley, pp. 435–472.

Miller, N. E. (1959). Liberalization of basic S-R concepts: Extensions to conflict behavior, motivation and social learning. In *Psychology: A Study of a Science,* Study 1, Vol. 2 (S. Koch, Ed.). New York: McGraw-Hill, pp. 196–292.

Miller, N. E. (1960). Learning resistance to pain and fear: Effects of overlearning, exposure and rewarded exposure in context. *J. Exp. Psychol. 60,* 137–145.

Miller, N. E. (1969). Learning of visceral and glandular responses. *Science 163,* 434–445.

Miller, N. E. (1971). *Neal E. Miller: Selected Papers.* Chicago: Aldine-Atherton.

Miller, N. E. (1972). Interactions between learned and physical factors in mental illness. *Semin. Psychiat. 4,* 239–254.

Miller, N. E. (1975). Applications of learning and biofeedback to psychiatry and medicine. In *Comprehensive Textbook of Psychiatry/II* (A. M. Freedman *et al.,* Eds.). Baltimore: Williams & Wilkins, pp. 349–365.

Miller, N. E., and Carmona, A. (1967). Modification of a visceral response, salivation in thirsty dogs, by instrumental training with water reward. *J. Comp. Physiol. Psychol. 63,* 1–6.

Miller, N. E., and Dworkin, B. R. (1974). Visceral learning: Recent difficulties with curarized rats and significant problems for human research. In *Cardiovascular Psychophysiology* (P. A. Obrist *et al.,* Eds.). Chicago: Aldine, pp. 312–331.

Miller, N. E., and Weiss, J. M. (1969). Effects of the somatic or visceral responses to punishment. In *Punishment and Aversive Behavior* (B. A. Campbell and R. M. Church, Eds.). New York: Appleton-Century-Crofts, pp. 343–372.

Mowrer, O. H. (1939). A stimulus-response analysis of anxiety and its role as a reinforcing agent. *Psychol. Rev. 46,* 553–565.

Mowrer, O. H., and Mowrer, W. M. (1938). Enuresis–a method for its study and treatment. *Am. J. Orthopsychiat. 8,* 436–459.

Myers, A. K. (1956). The effects of predictable vs. unpredictable punishment in the albino rat. Ph.D. Thesis, Yale University.

Pavlov, I. P. (1927). *Conditioned Reflexes* (G. V. Anrep, trans.). London: Oxford University Press. Reprinted, New York: Dover (1960).

Rabkin, J. G., and Struening, E. L. (1975). Social change, stress and illness. Paper presented at AAAS Annual Meeting, New York, New York.

Sargent, J. D., Green, E. E., and Walters, E. D. (1972). The use of autogenic feedback graining in a pilot study of migraine and tension headaches. *Headache 12,* 120–124.

Stein, M., Schiavi, R. C., and Camerino, M. (1976). Influence of brain and behavior on the immune system. *Science 191,* 435–440.

Schildkraut, J. J. (1969). *Neuropsychopharmacology and the Affective Disorders.* Boston: Little, Brown.

Selye, H. (1956). *Stress and Disease.* New York: McGraw-Hill.

Shapiro, A. K. (1960). A contribution to a history of the placebo effect. *Behav. Sci. 5,* 109–135.

Shapiro, D., Tursky, B., and Schwartz, G. E. (1970). Differentiation of heart rate and systolic blood pressure in man by operant conditioning. *Psychosom. Med. 32,* 417–423.

Shapiro, M. M., and Herendeen, D. L. (1975). Food-reinforced inhibition of conditioned salivation in dogs. *J. Comp. Physiol. Psychol. 88,* 628–632.

Vertes, R. P., and Miller, N. E. (1976). Brainstem neurons that fire selectively to a conditioned stimulus for shock. *Brain Res. 103,* 229–242.

Weiss, J. M. (1968). Effects of coping responses on stress. *J. Comp. Physiol. Psychol. 65,* 251–260.

Weiss, J. M. (1970). Somatic effects of predictable and unpredictable shock. *Psychosom. Med. 32,* 397–408.

Weiss, J. M. (1971). Effects of coping behavior in different warning signal conditions on stress pathology in rats. *J. Comp. Physiol. Psychol. 77,* 1–13.

besides "inhibiting drives" there might also be other functions. The most likely ones would be to stamp in sensory-motor connections (possibly in the cerebellum), to motivate the "replay" of sequential behavior memories (possibly in the hippocampus), and to "charge" (or connect drives to) cell assemblies active at the time of reward (possibly in the neocortex).

A current vogue emphasizes genetic differences between peoples by pointing to their inborn behaviors and capabilities and deemphasizes the motivation-reward-learning processes that were the main interest of older psychologists. It is therefore a matter of some optimistic importance to remember, in these dark days when we worry so much about the dangers of behavior, that a great deal of the specific content of behavior particularly in human beings (but also in all mammals including laboratory rats) is created and modified by training. This is because a large part of the inherited nervous system may be conceived as something of a reward-and-learning machine, almost a tabula rasa upon which the environment (physical and social) writes particular sensory-motor specifications at some time after birth. It is the working of this particular aspect of the adaptive system that has concerned psychologists for years. It is the brain mechanisms behind these that are sought by many of the current programs of neuropsychology. One or several structures of the CNS have this plastic and absorptive character. Each of these is ready to pick up programs from the world. It can be conceived as a delicately wired instrument like an oscilloscope or even better a computer waiting to be programmed, ready to listen to what is at first a very simple language which it has in common with the programmer. In this case the programmer is the physical environment and the social world.

The key factors in the programming of mammal behaviors go under names like motivation, reward, and learning. Subjects *learn* to do things, *if* they are rewarded for doing them, *and if* they are motivated to do them. There is of course a continuum from higher to lower motives. The lower end of the continuum is the one we have most in common with the laboratory rat. At this end there are four key factors. First there are drives, that is, special states created by alarming or dangerous deficits. Second there are incentive mechanisms, that is, reactions to promising stimuli which guide behavior even though deficits are not alarming. Third there are rewards, that is, targets that become objects of pursuit under either of the two kinds of motivating conditions, and which modify behavior repertories a little or a lot when they are achieved (or when they are brought to bear as stimuli). Fourth and finally, there are the learning mechanisms, that is, the set of built-in rules for modifying the repertory with or without rewards.

The episodes of behavior involved are triggered into action by either alarming shortages, or promised rewards, and the triggered behaviors may look much the same in either case.

Whichever way the behavior gets started it is "steered" by rewards. By steering I mean the rewards when they happen serve to shape the response repertory and in the long run this can be reflected in radically altered response probabilities. One way or another responses closer to rewards in a series get their probabilities elevated, and sensory signals close to rewards in a series are rendered attractive so that they become secondary targets of pursuit. Neither the new behaviors nor the new goals however are always probable or always pursued after the elevation by reward. The pursuit only happens if and when the appropriate drive or incentive state exists. It is as if the reward did not attach *itself* to contiguous behaviors and objects, nor just raise them in the repertory either. Instead it seems to attach drives to them. Much of the learning that goes on under the influence of drives or incentives and rewards is thus conceptualized as mainly involving the hookup of a drive to a sensory motor process. The hookup is caused by the application of a reward during or shortly after a sensory-motor event.

Besides being steered, behavior is punctuated and eventually terminated by rewards. These functions are equally as important to the animal as the steering function, even if not as interesting to the psychologist. When the animal is suddenly given access to a quantity of food, this immediately (though possibly temporarily) halts the instrumental behavior whether it was sustained by an incentive or a more alarming condition. The searching and the working stop and there is the end of the urgent and driven look in the animal's behavior. Consummatory responses which look more automatic than driven take their place. If the supply of food is small or if it is removed the animal may switch back to driven behavior; and in a Skinner box there may be many alternations so the behavior is punctuated by each reward. But if the supply is sufficient, the alarm state will give way to the incentive state, and then this will disappear. This makes way for other drives or dissimilar character, or of similar character but different direction.

Data from a number of brain experiments seemed at first to have relatively direct bearing on central questions of motivation, particularly those related to the steering of behavior toward some things and away from others, and the role of rewards and drives in the problem of learning. The brain experiments promised interesting advances just over the horizon; but so far they have added to the puzzles. Considerable thought is still needed to get a sensible thread of meaning from them and to decipher some of the most tractable directions they point toward further understanding. I want to set the outlines of this data on the table and try again to follow some of the best threads.[1]

[1] To do this I shall review work in which I participated on (1) behavioral features of "brain reward," and (2) "unit responses" of hypothalamic neurons; and work of others on (3) stimulation and lesions affecting basic drive behavior, and (4) the amine neurotransmitters and their maps. Although my wife, Marianne E. Olds, has worked a great deal on the latter problem, I have had no involvement in it (as little in fact as if it had been done in another laboratory). The advances made in the latter two fields, however, have done much to form or modify my views.

BRAIN STIMULATION

The first body of data came from electric stimulation of the brain in behaving animals. Experiments employing this method have resulted in a map of brain locations where electric stimulation caused animals to behave as if the stimulus itself were the goal object of an active drive or caused a condition so hedonically gratifying that no drive was necessary to get behavior going (Olds, 1962). These abnormal brain rewards motivated not only pedal behavior but also maze running and the crossing of aversive obstructions (Fig. 1).

Rats, cats, and monkeys had much the same map (Fig. 2). Even humans would perform nonsense tasks to stimulate analogous brain centers although they often seemed confused as to why they were doing it. In the rat, the olfactory bulb and much of the floor of the brain connected to it were implicated as areas where stimulation was rewarding.

On the floor of the rat brain a large central region is the hypothalamus. The borders are its outposts. Some of the outposts are in the olfactory parts of the forebrain. Others are toward the back of the brain. Through this back area come

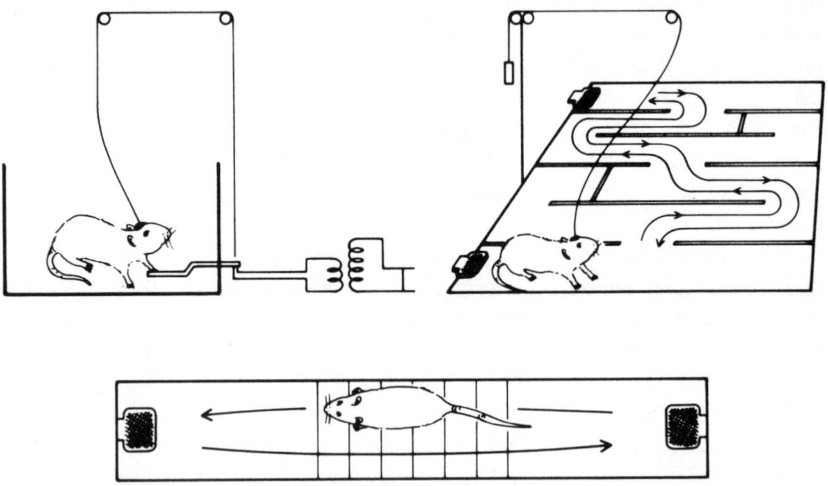

Fig. 1. Behaviors motivated by application of electric stimulation in the medial forebrain bundle after each pedal response (0.25-sec trains, 60-Hz alternating current, 50-mA rms). In the maze and the obstruction box, three pedal responses were rewarded at one pedal and then the animal was required to shuttle to the other for three more, and so forth. In the obstruction box, current of about 60 mA applied through the grid floor stopped hungry rats running for food and these also stopped rats running for 50-mA brain stimuli. However when the brain "reward" was increased to 200 mA these animals crossed obstructions applying more than 400 mA to the feet.

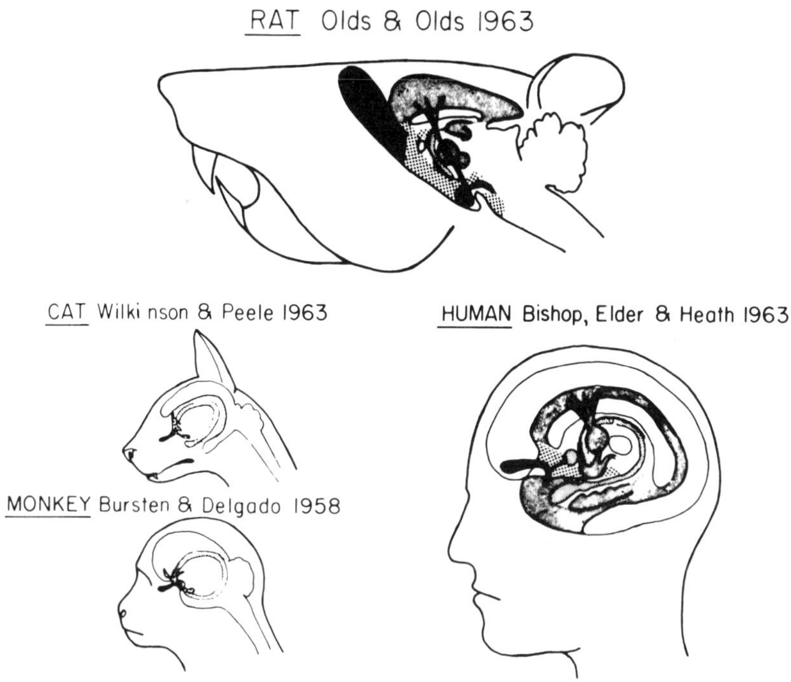

Fig. 2. Schematic pictures of the parts of the brain yielding brain reward behavior in different species. The map for the rat has been done carefully. Those of other species are estimated from a smaller number of tests, with extrapolation based on anatomical analogies.

messages from the visceral and gustatory receptors; and more directly into the hypothalamus from blood-brain windows come chemical messages from the circulation. These are the hormones.

The reward map covered most of the hypothalamus and its satellites (Fig. 3). In the hypothalamus it ran from far anterior to far posterior and from far lateral to the midline. A paradox of the map was that this same region was the home of aversive effects of electric stimulation (Roberts, 1958). If opposed aversive effects were not immediately obvious, they could usually be demonstrated by careful behavioral analysis. Because the whole hypothalamus was covered by a reward map and aversive countereffects were always in evidence, you may guess that the hypothalamus was homogeneous with respect to these maps. It was not.

In the far lateral parts of the hypothalamus and in some parts of the far medial hypothalamus there were locations where the rewarding effects of stimulation predominated. In these cases the animal was apparently at home with self-stimulation. No negative signs were seen during brain pedal behavior, and careful methods were required to reveal them.

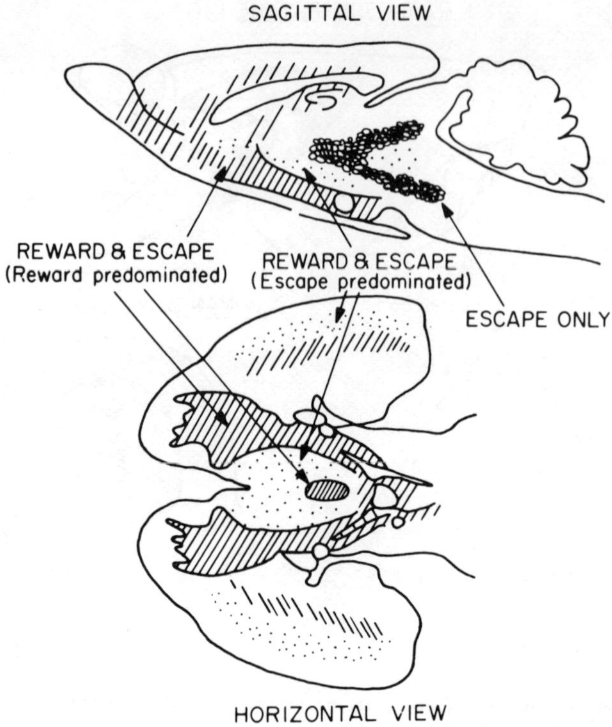

Fig. 3. More detailed map of main areas in the rat brain yielding reward and escape behavior.

In a large "in between" area, there was an obvious mixture of positive and aversive effects. The animal would pedal regularly and fast if closeted with an electric stimulus, but if there was a way out, the animal would escape. There was no amount of stimulation in these areas that seemed just right. The animal behaved as if it could not stand the stimulation but could not resist it (Olds and Olds, 1962).

In this same middle area there was a second paradox of the reward maps. Here the same electric stimulus often provoked drives as well as rewards. The drives depended partly on the location of the stimulus (Fig. 4). With probes in anterior hypothalamus there were sex responses and responses that adjusted the body temperature (Roberts *et al.,* 1967). In the anterior part of the middle hypothalamus there were both eating and drinking responses but the drinking responses predominated. In the posterior part of the middle hypothalamus there were more eating and drinking responses but here the eating responses predominated (Valenstein *et al.,* 1970). In the posterior part of the hypothalamus sex responses were evoked again (Herberg, 1963). Because there were many overlapping effects and sex responses were evoked in areas on both

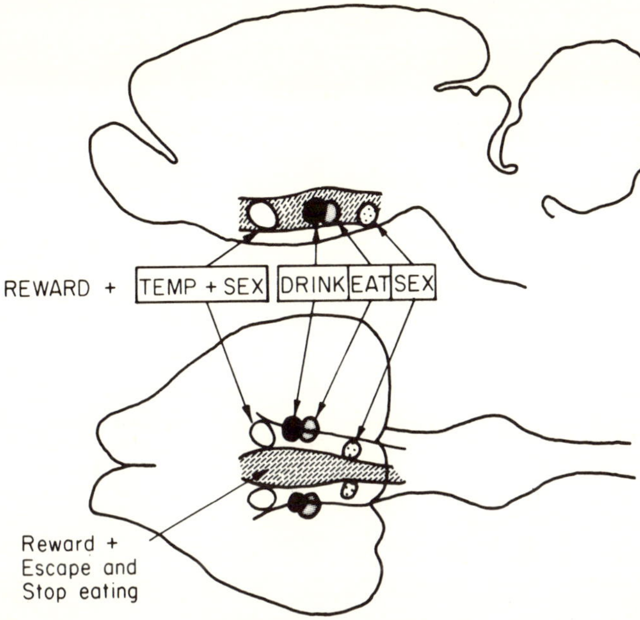

Fig. 4. Map of areas in rat brain yielding instrumental and consummatory behaviors aimed at different drive-object targets.

sides of the feeding and drinking areas, the idea of sharp localizations was rejected. But because this area could be mapped into four successive regions where stimulation caused temperature, drinking, eating, and sex responses, respectively, as most likely, the idea of totally unlocalized drive systems was rejected. Obviously the truth lay somewhere in between.

One feature of the "in between" answer was discovered. The goal objects of the "drives" caused by these stimulations were often changed by training. If the animal was stimulated regularly in the presence of a drive object, after a while the stimulus began to evoke an appropriate drive, that is, one with the available drive object as its target (Fig. 5).

For example, in one test, probes were placed in what originally seemed to be a feeding point (i.e., the stimulation evoked feeding as opposed to drinking in original choice tests). Then with only water present, 30-sec trains of stimulation were applied every 5 min for many days. Under the impetus of this, the feeding point changed. In the end it was a drinking point! This can be called the Valenstein effect after its discoverer (Valenstein *et al.*, 1968).

Because drives were mapped into different areas of the brain to begin with, it seemed wrong that they could be modified by training. One possible answer to this puzzle was that a family of drive neurons might be prerelated to a specific drive by their sensitivity to particular visceral inputs or to particular hormones,

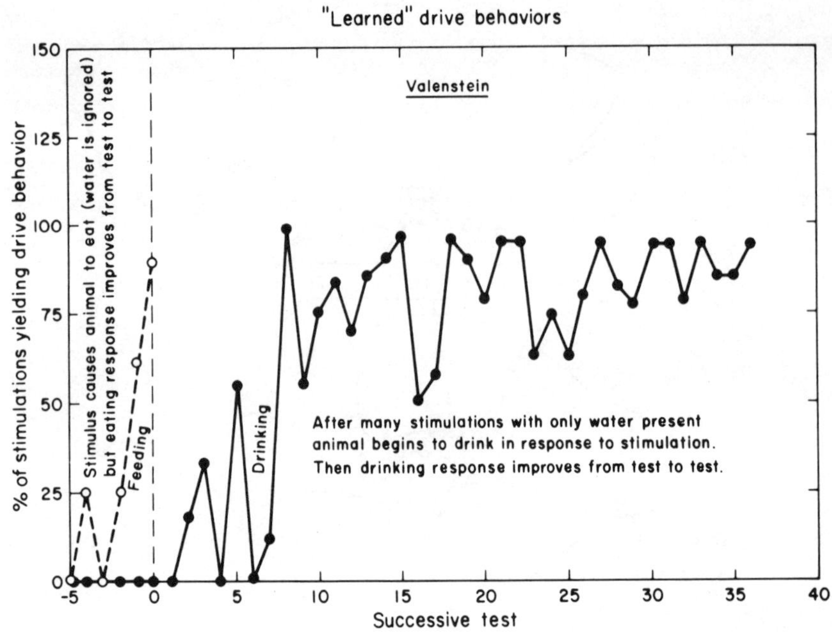

Fig. 5. Valenstein effect: modification of drive-object by training (Valenstein *et al.*, 1968). Prior to "training" the electric brain stimulus caused first little drive behavior and then after some repetition of the stimulus in the presence of food and water the stimulus evoked feeding. Then after a long period of stimulation in the presence of water only, the electric stimulus evoked drinking.

but their axons might become attached during development or learning to appropriate drive objects. The Valenstein effect might be evidence that electric stimulation could pervert this normal learning mechanism. It is possible to assume that the stimulus *was* applied in a "hunger center" but that gradually the training artificially caused the animal to respond as if water were a hunger-drive object.

To recapitulate the picture developed by hypothalamic stimulation—there were far lateral and far medial areas where reward predominated and in-between areas where aversive effects and drive effects were overlapped with reward.

LESIONS

A second body of data came from restricted destruction of small and deep "brain centers." These studies have divided the hypothalamus and its neighbors into a focus where lesions had one kind of effect, and a set of three surrounding areas where different kinds of opposed effects were observed (Fig. 6). Anatomically, the focus was the same lateral hypothalamus where electric stimulation had predominantly positive effects. It included also the boundary regions of the hypothalamus among which the substantia nigra is one we will be most interested in later. Lesions in lateral hypothalamus or along its boundary regions cause the "lateral hypothalamic syndrome," i.e., a loss of positive drive-reward behaviors and other operant behaviors (even ones aimed to avoid noxious stimulation; Teitelbaum and Epstein, 1962).

If, however, animals were kept alive for a few days after these lesions there was often good recovery of some of the reward behaviors. Animals died if not force-fed at first but they recovered in 1 to 3 weeks if kept alive by force feeding or other methods. After recovery the animals were dependent in a surprising way on the cortex for drive behavior (Teitelbaum and Cytawa, 1965). This was shown by use of a damaging manipulation of cortex. The application of KCl causes in normals a 4- to 8-hr period during which all instrumental behavior is abolished (and strange electric activity is recorded from the head). It is thought of as causing a temporary shutdown of cortex. In normals there appeared to be quite full recovery after several hours. In animals with lateral hypothalamic damage but apparently recovered from that, the cortex manipulation had a

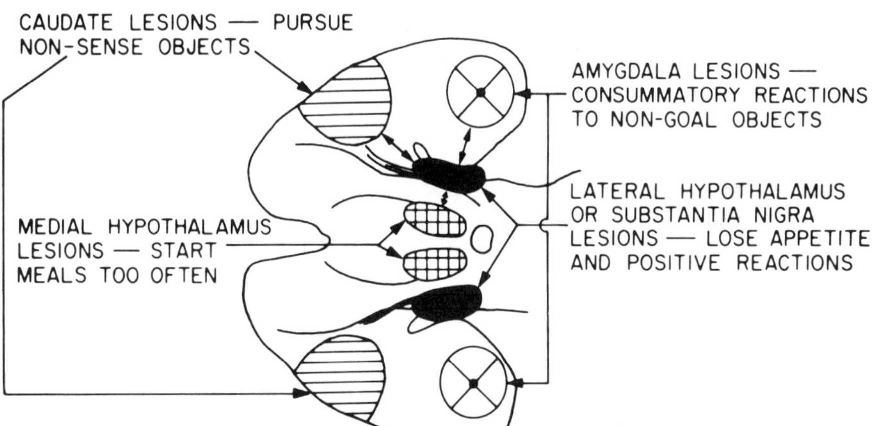

Fig. 6. Map of lesions in and near the hypothalamus affecting targeted instrumental-approach and consummatory behaviors. In the lateral-hypothalamic-substantia nigra focus, lesions halted such behavior temporarily and modified it permanently. In three neighbor areas there were opposed effects. Lesions in medial hypothalamus caused episodes of approach and consummatory behavior aimed at food to occur too frequently. Lesions in the caudate nucleus caused compulsive instrumental behavior aimed at moving "nonsense" objects. Lesions in the amygdaloid region caused attempts to perform consummatory behavior with nongoal objects.

much more devastating effect causing the full 3-week recovery to need redoing. This dependence of drive behavior on cortex after lateral hypothalamic damage apparently showed that recovery was not really complete.

There were other signs pointing in the same direction. The animals after recovery did not respond to cellular water deficits by drinking. They drank to wet their mouths and for a number of reasons that would seem irrelevant (if they did not serve luckily for the animal to keep it alive). Similarly it did not respond to glucose deficits by eating. And it failed to respond appropriately to sodium deficits. However with its repertory of redundant hunger controllers or learned feeding behaviors it got along. It ate well and looked robust. It was possible that a learned cortex repertory of drive behaviors recovered although a hypothalamic originator or starter of these was gone.

Fitting this view, a most important food-learning mechanism was also gone. The animal never more learned to exclude foods on the basis of poisoning or illness. A normal rat responds to foods that preceded illness as if they were aversive. This is called the Garcia effect after its discoverer (Garcia and Ervin, 1968). It may be thought of as the learning of aversive reactions to poisons. This learned aversion was gone after lateral lesions. In normal rats there is also learning of special positive reactions to foods that are correlated with recovery from illness. These learned positive reactions were gone after lateral lesions (Roth *et al.,* 1973). Because these food-learning phenomena matched the drive-target learning of the Valenstein experiments, it seemed doubly likely that the hypothalamus might be involved in the learning of drive targets, that is in the learned attachments of animals to objects.

There were three areas surrounding the lateral hypothalamus where lesions had opposed effects. The first set of opposed lesions was in the medial hypothalamus. These lesions caused the opposite of starvation. The animals ate too often because the beginning of meals was too early (as if no visceral or chemical trigger was needed; Le Magnen *et al.*, 1973).

The second set of opposed lesions was in the caudate nucleus. This is a motor center which is the reciprocal inhibitor of the substantia nigra. Lesions here caused nonsensical instrumental behavior directed at anything that moved (Villablanca, 1974). This looked like the other side of a coin, or at least like the inversion of the *loss* of pursuit behavior that occurred with lateral hypothalamus lesions.

The third set of opposed lesions was in the amygdala. This is an outpost in the olfactory forebrain. These lesions caused consummatory behavior toward dangerous objects or toward untested foods, or toward "wrong" objects. For example, there were attempts to eat or mate with nonfood or nonsex objects (Klüver and Bucy, 1937).

The fact that lesions in a central area stopped reward behaviors and lesions in three surrounding areas caused different kinds of excessive approach behavior suggested a multiple opponent process system: a central positive region in lateral hypothalamus and substantia nigra inhibiting and being inhibited by three surrounding regions. In such a system a shifting balance of excitation and inhibition would determine the presence or absence of approach behavior, and electric stimulation along the communication links might well have double or mixed effects.

Besides these, one other set of lesion studies more specifically related to self-stimulation deserves special mention. Many lesions failed to halt this behavior. Quite dramatic damage ablating all of cortex, paleocortex, and the basal ganglia except for the thalamus failed to halt self-stimulation (Huston and Borbely, 1974). Animals with these lesions behaved grossly, having lost all nuances of behavior. Gross behaviors like rearing on the hindlegs however could still be reinforced. When such behaviors were clearly characterized they could be greatly increased in frequency by electrical brain rewards. And after the reward was withdrawn, the behavior showed no signs of extinguishing leaving the impression that though the telencephalon was not required for operant learning of gross behaviors, it was required for normal extinction. The only way the behavor could be stopped in these animals was to reinforce its opposite. This study has many important implications for questions about why brain reward often extinguishes very rapidly. However, one of its main imports is that at least some of the "reward" neurons do not reside in the front end of the medial forebrain bundle system.

THE BRAIN AMINES

A third body of data came from neurochemical maps showing where certain brain chemicals reside and what neurons and fiber pathways carry them. These maps have seemed to identify a set of brain reward neurons and thus to give the beginnings of an interpretation of brain reward behavior and possibly the beginnings of an explanation of other reward behaviors. Small clumps of neurons in focal centers of the hindbrain, midbrain, and the boundaries of the forebrain send axons broadcast from the focal centers to the farthest reaches (Ungerstedt, 1971*a*). There are several similar clumps and several overlapping sets of broadcast fibers. The small origins and the widely diffusing fibers make these look like command centers that could send YES-NO messages to the whole brain (Fig. 7). The chemical messengers used by these neurons to send signals are noradrenaline, dopamine, and serotonin.

The noradrenaline fibers started farthest back and went farthest forward. They ran from a crossroads of the brain in the medulla to all of its outposts, the cerebellum, thalamus, paleocortex, and neocortex. The serotonin fibers started

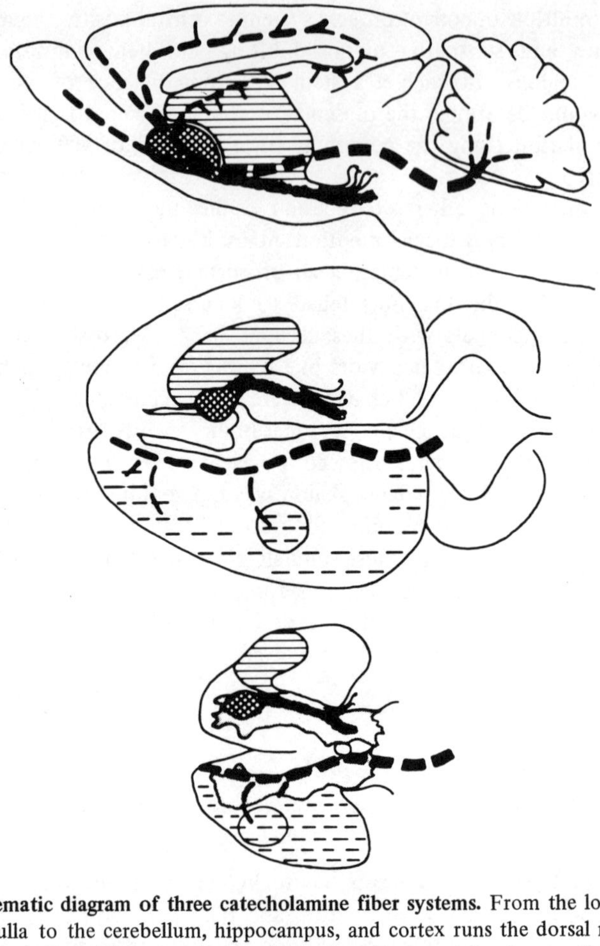

Fig. 7. Schematic diagram of three catecholamine fiber systems. From the locus coeruleus of the medulla to the cerebellum, hippocampus, and cortex runs the dorsal noradrenaline bundle (dashes). From the ventral midbrain to the olfactory tubercle runs the meso-limbic dopamine system (scrambled markings and cross hatch). From the substantia nigra to the caudate nucleus runs the nigro-striatal dopamine system (black lines). It is questionable whether these last two should be separated; and it is likely that at least one of them runs beyond the diagrammed targets because dopamine is found in the cortex. The raphe-paleocortex serotonin system is not shown; it starts between the noradrenaline and the dopamine systems and runs to the paleocortex (and likely also to parts of the neocortex).

in the middle of the midbrain and ran a less well-defined course to many parts of the forebrain. The dopamine fibers started in the front part of the midbrain and back part of the forebrain. They ran a shorter course ending mainly in structures below the cortex, i.e., in parts of the extrapyramidal motor system (which might be the main control system for purposive instrumental behavior) and in some poorly understood centers of the olfactory forebrain. Most likely they did not all end at their main subcortical stations, for dopamine itself was found along with noradrenaline in parts of the cortex.

For all amine fiber systems one property stood out, namely, there was a very small source of origin, and a very wide spread of influence. The noradrenaline, serotonin, and dopamine systems seemed almost like a small triumvirate—three little men deep in the brain making its command decisions.

Chemical studies showed the amines and particularly dopamine and noradrenaline to have special properties. First was the close similarity between them and the common stuff they were made of; more of one might mean less of the other (de la Torre, 1972). Second, they acted by means of a gate or switching chemical that was in common between the two of them and also in common with serotonin and many peptide hormones (Siggins *et al.,* 1969); as if they possibly acted in consort with or instead of peptide hormones. Third, they sometimes seemed to inhibit and excite and at the same time; that is, while they appeared to inhibit neuron activity (Phillis, 1970), this action often seemed to excite the animal. Fourth, there was a slow onset and a long duration of action with a time constant of $\frac{1}{3}$ sec to 1 sec like the minimum episodes of behavior itself (Fig. 8; Segal and Bloom, 1974*a,b*). Fifth, there was a process whereby the neurons involved seemed to pump these neurohumors back up after use as if supplies were scarce (Iverson, 1967). But sixth, there were special counteracting chemicals which provided a negative feedback on available supplies (a planned scarcity—like the Federal Reserve Board). All these fitted the catecholamines for a special function.

While it was by no means clear what functions they mediated, their properties would fit them for controlling behavioral priorities. This is because the repeating theme was competition for a limited resource, and the time constants involved were in the order of magnitude of behavioral episodes rather than of

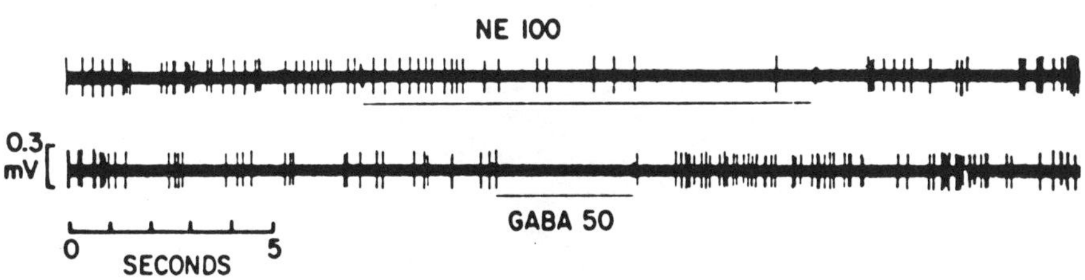

Fig. 8. Slow onset and long duration of the inhibitory response caused by electrophoretic application of noradrenaline in the hippocampus. In the upper trace the horizontal bar denotes the application of noradrenaline; the onset of effect requires more than a second and the effect lasts for several seconds. In the lower trace the bar denotes the application of the inhibitory neurohumor, gamma amino butyric acid; the onset and offset of effect are relatively immediate. (From Segal and Bloom 1974*a.*)

neurophysiological events. Furthermore, they were broadcast through the brain so they could exercise a controlling influence over all of its activities.

Because their properties seemed to suit them for controlling behavioral priorities, it was interesting that experiments linked them to drive and reward systems in strong and intricate ways.

These fibers pervaded the drive-reward systems in such a way as to match the drive-reward maps (German and Bowden, 1974). New maps based on the theory that these were reward neurons showed new rewarding locations that tracked the noradrenaline pathway toward the medulla and the dopamine pathway toward the substantia nigra (Fig. 9; Crow, 1971, 1972*a*; Ritter and Stein, 1973).

While stimulation at the sources of the two different catecholamines both appeared to reward behavior, interesting differences appeared. When stimulation was applied at the source of the noradrenaline system in locus ceruleus, the behavior caused was quiet and paced. When the stimulation was applied instead at one source of the dopamine pathways in the substantia nigra, the behavior was much more frenzied; it appeared highly motivated but also possibly conflicted (Crow, 1972*a,b*).

Furthering the view that these could be reward neurons were drug studies far

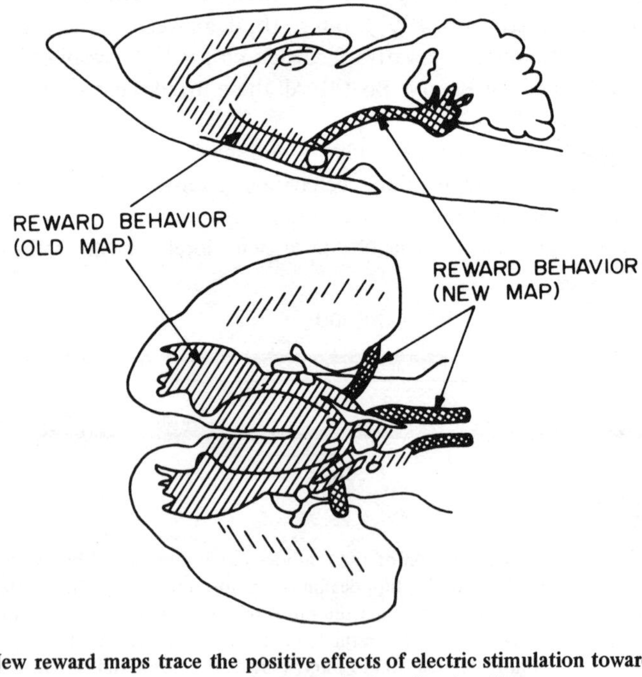

Fig. 9. New reward maps trace the positive effects of electric stimulation toward the origin of the dopamine and noradrenaline bundles.

too numerous to mention. Chemicals which liberated catecholamines from their inactive "capsules" promoted hedonic states in humans and added to self-stimulation behavior in animal experiments. Chemicals which blocked the degradation of catecholamines, and others which blocked inactivation by reuptake added to these positive effects. Chemicals which prevented the formation of catecholamines or blocked their receptors caused depressed conditions in humans and animals, and blocked brain reward behavior. In sum, the drug studies added considerably to the view that catecholamine neurons could be reward neurons.

Another source of support came from studies of lesions and poisons which damaged the CNS and were supposed to have specially damaging effects on catecholamine neurons, either destroying them or their active endings (Fig. 10). When lesions of this kind were placed in the dopamine bundles (or when catecholamine poisons were applied in the ventricles where they presumably had wide effects) the results were to abolish reward and drive behaviors, almost equaling exactly the lateral hypothalamic syndrome referred to earlier (Ungerstedt, 1971*b*; Stricker and Zigmond, 1974). Brain reward behavior was abolished by this kind of ventricular poisoning. In other experiments it was also greatly set back or altogether absent after electrolytic lesions applied to the origin point of the main ascending noradrenaline pathway (M. E. Olds, private communication; S. J. Ellman, private communication). When poison lesions were placed so as to directly affect only the secondary noradrenaline pathway (one directed at the medial hypothalamus) there were quite opposed effects which apparently mimicked the medial lesions that caused animals to overeat (Ahlskog and Hoebel, 1972). It was surprising that this catecholamine lesion had such a different effect from the one in the dopamine bundle, but this was thought to be due to subtle differences in function.

In other experiments (Pickel *et al.,* 1974), electrolytic lesions in the main noradrenaline fiber system caused a great regrowth and proliferation of the damaged fibers (Fig. 11). Catecholamine neurons are thought to be quite special in their tendency to regrow this way. This brought them in quite a different way into accord with the self-stimulation data because the time course of this remarkable regrowth matched well the time course of a well-known (but rarely reported) behavioral change. This was that self-stimulation behavior improved (thresholds declining and rates increasing) for a period of about 2 to 3 weeks after probe implantation. The improvement progressed steadily from the time of surgery whether or not the animals were provided with any stimulation during the 3-week period. The surprising proliferation of CA fibers during 3 weeks after surgery, matching as it did the improvement of self-stimulation (whose probes must have damaged CA fibers), may thus explain what has heretofore been a mystery to me (Olds, 1958) and may forge another important link connecting catecholamine systems to reward behaviors.

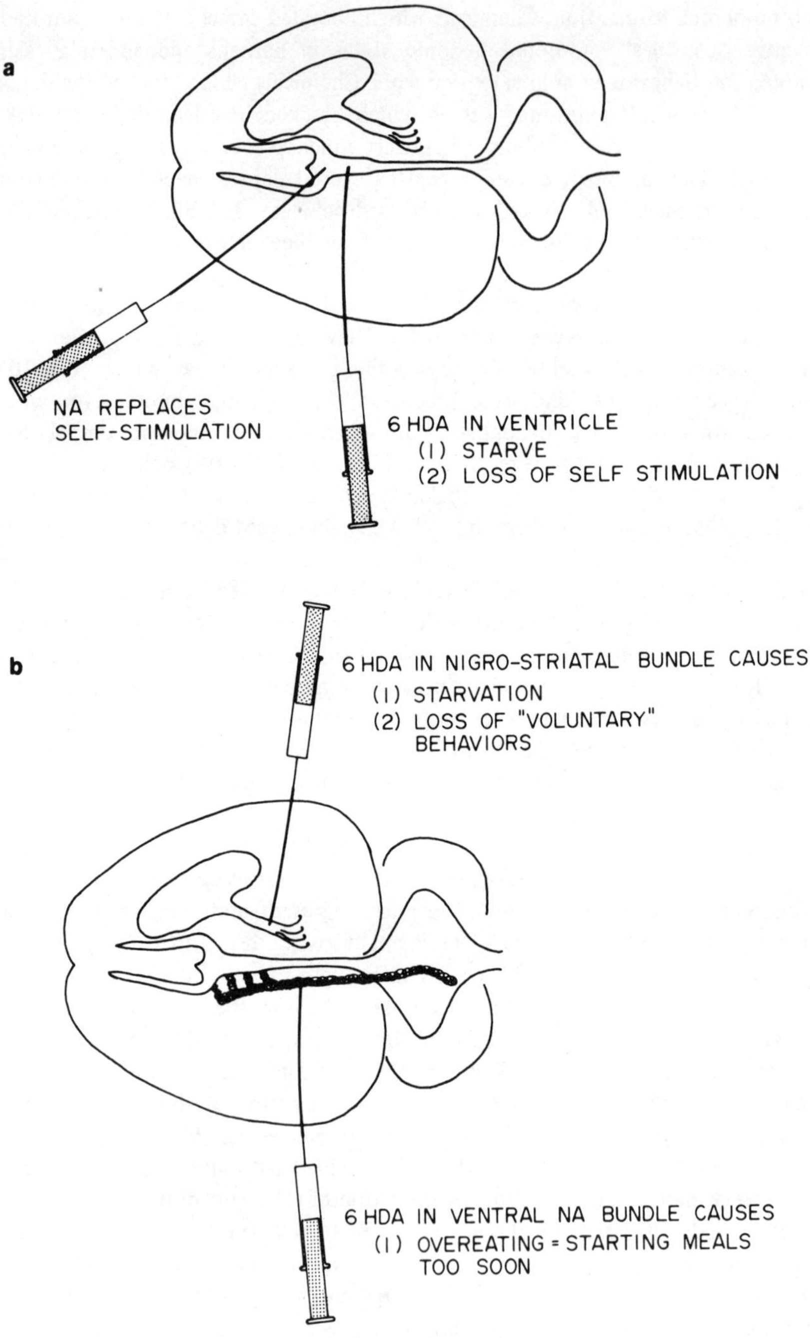
a
NA REPLACES
SELF-STIMULATION
6 HDA IN VENTRICLE
(1) STARVE
(2) LOSS OF SELF STIMULATION
b
6 HDA IN NIGRO-STRIATAL BUNDLE CAUSES
(1) STARVATION
(2) LOSS OF "VOLUNTARY"
BEHAVIORS
6 HDA IN VENTRAL NA BUNDLE CAUSES
(1) OVEREATING = STARTING MEALS
TOO SOON

c

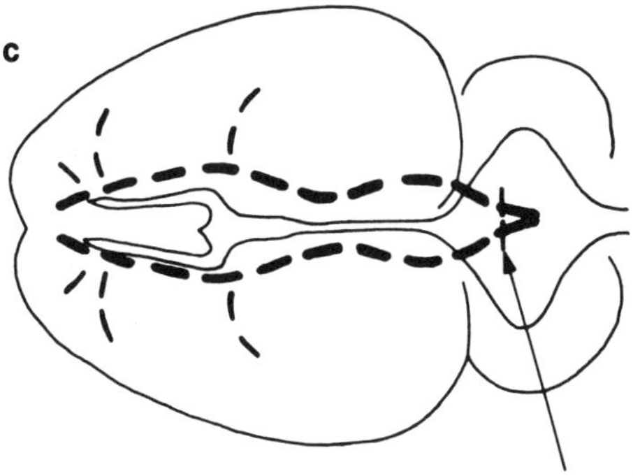

Fig. 10. Effects of the catecholamine neuron poison 6-hydroxydopamine (6-HDA) and of lesions in the locus coeruleus. Applied in the lateral ventricle (a) or in the substantia nigra (b), 6–HDA caused a loss of reward and drive behaviors similar to that of the "lateral hypothalamic syndrome." Ventricular application also caused a loss of self-stimulation behavior which could be at least partially restored by ventricular application of noradrenaline (a). Application of 6-HDA in the ventral noradrenaline bundle (b), which runs from medulla to hypothalamus caused excessive eating, matching what could be called the "medial hypothalamic syndrome." (c) Lesions in the locus coeruleus (the origin of the dorsal noradrenaline bundle which is aimed at the higher centers of the brain) caused a loss of self-stimulation behavior which could be restored at least to some degree by administration of amphetamine.

Even stronger support for the catecholamine theory of reward was the data that direct application of catecholamines into the ventricle had positive effects on brain reward behavior, either restoring it if it had been blocked by drugs or lesions, or promoting it if stimulation was applied at or near threshold levels (Wise *et al.*, 1973).

Many of the experiments that pointed strongly toward dopamine or noradrenaline involving them in brain reward or drive behavior pointed ambiguously at the third important amine, serotonin (Poschel and Ninteman, 1971). Drugs that manipulated this neurohumor could be positive or negative on brain reward behavior depending on other ill-defined aspects of an experiment. Quite different researches were more clear in pointing to an involvement of serotonin in quieting the animal for sleep or suppressing pain (Jouvet, 1974; Yunger and Harvey, 1973). It seemed possible therefore that some self-stimulation aimed at pain reduction might be promoted by serotonergic drugs, while other behavior aimed at producing a euphoric state might well be damped by the same drugs.

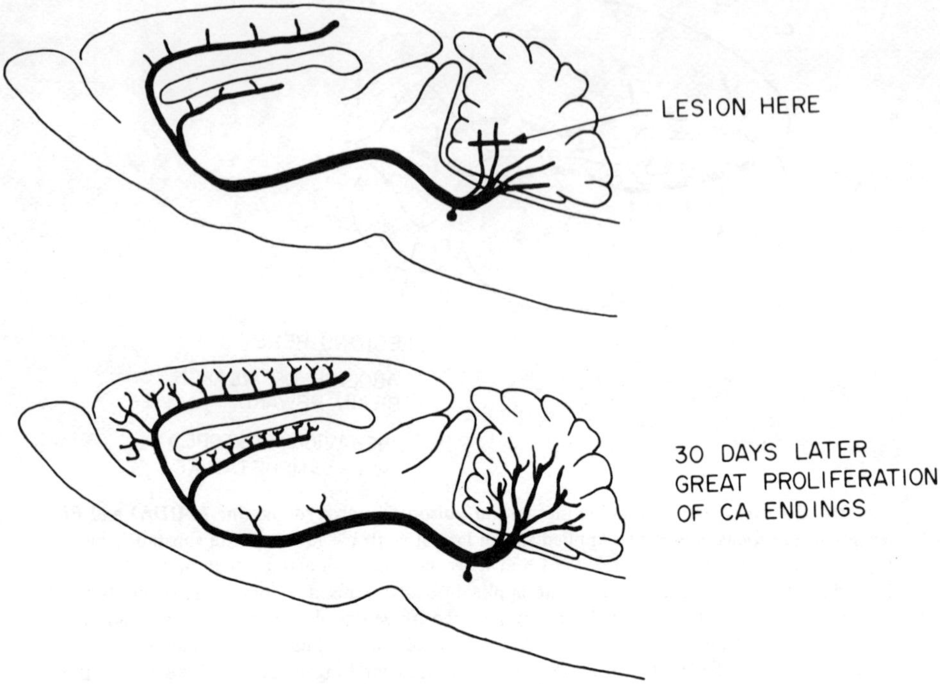

Fig. 11. Regrowth and proliferation of noradrenaline axons after damage to one branch. Lesions cutting half of the bundle from locus coeruleus to cerebellum caused a doubling of the axon terminals from this bundle in cerebellum and a 6-fold increase in endings in other parts of the brain coming from the same source neurons (Pickel *et al.*, 1974).

In any event, serotonin was related to the attenuation of arousal, and the two catecholamines were strongly related to the basic drives and to rewards.

There were some counterarguments. The main flaw in the argument relating catecholamines to brain reward was that many other behaviors besides reward ones were blocked or promoted by the same catecholamine manipulations that had effects on self-stimulation. However, it was argued with some grounds that a reward system might well be a common factor required for all kinds of operant behavior (even that aimed to avoid aversive conditions). Therefore if self-stimulation amounted to tapping directly into this common denominator line, then chemicals blocking it might well block a great deal of operant behavior. A second flaw was technical and unclear. It was that animals appeared to find self-stimulation even more rewarding when a great deal of the catecholamine from the nerve endings had already been released and become active so that each electric stimulation could not have added much to the active pool. A similar problem arose when damage to CA neurons by poisoning or lesions could be restored by administration of amphetamine in the circulation or noradrenaline in the ventricles. If the catecholamine fibers were damaged or ineffective and the electric stimulation was still effective, it might have been stimulating something else. While these arguments were puzzling, they had many sides; they could become damaging to the catecholamine view if we knew much more about the drugs involved than we do. But we do not know enough. Therefore the catecholamine theory of reward is not only viable but substantial.

To recapitulate once again, brain stimulation and lesion studies have shown that a bundle of pathways extending bidirectionally from the medulla to the telencephalon through the lateral hypothalamus probably contains the axons of a set of reward neurons. In self-stimulation experiments, stimulation of these pathways caused reward behavior; in lesion studies of the "lateral hypothalamic syndrome," cutting them suspended it temporarily and modified it permanently. Very large lesions removing neocortex and paleocortex and the basal ganglia except for the thalamus showed that at least some of the "reward neurons" did not have their cell bodies in the front end of this system. Other experiments have given strong but not yet compelling evidence that two or three families of catecholamine-containing neurons with cell bodies at the other end of the system and widely broadcast axons might be the reward neurons. Indole amine neurons were ambiguously involved in these effects and it seemed possible to view them as mainly involved in suppression of arousing and aversive states and only secondarily in reward and punishment.

UNIT RECORDINGS

A fourth set of experiments was aimed at explaining some of the others by recording directly from neurons in and near the lateral hypothalamic centers that seemed from stimulation studies to be reward centers, from lesion studies to be drive centers, and from catecholamine studies to be important targets for all the amine fiber systems.

Neurons near the so-called lateral hypothalamic feeding center and the substantia nigra (origin of main dopamine systems) were recorded in chronic animals during reward behavior. These experiments inidcated a set of resident neurons that were regularly inhibited by rewards. Their firing rate was suppressed or attenuated by three different kinds of rewarding stimulation. In my laboratory, Hamburg (1971) showed attenuation with food reward (Fig. 12), and Ito (1972) showed it with brain reward (Fig. 13). Elsewhere, Kerr *et al.*, (1974) showed it with morphine reward." All of these experiments suggested that these were possibly "drive" neurons that monitored and were inhibited by medial forebrain bundle "reward" axons.

Fig. 12. Silencing of lateral hypothalamic spikes during feeding (Hamburg, 1971). The two traces are recordings from the same probe. Spikes appeared at about 20 per sec in trace No. 1 while the animal was hungry awaiting food. These spikes disappeared in trace No. 2 while the animal was eating. The bursts in 2 are "chewing artifacts" not neuron action potentials. Traces are about 1 sec in duration.

Other data suggested that if these were "drive neurons" in the sense of being prewired on the input side to detectors of physiological drive conditions, they were also amenable to connective changes during training. One kind of evidence pointing this way came from conditioning studies (Linseman and Olds, 1973); these indicated that lateral hypothalamic neurons were often excited after

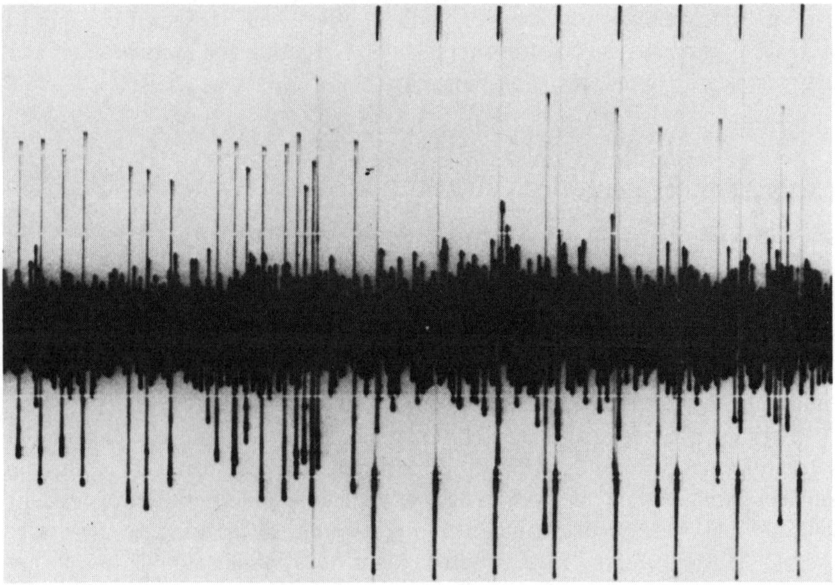

Fig. 13. Silencing of lateral hypothalamic spikes during brain reward stimulation. The very large spikes (going off the figure) in the second half of the trace are shock artifacts from the self-administered train of rewarding brain stimulation. The large spikes in the first half of the trace are action potentials recorded from a large lateral hypothalamic neuron. The interval between shock artifacts is 25 msec (shocks at 40 per sec).

training by food-related conditioned stimuli that were ineffective prior to training (Fig. 14).

THEORIES

Stimulation and lesion studies have suggested the possibility of drive neurons in or near the lateral hypothalamus. The map localizing the different drives to some degree suggested that there were most likely some genetically determined input connections. The conditioned responses recorded from neurons here suggested that there were also some modifiable input connections. The silencing of lateral hypothalamic units by rewards suggested that there were also some inhibitory input connections, possibly coming from reward neurons (Fig. 15). The methods of Ungerstedt suggested that some of the drive neurons might be dopaminergic. The experiments of Crow and Ritter and Stein suggested that some of the reward neurons might be noradrenalinergic.

Might some dopamine neurons also be thought of as reward neurons? Most likely yes, but in a subtle way. Because there were different effects of stimulation and lesions in the paths of the two catecholamines, Crow (1972*b*, 1973) guessed that one of them, noradrenaline, might be involved in those rewards that come toward the end of a consummatory episode and carry the seeds of satiety.

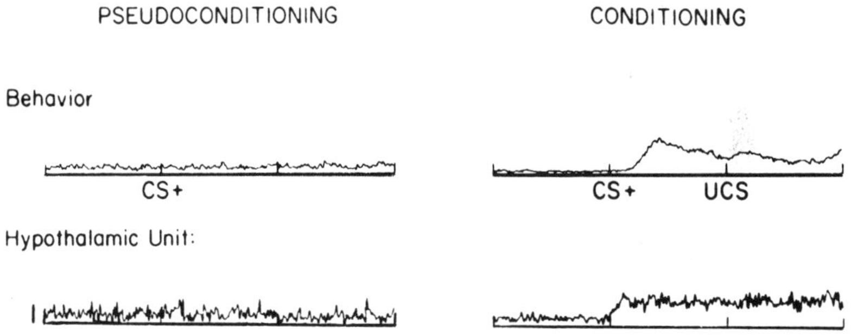

Fig. 14. Behavior and hypothalamic unit responses before and after conditioning. The upper trave portrays the average output of a detector attached to the head that measured head movements in arbitrary units. The lower trace represents the spike frequency of a lateral hypothalamic unit, the vertical bar at the left representing a rate of 5 spikes per second. The traces represent 3 sec. At the end of the first sec a tone (CS+) was started which then continued to the end. During conditioning a pellet dispenser (UCS) was triggered at the end of 2 sec. Prior to conditioning (pseudoconditioning) the tone caused minor changes in the unit and behavior responses. After conditioning it caused behavior changes with a 170-msec latency and unit changes with a 20- to 40-msec latency (Linseman and Olds, 1973).

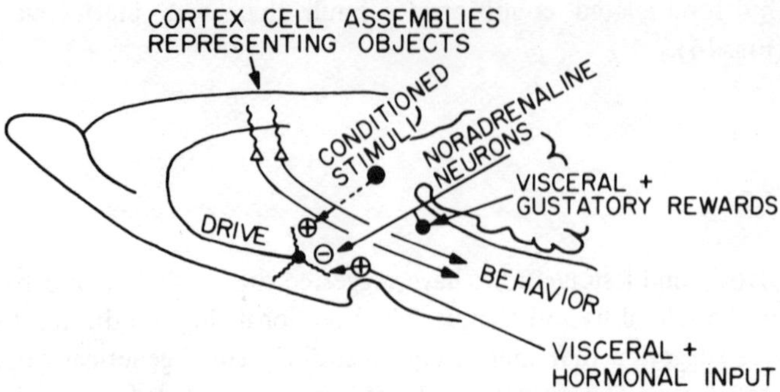

Fig. 15. A new "drive-reduction theory of reward." Noradrenaline neurons from the medulla would be triggered by rewarding gustatory and visceral inputs. They would act to silence drive neurons housed in or near the lateral hypothalamus (some of which could be dopamine neurons). Silencing of the drive neurons would cause them to become coupled to cortex cell assemblies active at the time.

And that the other one, dopamine, might be involved in those rewards that come at the beginning of a consummatory episode or in the promising phases of an instrumental one which are therefore involved in getting things going. This good idea seems equally interesting if modified slightly to suggest that dopamine neurons might be some kind of learned drive neurons involved in energizing the positive travail on the pathway toward a reward cache, and the noradrenaline ones might be inhibitors of these that mediate a change from operant to respondent behavior when a hoard is at hand.

Crow's argument fits because self-stimulation was slow and paced in a mainly noradrenaline region, and frenzied in a dopamine one. Actually the frenzied behavior occurred when probes were in areas populated by both catecholamines, and there was a third step in the series: frenzied and highly conflicted self-stimulation was regularly observed when probes were in a third region where histochemistry pointed to an intersection of acetylcholine and noradrenaline systems. The three-step series has led me to suppose that the concept of drive like that of reward should have at least two parts. In this case, one would be an incentive part identical with the incentive part of reward (i.e., neurons that would become active during promising conditions and motivate operant behavior). The other would be an alarm part (i.e., neurons that would become active when supplies were dangerously low). If so, the alarm kind would be inhibited by reward elements insofar as they signaled the end of the danger. This might be correlated with an acetylcholinergic system countered by a noradrenaline one. The incentive kind would also be inhibited by rewards insofar as they triggered consummatory responses so that operant behavior was suppressed. This might be correlated with a dopamine system countered by a noradrenaline one. In other words, two different kinds of drive neurons in the hypothalamus could be countered for different reasons by differently functioning noradrenaline neurons.

The opponent processes interaction between the noradrenaline systems and the drive neurons might well be the main target of lesions cutting between medial and lateral hypothalamus. Such lesions would leave drive or incentive motivation to overrun. The other two lesions affecting approach behavior (in the caudate nucleus and in the amygdala) had to do with a different problem. This was the channeling of drives onto specific targets.

The experiments of Garcia and Valenstein pointed to changeable targets of brain-stimulated drives and also of naturally occurring ones. The locus of Valenstein's probes and of lesions countering Garcia's effects pointed to the lateral hypothalamic region. One interpretation was that the same drive or learned drive neurons that we have been talking about had not only some modifiability on the input side but also some at the other end. That is, they might have changed output connections depending on good and bad aftereffects of consummatory behavior.

The "law of learning" for this changing of drive targets might be that the drive axons would become attached to basal ganglia and cortex cell assemblies active at the time of drive inhibition. The long-imagined hypothalamo-cortex axons to mediate this have recently been amply demonstrated with horseradish peroxidase (Fig. 16; Kievit and Kuypers, 1975).

Thus the possibility exists of drive neurons excited by visceral inputs and hormones and by conditioned signals, silenced by rewards, and connected during development or learning to basal ganglia and cortex cell assemblies active at the time of their "silencing" (Fig. 15).

The widespread ramifications of the noradrenaline axons suggested that, if these were reward neurons, there must be other functions besides inhibiting lateral hypothalamic neurons. This brings me to my final set of points. There are several different ways that reward neurons should be involved in more specialized behavioral steering mechanisms.

At least three different ways would be appropriate to different levels of CNS organization (see Table 1). At the level of motor skills, reward might directly cause connections to form between a warp of sensory fibers and a woof of motor dendrites. This kind of process would fit into the cerebellum where the parallel fibers form a warp and the Purkinje dendrites form a woof (see Fig. 17).

At a second level there would be sequential memory as when a rat runs a maze or a human remembers how he got somewhere and takes the same route a second time. In these cases, recordings would be made of successive events on the first occurrence of a behavior sequence. These would have some characteristics of a movie film, or a magnetic tape recording; but a better metaphor would be the sequential memory locations in a computer. Successive small episodes with their sensory, motor, and reward components would be recorded in

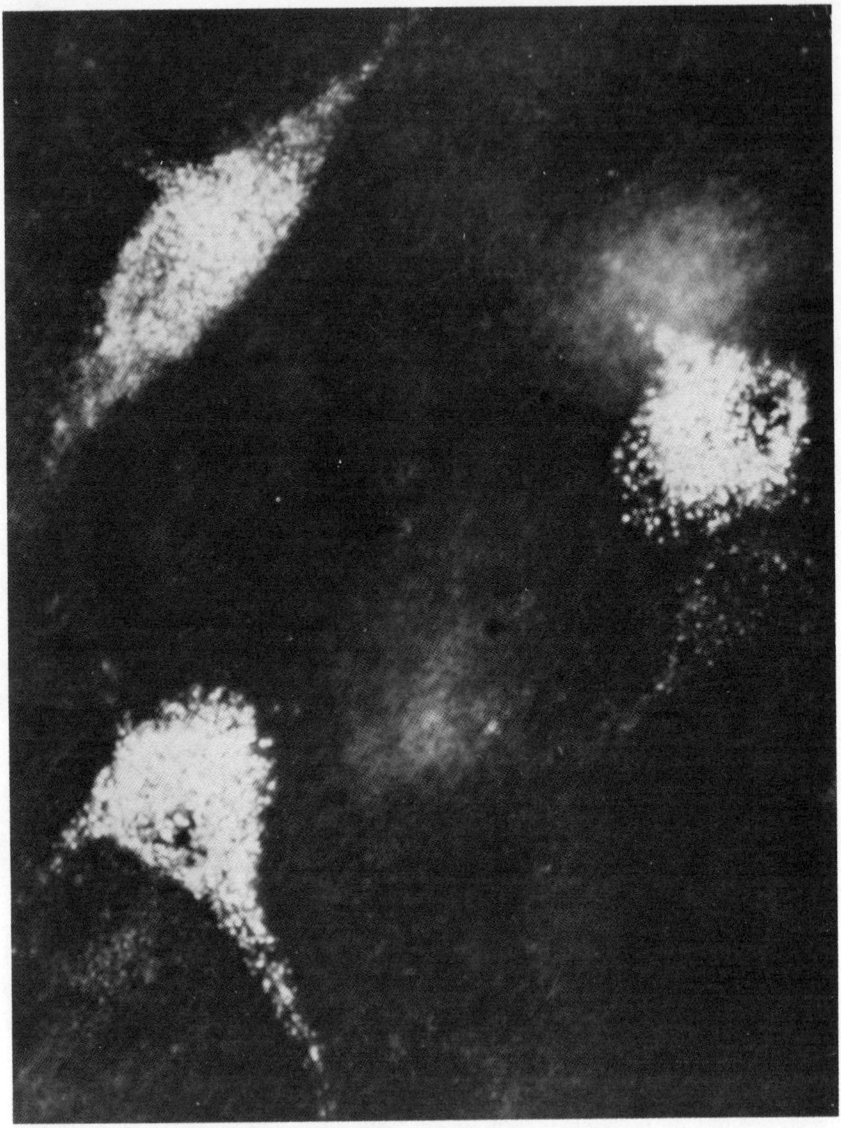

Fig. 16. Drive neurons? These are neurons in or near the lateral hypothalamus stained with horseradish peroxidase that was injected into the cortex giving direct evidence for neurons ascending from this area to the cortex. (From Kievit and Kuypers, 1975.)

Table 1

Function	Possible anatomical correlate	Mechanism	Role of reward
Learning motor skills	Cerebellum	Sensory-motor connections	"Stamp-in" connections
Learning behavior sequences	Hippocampus	Convert temporal sequences to spatial memory arrays	Cause rewarded behaviors to recur
Learning maps and objects	Cortex	Make behavior-topographic maps of external objects	Cause object to be pursued

sequential memory addresses. Later, near-matching sensory inputs would rearouse a memory and this would rearouse its successors. A "dry run" (without behavior) would occur first. If an appropriate reward memory was discovered among the near successors, this would cause a vector at least toward expression of some of the recorded behaviors. From observations in the hippocampus of highly organized axonal systems that run like the digit lines of a computer

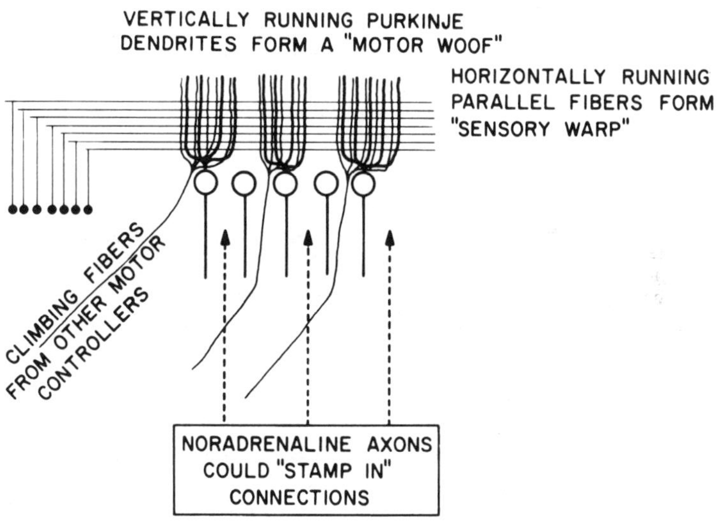

Fig. 17. A motor-skill learning mechanism that could be housed in the cerebellum. Each output pattern would be represented by a pattern of Purkinje cell activity; climbing fibers might well force Purkinje cells into appropriate patterns when behavior was driven from other sources. The sensory field would be represented by a pattern of parallel fiber activity. Each sensory-motor correlation would leave a very temporary trace (at the parallel-to-Purkinje synapses of the active elements). Reinforcement through the noradrenaline inputs from locus ceruleus might stamp in or prolong this trace.

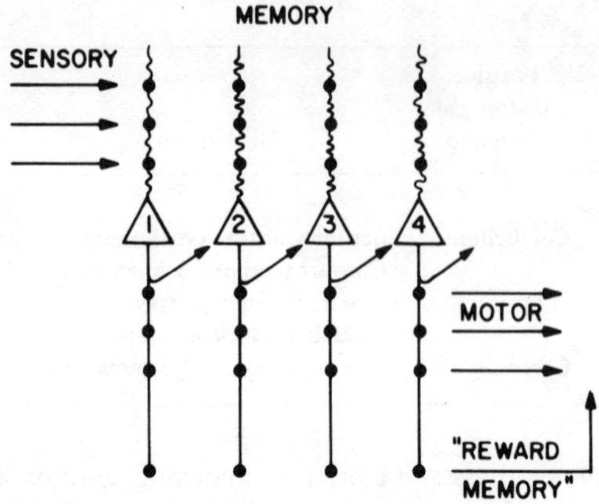

Fig. 18. A sequential recording mechanism that could be housed in the hippocampus. Each episode (170–300 msec) would be recorded on a set of memory elements, e.g., the CA3 pyramids. The sensory side of the episode would be coded in the activity pattern of a major input path, e.g., the perforant pathway or the pathway from cingulate cortex. The motor side of the episode would be coded in a set of dendrites crossed by all memory axons (as the CA1 dendrites are crossed by Schaeffer collaterals). The reward side would be coded in other targets of CA3 axons), e.g., targets of the descending fornix. During recording the memory elements would be successively activated (possibly by the theta rhythm) and they would become coupled by a principle of coincident axo-dendritic firing. Later near-matching sensory inputs would rearouse memory elements and these would arouse their memory successors. If a near successor was coupled to a "reward-output" this would have some tendency to release the other behaviors.

core-memory through oriented dendrite systems that look like the memory lines of such a device, it has seemed possible that this kind of process might occur there (Fig. 18).

At a third level, there would be representation of objects and object-arrangements. These would be sensory-motor or cognitive maps, with control elements at any one node pointing toward a behavior with one hand and toward a second node (the representation of the expected outcome) with the other. These control elements could be the layer 5 pyramidal cells and the nodes could be the columns of neocortex. Motive cells at a second node would need to "point back" to motivate the control elements pointing toward them. The motive cells could be other pyramidal cells or cells of a different type (Fig. 19). Reward in this kind of system would serve to "charge" in some way the motive elements of those columns or cell assemblies which were active at the time of the rewarding events. This might be the same function indicated earlier by saying drive neurons might become attached to cell assemblies active at the time of their inhibition.

The suggestions contained here, therefore, are that there may be reward

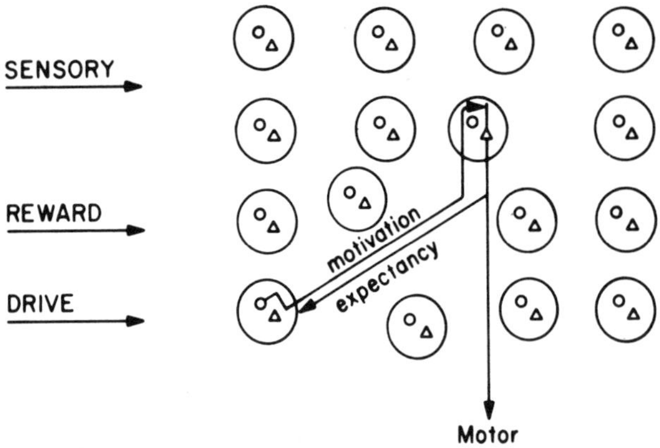

Fig. 19. A behavioral–topographic coding mechanism that could be housed in the cortex. Objects and object-arrangements could be represented by links between feature-detecting or object-detecting columns. The motor outputs, e.g., layer 5 pyramids, from a column would point to a behavior with a descending axon and to another column with an axon collateral; this would be to arouse an expectancy while initiating a behavior. Motive elements in the other column would point back to motivate the layer 5 element. The likelihood of the behavior would thus depend partly on the amount of motivational "charge" characterizing the motive elements in the second column. Rewarding the animal would increase the charge of active columns (possibly by connecting drive neurons to them).

neurons, and these may have four functions: (1) to inhibit drive neurons in the lateral hypothalamus, (2) to stamp in sensory-motor connections in the cerebellum, (3) to plant emotional codes on certain sequential memories in the hippocampus, and (4) to charge (or connect drives to) cell assemblies in neocortex which were active at the time of the reward. It was also suggested that catecholamine neurons might be the reward neurons. If one neurohumor were involved in these disparate functions, it would not be the first time that nature has pressed an active biochemical into multiple roles.

REFERENCES

Ahlskog, J. E., and Hoebel, B. G. (1972). Overeating and obesity from damage to a noradrenergic system in the brain. *Science 182,* 17–27.

Bishop, M. P., Elder, S. T., and Heath, R. G. (1963). Intracranial self-stimulation in man. *Science, 140,*394–396.

Bursten, B., and Delgado, J. M. R. (1958). Positive reinforcement induced by intracranial stimulation in the monkey. *J. Comp. Physiol. Psychol. 51,* 6–10.

Crow, T. J. (1971). The relation between electrical self-stimulation sites and catecholamine-containing neurones in the rat mesencephalon. *Experientia 27,* 662.

Crow, T. J. (1972*a*). A map of the rat mesencephalon for electrical self-stimulation. *Brain Res. 36,* 265–273.

Crow, T. J. (1972*b*). Catecholamine-containing neurones and electrical self-stimulation: 1. A review of some data. *Psychol Med. 2,* 414–421.

Crow, J. T. (1973). Catecholamine-containing neurones and electrical self-stimulation: 2. A theoretical interpretation and some psychiatric implications. *Psychol. Med. 3,* 66–73.

de la Torre, J. C. (1972). *Dynamics of Brain Monoamines.* New York: Plenum.

Garcia, J., and Ervin, F. R. (1968). Gustatory-visceral and telereceptor-cutaneous conditioning–Adaptation in internal and external milieus. *Commun. Behav. Biol., Part A, 1,* 389–415.

German, D. C., and Bowden, D. M. (1974). Catecholamine systems as the neural substrate for intracranial self-stimulation: a hypothesis. *Brain Res. 73,* 381–419.

Hamburg, M. D. (1971). Hypothalamic unit activity and eating behavior. *Am. J. Physiol. 220,* 980–985.

Herberg, L. J. (1963). Seminal ejaculation following positively reinforcing electrical stimulation of the rat hypothalamus. *J. Comp. Physiol. Psychol. 56*, 679–685.

Huston, J. P., and Borbely, A. A. (1974). The thalamic rat: general behavior, operant learning with rewarding hypothalamic stimulation, and effects of amphetamine. *Physiol. Behav. 12,* 433–448.

Ito, M. (1972). Excitability of medial forebrain bundle neurons during self-stimulating behavior. *J. Neurophysiol. 35,* 652–664.

Iversen, L. L. (1967). *The Uptake and Storage of Noradrenaline in Sympathetic Nerves.* Cambridge, England: The University Press.

Jouvet, M. (1974). Monoaminergic regulation of the sleep-waking cycle in the cat. In: *The Neurosciences, Third Study Program,* F. O. Schmitt and F. G. Worden (Eds.). MIT Press, Cambridge, Mass., pp. 49–508.

Kerr, F. W., Triplett, J. N., and Beeler, G. W. (1974). Reciprocal (push-pull) effects of morphine on single units in the ventromedian and lateral hypothalamus and influences on other nuclei; with a comment on methadone effects during withdrawal from morphine. *Brain Res. 74,* 81–103.

Kievit, J., and Kuypers, H. G. J. M. (1975). Basal forebrain and hypothalamic connections to frontal and parietal cortex in the rhesus monkey. *Science 187,* 660–662.

Klüver, H., and Bucy, P. C. (1937). Psychic blindness and other symptoms following bilateral temporal lobectomy in rhesus monkey. *Am. J. Physiol. 119,* 352–353.

Le Magnen, J., Devos, M., Gaudilliere, J.-P., Louis-Sylvestre, J., and Tallon, S. (1973). Role of a lipostatic mechanism in regulation by feeding of energy balance in rats. *J. Comp. Physiol. Psychol. 84,* 1–23.

Linseman, M. A., and Olds, J. (1973). Activity changes in rat hypothalamus, preoptic area and striatum associated with Pavlovian conditioning. *J. Neurophysiol. 36,*1038–1050.

Olds, J. (1958). Discussion. In *CIBA Foundation Symposium on the Neurological Basis of Behaviour.* G. E. W. Wolstenholme and C. M. O'Connor (Eds.). London: Churchill, p. 89.

Olds, J. (1962). Hypothalamic substrates of reward. *Physiol. Rev. 42*, 554–604.

Olds, M. E., and Olds, J. (1962). Approach–escape interactions in rat brain. *Am. J. Physiol. 203*, 803–810.

Olds, M. E., and Olds, J. (1963). Approach–avoidance analysis of rat diencephalon. *J. Comp. Neurol. 120*, 259–295.

Phillis, J. W. (1970). *The Pharmacology of Synapses.* New York: Pergamon Press.

Pickel, V. M., Segal, M., and Bloom, F. E. (1974). Axonal proliferation following lesions of cerebellar peduncles. A combined fluorescence microscopic and radioautographic study. *J. Comp. Neurol. 155,* 43–60.

Poschel, B. P. H., and Ninteman, F. W. (1971). Intracranial reward and the forebrain's serotonergic mechanism: Studies employing para-chlorophenylalanine and para-chloroamphetamine. *Physiol. Behav.* 7,39–46.

Ritter, S., and Stein, L. (1973). Self-stimulation of noradrenergic cell group (A6) in locus coeruleus of rats. *J. Comp. Physiol. Psychol. 85,* 443–452.

Roberts, W. W., Steinberg, M. L., and Means, L. W. (1967). Hypothalamic mechanisms for sexual, aggressive, and other motivational behaviors in the opossum, *Didelphis virginiana. J. Comp. Physiol. Psychol. 64,* 1–15.

Roberts, W. W. (1958). Both rewarding and punishing effects from stimulation of posterior hypothalamus of cat with same electrode at same intensity. *J. Comp. Phychol. 51,* 400–407.

Roth, S. R., Schwartz, M., and Teitelbaum, P. (1973). Failure of recovered lateral hypothalamic rats to learn specific food aversions. *J. Comp. Physiol. Psychol. 83,* 184–197.

Segal, M., and Bloom, F. (1974*a*). The action of norepinephrine in the rat hippocampus. I. Iontophoretic studies. *Brain Res. 72,* 79–97.

Segal, M., and Bloom, F. (1974*b*). The action of norepinephrine in the rat hippocampus. II. Activation of the input pathway. *Brain Res. 72,* 99–114.

Siggins, G. R., Hoffer, B. J., and Bloom, F. E. (1969). Cyclic adenosine monophosphate: possible mediator for norepinephrine effects on cerebellar Purkinje cells. *Science, 165,* 1018–1020.

Stricker, E. M., and Zigmond, M. J. (1974). Effects on homeostasis of intraventricular injections of 6-hydroxydopamine in rats. *J. Comp. Physiol. Psychol. 86,* 973–994.

Teitelbaum, P., and Cytawa, J. (1965). Spreading depression and recovery from lateral hypothalamic damage. *Science 147,* 61–63.

Teitelbaum, P., and Epstein, A. N. (1962). The lateral hypothalamic syndrome: recovery of feeding and drinking after lateral hypothalamic lesions. *Psychol. Rev. 69,* 74–90.

Ungerstedt, U. (1971*a*). Stereotaxic mapping of the monoamine pathways in the rat brain. *Acta Physiol. Scand. (Suppl. 367),* 1–48.

Ungerstedt, U. (1971*b*). Aphagia and adipsia after 6-hydroxydopamine induced degeneration of the nigro-striatal dopamine system. *Acta. Physiol. Scand. (Suppl. 367),* 95–122.

Valenstein, E. S., Cox, V. C., and Kakolewski, J. W. (1968). Modification of motivated behavior elicited by electrical stimulation of the hypothalamus. *Science 157,* 552–554.

Valenstein, E. S., Cox, V. C., and Kakolewski, J. W. (1970). Reexamination of the role of the hypothalamus in motivation. *Psychol. Rev. 77*:16–31.

Villablanca, J. (1974). Presentation of films of kittens and cats with bilateral ablations of the caudate nuclei. Conference on Brain Mechanisms in Mental Retardation. Oxnard, California, Jan. 13–16. (Sponsored jointly by MRRC of UCLA and NICHHD, Washington, D. C.)

Wilkinson, H. W., and Peele, T. L. (1963). Intracranial self-stimulation in cats. *J. Comp. Neurol. 121,* 425–440.

Wise, C. D., Berger, B. D., and Stein, L. (1973). Evidence of alpha-noradrenergic reward receptors and serotonergic punishment receptors in the rat brain. *Biol. Psychiatry 6,* 3–21.

Yunger, L. M., and Harvey, J. A. (1973). Effect of lesions in the medial forebrain bundle on three measures of pain sensitivity and noise-elicited startle. *J. Comp. Physiol. Psychol. 83*, 173–183.

Constitutional Differences in Physiologic Adaptation to Stress and Distress

SAMUEL A. CORSON
and ELIZABETH O'LEARY CORSON

DELINEATION OF THE PROBLEM

Because of the widespread confusion about the definition of the term "stress," we shall use the definition as stated by Selye in his latest writings: "Stress is the non-specific response of the body to any demand made upon it" (Selye, 1971). In this sense, stress represents a reaction of a living organism to any stimuli which would tend to disturb the homeokinetic state or which would tend to satisfy the particular drives (needs) of the organism at a particular time.

The aim of this paper is to point out that in addition to these nonspecific Selye-type stress responses, there are *specific patterns* of physiologic and behavioral reactions, i.e.:

1. The response patterns to biologically positive unconditional or conditional stimuli (e.g., conditional responses to alimentary reinforcement) are different from those evoked by biologically negative (aversive) stimuli (e.g., conditional defense reactions).

SAMUEL A. CORSON and ELIZABETH O'LEARY CORSON · Laboratory of Cerebrovisceral Physiology, Division of Behavioral and Neurobiological Sciences, Departments of Psychiatry and Biophysics, The Ohio State University, Columbus, Ohio.

2. There are marked constitutional differences in the stress reactions of organisms, particularly in regard to aversive stimuli.
3. These constitutional differences become especially apparent in *longitudinal studies,* particularly on exposure to chronic or repetitive aversive stimuli.
4. The stress reactions of organisms to unavoidable aversive stimuli are different from those resulting from exposures to avoidable aversive stimuli, that is, when the organisms can develop control over the aversive situations. It is the exposure to unavoidable (uncontrollable) aversive stimuli that leads to *distress* in susceptible individuals.
5. Research designs utilizing a systems approach are more productive in elucidating specific adaptive psychophysiologic and behavioral patterns than are reductionist methods of recording isolated or single parameters.

WHY A SYSTEMS APPROACH?

A systems approach in psychobiology is suggested by the recognition of the inadequacy of the reductionist-mechanistic approach in the elucidation of the nature of psychophysiologic integration and adaptation. For many years physicians and researchers have been aware in a general way of the importance of integrative physiology and the need to consider the patient as a whole. However, in the actual design of experiments, many researchers and clinicians have generally focused their attention on, and drawn conclusions from, measurements of a few isolated variables (often of just a single parameter) without due regard to the place of these variables in the total integrative adaptive activity of the organism.

Recognition of the inadequacy of reductionism does not imply disparagement of the fractional-analytic approach per se in psychobiologic and biomedical research. Such investigations have made, and will continue to make, important contributions to behavioral and biological sciences. The argument is not to eliminate reductionist-analytic studies. Rather, it is to assert that the functional and behavioral characteristics of biologic systems cannot be understood merely by studying individual isolated components of the living machinery or by limiting our observations to a particular parameter.

The reductionist approach is based on the assumption that any given phenomenon or event can be considered as a closed system involving linear cause-effect relationships. The main trouble with reductionist science is the failure to appreciate the limitations of reductionism and the failure to recognize the fact that in nature, and particularly in living systems, we are dealing with open systems involving nonlinear complex patterns of interrelationships.

The predominance of reductionism in biomedical and psychologic sciences has led to some serious philosophic blind alleys and therapeutic errors and has contributed to development of iatrogenic disorders. In the behavioral sciences, reductionism reached the ultimate reductio ad absurdum by a failure to distinguish between the needs of relatively homogeneous inbred, captive laboratory rats and pigeons and the aspirations, dreams, and strivings of the rich heterogeneous tapestry of mankind.

The essence of the systems approach can perhaps be best represented by a pithy quotation from P. A. Weiss (1969, p. 7): "We are concerned with *living organisms,* and for those, we can assert definitely, on the basis of empirical investigation, that the mere reversal of our prior analytic dissection of the Universe by *putting the pieces together* again, whether in reality or just in our minds, *can yield no complete explanation of the behaviour of even the most elementary living system.*"

EXPERIMENTAL DATA

In this brief presentation, we shall try to illustrate the usefulness of the systems approach in psychobiology by offering a review and critical analysis of some of our experiments involving conditional renal responses and the effects of emotional stress on renal function.

An association of water and electrolyte disturbances with emotional factors has been reported by many clinicians. Since we recently reviewed this literature (Corson and Corson, 1969, 1971), we shall mention here only a few representative studies.

Probably one of the oldest and best-known clinical syndromes, premenstrual tension, has long been coupled with the occurrence of edema (Frank, 1931; Thomas, 1933; Thorn *et al.*, 1938). Schottstaedt *et al.* (1955), Coppen (1965), Crammer (1959), and Jenner *et al.* (1967) reported water and electrolyte disturbances in patients with affective and behavioral disorders.

There are many references in the literature to the effect that higher organisms, including man, often respond with antidiuresis to noxious or emotionally aversive stimuli. The reports regarding the mechanisms involved in these antidiuretic reactions have been rather confusing. Some authors claimed that these antidiuretic responses are due to the release of vasopressin, the pituitary antidiuretic hormone (ADH). Others ascribed the antidiuresis to changes in renal hemodynamics. The problem was further compounded by reports of *diuretic* reactions to emotional stress in human subjects (Miles, 1953; Miles *et al.*, 1952).

Pavlovian conditioning experiments on several breeds of dogs in our laboratory (Corson, 1966*a,b*; Corson and Corson, 1966, 1971) resolved these contradictory reports by demonstrating that there are marked and stable constitutional differences in renal responses to psychologic stressors.

Our experimental design involved studying sequentially the same dogs for prolonged periods in three distinct environments maintained at a temperature of 21–23°C:

1. A neutral control room where baseline data were recorded.
2. An aversive room where Pavlovian motor defense reflexes were developed by reinforcing neutral tones with unavoidable electrocutaneous stimuli.
3. An operant conditioning room where similar neutral tones were reinforced with electrocutaneous stimuli but where dogs were permitted to develop discriminated avoidance responses.

It *was essential to have a different experimenter involved in each of the experimental rooms* in order to secure reproducible data.

We observed three types of reactions to the aversive Pavlovian room:

1. Some dogs (some border collies, wirehair fox terriers, cocker spaniels, and German shepherd dogs) exhibit marked and persistent antidiuretic responses with high urine osmolality, suggesting the involvement of vasopressin. Subsequently we demonstrated that the antidiuresis was due primarily to vasopressin release.
2. Other dogs (some beagles and other hounds) exhibit only temporary antidiuretic responses which are occasioned chiefly by decreased glomerular filtration. After several conditioning sessions, these antidiuretic responses disappear. We referred to these dogs as the pseudoextinction type (see Corson, 1966*a,b*).
3. Some dogs (some beagles and other hounds) may never show any antidiuretic responses to the aversive room. Thus, the contradictory reports in the literature are most probably due to the fact that the authors were dealing with constitutional differences but were not aware of this.

For purposes of identification we referred to the dogs in category 1 as *antidiuretic* dogs (AD dogs), whereas categories 2 and 3 we called *diuretic* dogs (D dogs).

We were able to resolve the above-mentioned problem by avoiding the pitfalls of reductionist behaviorism, which neglects genetic psychobiologic differences and assumes the existence in all organisms of a mythical standard behavioral machinery which responds in a uniform manner to psychologic stressors. It should also be emphasized that the constitutional differences in renal responses to emotional stressors were brought out primarily in *longitudinal studies* involving *repeated exposure to psychologic stressors.*

What remained perplexing to us and to other investigators was what is the function of antidiuresis in the adaptation of higher organisms to emotional stressors? Moreover, if the antidiuresis does serve some adaptive function, then how do the D types manage adaptation to psychologic stressors without including an antidiuretic component?

The answers to these questions came from a systems approach investigation involving simultaneous recording of several parameters in an attempt to assess the total integrative physiologic pattern of adaptation to repeated exposure to psychologic stressors.

Our methods have been described in previous publications (Corson, 1966*a,b*; Corson, 1971), so that it will suffice in this presentation to summarize briefly our data and conclusions.

An intimation of some of the answers came from analyzing the respiratory and other physiological responses of the two types of dogs to psychologic stressors. Figure 1 illustrates such responses in an AD dog, Harry, and a D dog, David. Figure 2 shows the reactions exhibited by an AD dog, Penny, and a D dog, Nicky. It turned out that the AD dogs exhibited intense and persistent polypnea. On the other hand, the D dogs under the same experimental conditions failed to exhibit any evidence of panting.

This intense polypnea in the AD dogs appeared as soon as the animals were brought into the aversive conditioning room and persisted throughout the entire experiment. *No panting was observed when these dogs were repeatedly studied in the control room where the animals were never exposed to emotional stressors.* It must be emphasized that the ambient temperature in the conditioning and in the control rooms was maintained within a thermoneutral range of 21–23°C.

Further examination of Figs. 1 and 2 reveals that the AD dogs in the aversive conditioning room exhibit also persistent tachycardia and copious salivation, whereas in the control room there is no salivation and no increase in heart rate. In contrast, the D dogs exhibit no salivation or tachycardia in either of the rooms. The salivation of the AD dogs in the conditioning room has no relationship to conditional alimentary reflexes, since no food reinforcement was used with these dogs.

This tetrad of responses (antidiuresis, polypnea, salivation, and tachycardia) to the conditioning room suggested to us that the aversive environment initiated a shift of adaptive physiologic mechanisms from those maintaining osmotic homeostasis to those required for the maintenance of thermal homeostasis. It must be remembered that our experimental dogs received a large water load (25 ml/kg) at the beginning of each experiment. Under normal conditions the resulting dilution of body fluids would stimulate the hypothalamo-hypophysial osmoreceptors to inhibit the release of antidiuretic hormone and thus lead to a diuresis.

Our AD dogs behave in the conditioning room as though they were engaged in intense muscular work, comparable to Cannon's description of the "fight or

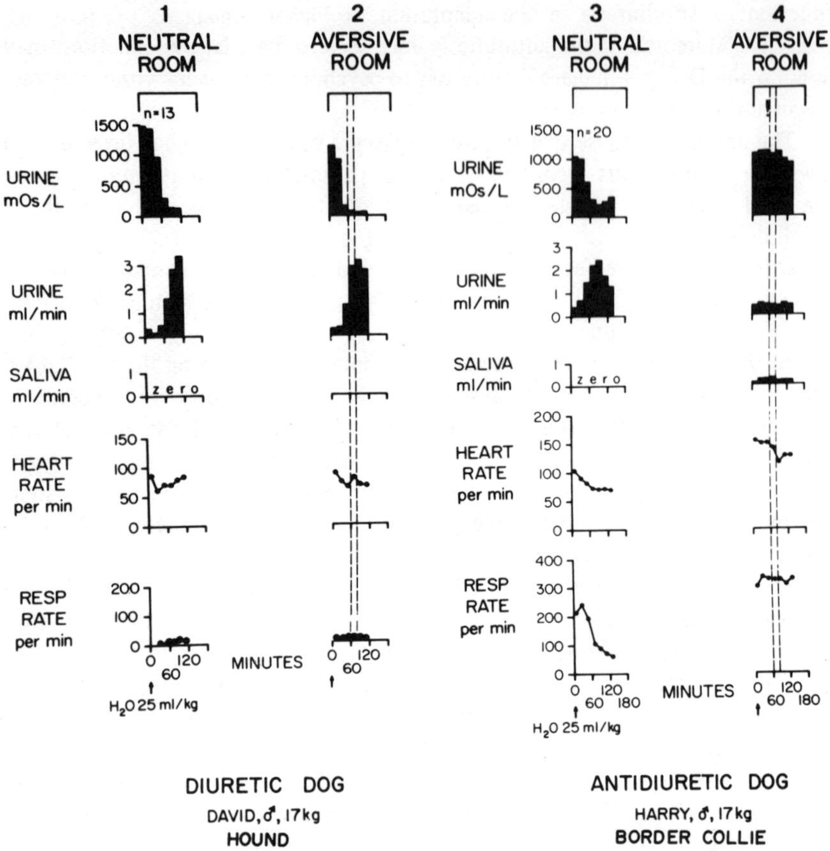

Fig. 1. Comparative psychophysiologic reactions of a high-adaptation (diuretic) dog, David, and a low-adaptation (antidiuretic) dog, Harry, to a neutral control room (columns 1 and 3) where no signals of any kind were presented, and to an aversive room (columns 2 and 4) where Pavlovian motor defense reactions were developed by means of unavoidable electrocutaneous reinforcement to a leg. The broken vertical lines in columns 2 and 4 indicate the 20-min period during which reinforced, unreinforced, or inhibitory tones were presented. Administration of water (25 ml/kg) at time zero is indicated by an arrow.

flight" reaction. We postulated that the primary response of the AD dogs to psychologic stressors is probably increased thermogenesis.

Continuous recording of oxygen consumption (Corson *et al.*, 1969; Corson, 1971) confirmed our hypothesis that the AD dogs in fact do show increased energy metabolism in the aversive environment. In contrast, the D dogs exhibit no increase in oxygen consumption in the aversive Pavlovian room.

Panting and salivation in the AD dogs therefore represent compensatory reactions designed to maintain thermal homeostasis. *The antidiuresis induced by*

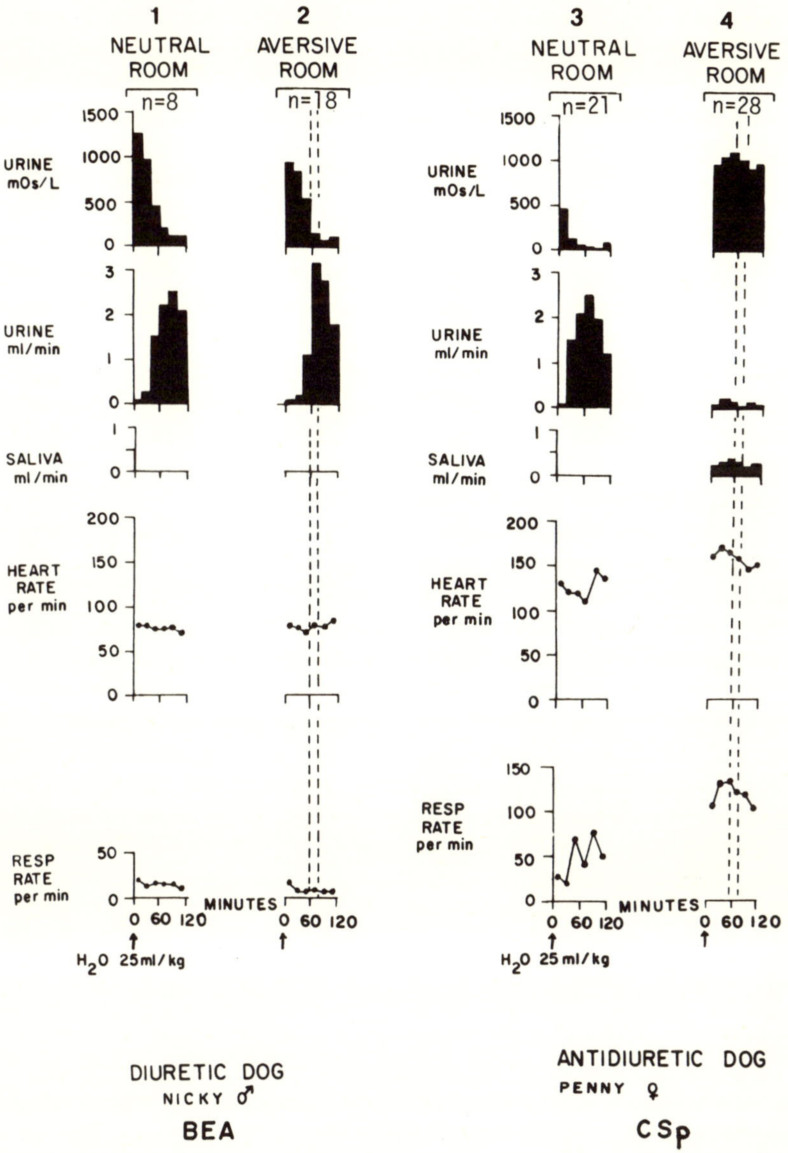

Fig. 2. Comparative psychophysiologic reactions of a high-adaptation (diuretic) dog, Nicky, and a low-adaptation (antidiuretic) dog, Penny, to a neutral control room (columns 1 and 3) where no signals of any kind were presented, and to an aversive room (columns 2 and 4) where Pavlovian motor defense reactions were developed by means of unavoidable electrocutaneous reinforcement to a leg. Designations are the same as in Fig. 1. Bea = beagle; C Sp = cocker spaniel.

vasopressin release thus serves the function of conserving body water, so that it will be available for thermoregulatory salivation. The diuretic dogs do not exhibit antidiuresis, since they do not show increased thermogenesis in the aversive environment and do not require the conservation of water for evaporative cooling.

It is important to point out that after several experimental sessions, the physiologic distress reactions (psychovisceral turmoil) appeared in the antidiuretic dogs as soon as they were brought into the aversive Pavlovian conditioning room, regardless of the conditional or unconditional stimuli presented. In contrast to the conditional motor defense responses, the conditional visceral reactions exhibited poor differentiation and were extremely difficult to extinguish in experiments in the Pavlovian conditioning room. We referred to this discrepancy between conditional Pavlovian somatic and visceral responses as *somatovisceral dichotomy.* This phenomenon is similar to Gantt's (1944, 1953, 1962) elegant description of schizokinesis and may represent a significant basis for the development of psychosomatic disorders.

Our hypothesis that the physiologic reactions of the antidiuretic dogs to the aversive room are comparable to reactions involved in intensive muscular work is supported by studies reported by Brod (1960, 1965) and by Brod *et al.* (1959*a*, 1962, 1964) in experiments on human subjects. Difficult arithmetical problems were presented at such a rate that the subjects were unable to carry them out. These tasks were "resented by most of the subjects as very unpleasant, exhausting, and producing in some of them a feeling of anger or frustration." Brod described the circulatory changes in these subjects as being comparable to those observed during strenuous muscular activity. It is interesting that similar circulatory patterns were reported by Brod *et al.* (1959*b*, 1962) as being characteristic for patients with essential hypertension and for patients with chronic cardiac failure (Brod *et al.*, 1964).

THE PSYCHOBIOLOGY OF STRESS AND DISTRESS

Concurrent measurements of several parameters made it possible to elucidate the adaptive function of psychogenic vasopressin release. In essence, the antidiuretic dogs exhibit a state of *distress* in the aversive room, i.e., *low psychobiologic adaptation,* at least as far as the cardiorespiratory systems are concerned. In contrast, the diuretic dogs show *high adaptation* to the same type of stressful situations.

We postulated that the persistence of the "fight or flight" quintet in the antidiuretic dogs in the Pavlovian conditioning room is associated with anticipatory triggering of the cardiovascular, respiratory, renal, and thermoregulatory control centers in the face of inability to achieve a consummatory adaptive response.

Exposure of an animal to new stimuli leads to an information deficit (Simonov, 1965) and the triggering of an orienting-investigative reflex. If the stimuli have a biological significance (positive or negative), the central nervous system initiates appropriate adaptive approach or avoidance responses. Anokhin (1935, 1974) postulated the operation in the central nervous system of control mechanisms for evaluating whether the response has actually achieved biological adaptation. Anokhin referred to such a coordinating evaluating center as an *action acceptor.*

The appropriate adaptive response of our experimental dogs in a Pavlovian paradigm with electrocutaneous reinforcement would be to fight or run. Since this is not possible in the Pavlovian experimental setup, the action acceptor will continue to signal that an adaptive response has not been achieved. This then would lead to the perpetuation of a chronic information deficit and a continuous internal (visceral) turmoil evidenced in our antidiuretic dogs by increased oxygen consumption, tachycardia, polypnea, salivation, and antidiuresis.

We reasoned that if this is true, then exposure of the antidiuretic dogs to similar conditional and unconditional stimuli under conditions where they could develop an avoidance response should tend to ameliorate the psychogenic visceral disquietude.

Figures 3 and 4 illustrate comparative visceral reactions in antidiuretic dog Mitzi in the control room (I), the Pavlovian conditioning room (II), and the operant avoidance room (III). In column II are plotted data derived from *extinction sessions in the Pavlovian conditioning room* when A tones were presented *without electrocutaneous reinforcement.* In column III are plotted data from experiments in another room (with a different experimenter) after the same dog reached 90–100% level of avoidance. This means that whereas the dog received no shocks at all in the Pavlovian extinction sessions, the dog did receive some shocks in the operant room.

As can be seen in Fig. 3, there was a significant decrease in the antidiuresis and in the urine osmolality (indicating a decrease in circulating vasopressin) in the operant room, as compared to the Pavlovian room, in spite of the fact that in the operant room the dog did receive occasional shocks. Note that the psychogenic thermoregulatory salivation also disappeared in the operant room.

Figure 4 depicts similar data in the same dog for heart rate, respiratory rate, and rectal temperature. All these parameters were attenuated in the operant room, so that the values began to approach the baseline magnitudes observed in the control room.

Thus, it appears that it is not the noxious stimuli per se that provoke inappropriate visceral-autonomic hyperactivity but the inability to achieve an adaptive consummatory response, or to develop control over the situation. It is

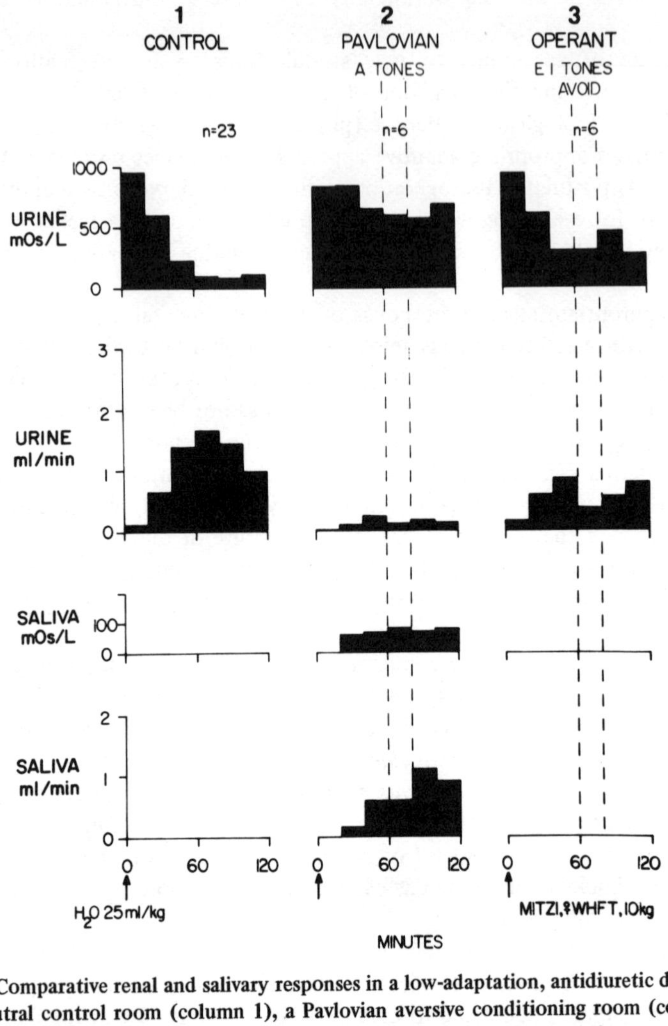

Fig. 3. Comparative renal and salivary responses in a low-adaptation, antidiuretic dog, Mitzi, in a neutral control room (column 1), a Pavlovian aversive conditioning room (column 2), and an operant conditioning room (column 3). In column 2 are depicted the mean values of renal and salivary responses in six extinction sessions conducted in the Pavlovian aversive room (where the dog had previously been exposed to unavoidable shock). In column 3 are depicted the mean values of six experiments in the operant room after the dog reached 90–100% level of avoidance. Note the amelioration of the antidiuresis and the decrease in urine osmolality in the operant avoidance room as compared to the values observed in the Pavlovian aversive room. Note also that in the avoidance experiments the psychogenic thermoregulatory salivary responses disappeared. WHFT = wirehair fox terrier. A tones are tones associated in previous experiments with electrocutaneous reinforcement; EI tones, a random sequence of excitatory and inhibitory tones used in connection with the development of discriminated avoidance responses.

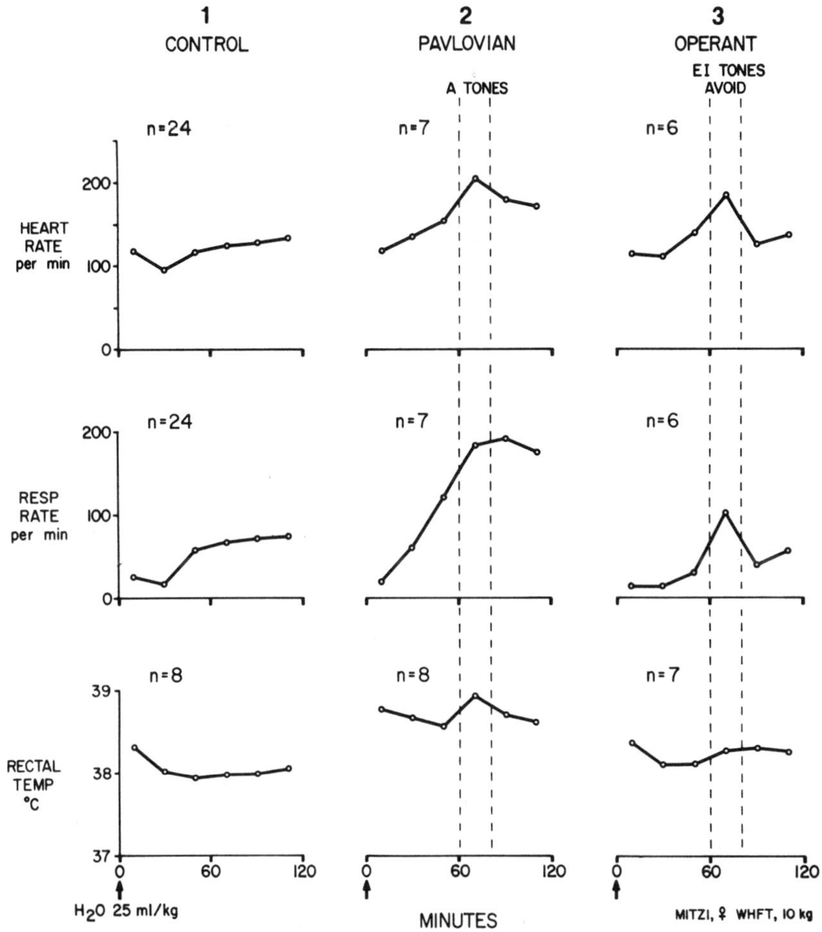

Fig. 4. Comparative values for heart rate, respiratory rate, and rectal temperature observed in a low-adaptation, antidiuretic dog, Mitzi, in a neutral control room and in Pavlovian aversive and operant conditioning rooms. The experiments are the same as those depicted in Fig. 3. Note the amelioration of the cardiac, respiratory, and thermal responses in the operant rooms as compared to the values observed in the Pavlovian aversive group.

such a psychosocial setting that may lead to the development of *distress reactions.*

It should be emphasized that distress to a particular type of aversive situation will develop only in individuals with appropriate psychobiologic programming. Our diuretic dogs do not appear to develop a distress response to the Pavlovian aversive room.

DISCUSSION

Our low-adaptation, antidiuretic dogs exhibiting psychophysiologic distress on exposure to unavoidable nociceptive stimuli in a Pavlovian paradigm may be considered a biologic model of anxiety and frustration.

A great deal of confusion has arisen because of vague definitions of the concept of anxiety. Mowrer (1939), for example, stated: "Anxiety is a learned response, occurring to signals (conditioned stimuli) that are premonitory . . . situations of injury or pain . . . anxiety is thus basically anticipatory in nature and has thus great biological utility in that it adaptively motivates living organisms to deal with traumatic events in advance of their actual occurrence . . . Anxiety, i.e., mere anticipation of actual organic need or injury, may effectively motivate human beings."

It appears to us that it serves no useful purpose to confuse anxiety with biological drive or motivation. The studies of Cattell (1964) and Cattell and Scheier (1961), using factor analysis techniques, demonstrated that there is only one type of anxiety and that it never serves as a motivational drive, but on the contrary always causes an impairment of performance.

Our data suggest that it is exposure to *unavoidable stressors* or *insoluble problems* that may lead to the development of a state of anxiety and distress in *organisms with certain constitutional makeup.*

The fact that the psychovisceral reactions of the antidiuretic dogs were markedly ameliorated when the animals were permitted to develop discriminated avoidance responses supported the proposition that it is the inability to cope with the stressful situation that provokes anxiety and psychopathology. This is in accordance with the views expressed so elegantly by Lazarus (1966, 1967) and in the well-controlled studies reported by J. M. Weiss (1972).

Mowrer and Viek (1948) reported that rats subjected to unavoidable shock (the authors referred to it as uncontrollable shock) exhibited a greater degree of conditional suppression than rats shocked but permitted to escape.

At first glance, our data on the comparative visceral responses of low-adaptation dogs to Pavlovian and operant conditioning to aversive stimuli appear to be at variance with the reports of Brady *et al.* (1958) on avoidance behavior and the development of gastroduodenal ulcers in the "executive" monkeys. However, our experimental design is different from that of Brady *et al.* The point is that the reinforcement schedule in our dogs in the operant situation was the type which permitted relatively easy mastery. The intervals between electrocutaneous reinforcements were 1 min in both the Pavlovian and operant rooms. Moreover, the total period for the presentation of stimuli was only 20 min during a 2-hr experiment. In the experiments of Brady *et al.*, the "executive" monkeys were exposed to an unsignalled avoidance task involving the pressing of a lever *every 20 sec.* (in order to avoid shock) *for 6 hr on and 6 hr off every*

24-hr day for a period of 6 to 7 weeks. This schedule represents a rather severe form of exposure to exhausting physical and psychological stressors. Such a schedule may be comparable to an extreme form of speedup on a factory assembly line and would be expected to lead to the development of distress reactions. Moreover, the "executive" monkeys were selected on the basis of being high-avoidance responders. Such a biased selection may have resulted in the assignment of "executive" roles to monkeys with a constitutional predisposition to psychovisceral disorders such as ulceration.

The importance of unavoidable psychologic stressors and of frustration in the development of psychosomatic disturbances was pointed out in the admirable studies of Wolf *et al.* (1948) on human subjects. These authors reported the case of a patient who actually left the interview office to beat up the person he resented. His resting diastolic pressure during the interview was 110 mm. After his return from this rather uninhibited physical aggression, his diastolic pressure decreased to 85 mm.

Wolf *et al.* (1955) concluded that the hypertensive patients were generally "fundamentally driving and often hostile, but not able fully to commit or assert themselves." These authors also had the impression that "freer or more fearless self-assertion, brought about by a variety of devices, has been associated among our subjects with a short or long-lasting lowering of arterial pressure."

Hambling (1959) reported cases of hypertensive patients who exhibited marked elevation of diastolic blood pressure whenever they were faced with a frustrating situation that was beyond their control. The same patients remained normotensive in stressful situations which permitted them appropriate response outlets.

Our experimental data suggest that Rahe's (1969) model of the relationship between any life change (the summation of unpleasant and pleasant changes) and morbidity may need some revision. In considering a life change, one should take into account how a particular life change is perceived by a given individual. A promotion may be considered a pleasant challenge by an individual whose ability matches the requirements of the new responsibilities. Another person may be distressed by a similar promotion because of a real or imaginary dichotomy between ability and the new responsibilities.

In a recent well-documented study, Rahe *et al.* (1974) present a most instructive model for life changes and illness research based on cross-cultural data from men in the U.S. Navy and in the Norwegian Navy. This model incorporates a number of intermediate steps that may influence the development of physical illness in relation to different life changes.

The importance of individual differences in responses to life changes is exemplified by the reactions of six Detroit automobile workers to a recently introduced group assembly system at the Saab automobile plant in Sweden (*New York Times,* Dec. 24, 1974, p. 25; *New York Times,* Jan. 5, 1975, Section E, p. 11).

According to the *New York Times* report, in the group assembly, fitters work in teams of three, each team assembling a complete engine at a pace determined by the group. According to the Saab officials, after the introduction of the group assembly plan, there has been less absenteeism and employee turnover than with the conventional production line. This would suggest a positive reaction of the Swedish automobile workers to the group assembly system. Yet, of the six Detroit automobile workers (who spent 4 weeks at the Saab plant), only one expressed a definite preference for the group assembly method. The other five workers "were willing to accept boredom and the freedom to lose themselves in their own thoughts while working, rather than accept a work situation demanding greater concentration." Thus the same situation or life change may be perceived as distressful by some individuals, whereas others may find such a life change as a pleasant and stimulating challenge.

Levi (1974) reported that urinary catecholamine output may be increased by unpleasant as well as by pleasant emotional stumuli. This would simply suggest that the *measurement of urinary catecholamines alone* does not represent an appropriate parameter for the differentiation between stress and distress. This again underscores the need for recording concurrent psychophysiologic responses. The extensive well-controlled studies by Mason (1968) on patterns of psychoendocrine reactions represent a most significant contribution in this area.

In our experiments on dogs, we have never seen the pattern of psychophysiologic distress reactions in Pavlovian conditioning experiments involving reinforcement with biologically positive (pleasant) stimuli.

It appears to us that in order to gain some understanding and control of psychosocial stressors, it is essential to embark on longitudinal studies involving concurrent measurements of many physiologic, neuroendocrine, and biochemical parameters in animals with controlled genetic and environmental histories exposed to different types of avoidable and unavoidable stressors. The inclusion of different species of animals in long-term studies may help considerably in our attempts to elucidate the mechanisms whereby psychosocial stressors may contribute to the development of specific disease entities. Data derived from such investigations may contribute to our ability to develop our social-economic and educational institutions and social interactions in such a manner as to lead to a more harmonious biological adaptation.

SUMMARY

Our experiments demonstrated consistent significant constitutional differences in the reactions of dogs exposed to an environment associated with unavoidable unconditional and conditional aversive stimuli.

These individual differences in reactions are not merely quantitative. These are differences in the *quality of responses,* i.e., the dogs exhibit *different patterns* of physiologic and endocrine reactions. Some dogs (e.g., many wirehair fox terriers) develop a persistent quintet of distress responses to the aversive room: increased energy metabolism, tachycardia, polypnea, copious salivation, and vasopressin release (low-adaptation, antidiuretic dogs). In other words, these dogs show persistent physiologic responses to symbols of danger comparable to those described by Cannon for a "fight or flight" reaction, even though no fight or flight is possible. Other dogs (e.g., many beagles), exposed to the same aversive environment, do not develop this pattern of reactions (high-adaptation, diuretic dogs).

A systems appraoch involving concurrent recording of several relevant parameters in longitudinal studies made it possible for us to elucidate the probable adaptive function of psychogenic vasopressin release.

Dogs which exhibit physiologic distress reactions to an unavoidable aversive situation showed marked amelioration of these distress reactions when these animals were studied in a different room by a different experimenter and were permitted to develop discriminated avoidance responses and thus gain control over the aversive stimuli.

When our low-adaptation, antidiuretic dogs were involved in Pavlovian conditioning experiments using reinforcement with biologically positive (pleasant) stimuli (e.g., food reinforcement), the above-described pattern of distress did not appear.

ACKNOWLEDGMENTS

Supported in part by USPHS grants MH 12089, MH 18098, The OSU Graduate School Biomedical Science Research Support Grant, and Grant Foundation, Inc. We wish to express our appreciation to Vladimir Kirilcuk for his expert participation in the Pavlovian and operant conditioning and energy metabolism experiments. Julius Hajek and Marta Hajkova made a major contribution to the development of our studies on continuous recording of oxygen consumption and CO_2 production during Pavlovian and operant conditioning experiments.

REFERENCES

Anokhin, P. K. (Ed.). (1935). The problem of center and periphery in the contemporary physiology of nervous activity. In *The Problem of Center and Periphery in the Physiology of Nervous Activity.* Gorki: Gosizdat.

Anokhin, P. K. (1974). *Biology and Neurophysiology of the Conditioned Reflex and Its Role in Adaptive Behavior* (Samuel A. Corson, Scientific and Translation Editor). Oxford: Pergamon Press.

Brady, J. V., Porter, R. W., Conrad, D. G., and Mason, J. W. (1958). Avoidance behavior and the development of gastroduodenal ulcers. *J. Exp. Anal. Behav. 1,* 69–72.

Brod, J. (1960). Haemodynamic response to stress and its bearing on the haemodynamic basis of essential hypertension. *WHO Symposium on the Pathogenesis of Essential Hypertension, Prague,* pp. 256–264.

Brod, J. (1965). Coordination of circulation during emotion. *XXIII International Congress of Physiological Sciences.* Excerpta Medica International Congress Series, No. 87, pp. 157–164.

Brod, J., Fencl, V., Hejl, Z., and Jirka, J. (1959*a*). Circulatory changes underlying blood pressure elevation during acute emotional stress (mental arithmetic) in normotensive and hypertensive subjects. *Clin. Sci. 18*(2), 269–279.

Brod, J., Fencl, V., Hejl, Z., and Jirka, J. (1959*b*). Haemodynamics in essential hypertension. *Nature 184,* 1643–1644.

Brod, J., Fencl, V., Hejl, Z., and Ulrych, M. (1964). Muscle blood flow in heart failure. *Fourth European Congress of Cardiology, Prague,* (abstract 47).

Brod, J., Fencl, V., Jirka, J., Hejl, Z., and Ulrych, M. (1962). General and regional haemodynamic pattern underlying essential hypertension. *Clin. Sci. 23,* 339–349.

Cannon, W. B. (1929). Organization for physiological homeostasis. *Physiol. Rev. 9,* 397.

Cannon, W. B. (1932). *The Wisdom of the Body.* New York: W. W. Norton and Co.

Cattell, R. B. (1964). Psychological definition and measurement of anxiety. *J. Neuropsychiat. 5,* 396.

Cattell, R. B., and Scheier, I. H. (1961). *The Meaning and Measurement of Neuroticism and Anxiety.* New York: The Ronald Press Co.

Coppen, A. (1965). Mineral metabolism in affective disorders. *Brit. J. Psychiat. 111,* 1133.

Corson, S. A. (1957). Review of *S. P. Botkin and the Neurogenic Theory of Medicine* by F. R. Borodulin, 2nd edition, 1953, 184 pp. (in Russian), Moscow: Medgiz. *Science 125*(3237), 75–77.

Corson, S. A. (1966*a*). Conditioning of water and electrolyte excretion. *Res. Publ. Assoc. Nerv. Ment. Dis. 43,* 140.

Corson, S. A. (1966*b*). Neuroendocrine and behavioral response patterns to psychologic stress and the problem of the target tissue in cerebrovisceral pathology. *Proc. Conference on Psychophysiological Aspects of Cancer, Ann. NY Acad. Sci. 125*(Art. 3), 890.

Corson, S. A. (1971). Pavlovian and operant conditioning techniques in the study of psychosocial and biological relationships. In *Society, Stress and Disease,* Vol. 1: *The Psychosocial Environment and Psychosomatic Diseases* (L. Levi, Ed.). London: Oxford University Press, pp. 7–21.

Corson, S. A., and Corson, E. O'L. (1969). The effects of psychotropic drugs on conditioning of water and electrolyte excretion: experimental research and clinical implications. In *Psychotropic Drugs in Internal Medicine* (A. Pletscher and A. Marino, Eds.). Excerpta Medica International Congress Series, No. 182, 147–164.

Corson, S. A., and Corson, E. O'L. (1971). Psychosocial influences on renal function–implications for human pathophysiology. In *Society, Stress and Disease,* Vol. 1: *The Psychosocial Environment and Psychosomatic Diseases* (L. Levi, Ed.). London: Oxford University Press, pp. 338–351.

Crammer, J. L. (1959). Water and sodium in two psychotics. *Lancet 1,* 1122.

Frank, R. T. (1931). Hormonal causes of premenstrual tension. *Arch. Neurol. Psychiat. 26,* 1053.

Gantt, W. H. (1944). *Experimental Basis for Neurotic Behavior.* New York: Hoeber Medical Division of Harper and Row.

Gantt, W. H. (1953). Principles of nervous breakdown–schizokinesis and autokinesis. *Ann. NY Acad. Sci. 56*(2), 143.

Gantt, W. H. (1962). Factors involved in the development of pathological behavior: schizokinesis and autokinesis. *Perspect. Biol. Med. 5*(4), 473–482.

Hambling, J. (1959). Essential hypertension. In *The Nature of Stress Disorder.* Springfield, Ill.: Charles C Thomas.

Jenner, F. A., Gjessing, L. R., Cox, J. R., Davies-Jones, A., Hullin, R. P., and Hanna, S. M. (1967). A manic depressive psychotic with a persistent forty-eight hour cycle. *Brit. J. Psychiat. 113,* 895.

Lazarus, R. (1966). *Psychological Stress and the Coping Process.* New York: McGraw-Hill, Inc.

Lazarus, R. (1967). Cognitive and personality factors underlying threat and coping. In *Psychological Stress* (M. H. Appley and R. Trumbull, Eds.). New York: Appleton-Century-Croft, pp. 151–181.

Levi, L. (1974). Psychosocial stress and disease: a conceptual model. In *Life Stress and Illness.* (E. K. E. Gunderson and Rahe, R. H., Eds.). Springfield, Ill.: Charles C Thomas.

Mason, J. W. (1968). Organization of the multiple endocrine responses to avoidance in the monkey. *Psychosom. Med. 30*(5), Part 2, 774–790.

Miles, B. E. (1953). Effect of emotion on renal function in normotensive and hypertensive women. *Lancet, 2,* 539.

Miles, B. E., De Wardener, H. E., and McSwiney, R. R. (1952). Renal function during emotional diuresis. *Am. J. Med. 12,* 659.

Mowrer, O. H. (1939). A stimulus-response analysis of anxiety and its role as a reinforcing agent. *Psychol. Rev. 46,* 553–565.

Mowrer, O. H., and Vieck, P. (1948). An experimental analogue of fear from a sense of helplessness. *J. Abnorm. Soc. Psychol. 43,* 193–200.

Rahe, R. H. (1969). Life crisis and health change. In *Psychiatric Drug Responses: Advances in Prediction* (P. R. A. May and J. R. Wittenborn, Eds.) Springfield, Ill.: Charles C Thomas.

Rahe, R. H., Fløistad, I., Bergan, T., Ringdal, R., Gerhardt, R., Gunderson, E. K. E., and Arthur, R. J. (1974). A model for life changes and illness research. Cross-cultural data from the Norwegian Navy. *Arch. Gen. Psychiat. 31,* 172–177.

Schottstaedt, W. W., Grace, W. J., and Wolff, H. G. (1955). Life situation, behavior patterns, and renal excretion of fluid and electrolytes. *J. Am. Med. Assoc. 157,* 1485.

Selye, H. (1971). The evolution of the stress concept–stress and cardiovascular disease. In *Society, Stress and Disease, Vol. 1: The Psychosocial Environment and Psychosomatic Diseases* (L. Levi, Ed.). London: Oxford University Press, pp. 299–311.

Simonov, P. V. (1965). The role of emotions in the adaptive behavior of living systems. *Vopr. Psikhol. 4,* 75–84.

Thomas, W. A. (1933). Generalized edema occurring only at the menstrual period. *J. Am. Med. Assoc. 101,* 1126.

Thorn, G. W., Nelson, K. R., and Thorn, D. W. (1938). Study of mechanism of edema associated with menstruation. *Endocrinol. 22,* 155.

Weiss, J. M. (1972). Influence of psychological variables on stress-induced pathology. In *Physiology, Emotion & Psychosomatic Illness.* Ciba Foundation Symposium 8 (New Series), pp. 253–279.

Weiss, P. (1925). Tierisches Verhalten als "Systemreaktion." Die Orientierung der Ruhestellungen von Schmetterlingen (Vanessa) gegen Licht and Schwerkraft. *Bio. Gen. 1,* 168.

Weiss, P. A. (1969). The living system: determinism stratified. In *Beyond Reductionism* (A. Koestler and J. R. Smythies, Eds.), New York: The Macmillan Co.

Wiener, N. (1948, 1961). *Cybernetics–or Control and Communication in the Animal and the Machine.* Cambridge, Mass.: The MIT Press.

Wolf, S., Cardon, P. V., Jr., Shepard, E. M., and Wolff, J. G. (1955). *Life Stress and Essential Hypertension.* Baltimore: The Williams and Wilkins Co.

Wolf, S., Pfeiffer, J. B., Ripley, H. S., Winter, O. S., and Wolff, H. G. (1948). Hypertension as a reaction pattern to stress: summary of experimental data on variations in blood pressure and renal blood flow. *Ann. Int. Med. 29,* 1056–1076.

Wolff, H. G. (1953). *Stress and Disease.* Springfield, Ill.: Charles C Thomas.

Motivation, Mood, and Mental Events

Patterns and Implications for Adaptive Processes

ERIC KLINGER, STEVEN G. BARTA, THOMAS W. MAHONEY, *et al.*

Psychological adaptation is a continuous, time-bound process. People take in information, process it in relation to their goals, build mental schemes for acting on the world around them, act, generate new information, and so on. How well people succeed in coming to terms with their situations depends to an important extent on the ways in which they execute these moment-to-moment steps: which information they choose to process, how they combine it, how accessible they keep it, what other materials and skills they bring to bear on it, and what decisions they arrive at and act upon on the basis of it. These activities will be adaptive insofar as the goals themselves correspond to the requirements of survival. If people notice, store, and think about matters that affect their major interests they can be expected to deal with their environments more effectively than if they do not. Therefore, understanding human adaptation would seem to require among other things understanding the interface between motivational states and cognitive operations. The research described here focuses on this

ERIC KLINGER, STEVEN G. BARTA, THOMAS W. MAHONEY, *et al.* • University of Minnesota, Morris, Minnesota.

crucial linkage between cognition and motivation, and it probes the disordering of this motivational guidance system during depressed moods.

Our key concept for thinking about motivation is the "current concern" (Klinger, 1971, 1975). This construct refers to the state of an organism between the time that it becomes committed to pursuing a goal and the time that it either attains the goal or abandons pursuit of it. Since concerns are considered specific to goals, an individual is normally involved in several concerns at the same time.

One important feature of this view is the assumption that a person committed to an incentive continues to respond to cues related to it without the need for specific drive states, and without needing the incentive itself to arouse a central excitatory state. This view therefore posits a continuing motivational process that serves as the basis for monitoring of the environment and for cognitive processing of information that might be related to each of the person's concerns. Whatever the mechanism for that continuing process may be, previous evidence suggested that it sensitizes people to notice and think about cues related to their concerns (Klinger, 1971, 1975), and there is good reason to believe that its effects vary according to the person's relationship to the incentive: how well or poorly the pursuit of it is going, how valuable it is, and how pressing. The purpose of the research reported here is to confirm that such a continuing process indeed governs cognitive processes, to quantify the effects, and to begin exploring their parameters.

ASSESSING CURRENT CONCERNS

The first task was to devise methods for assessing current concerns. The primary requirement for identifying a current concern is that there be an object or event which the subject is committed to pursuing, attaining, enjoying, retaining, etc., or avoiding, preventing, etc., such as completing a certain course, maintaining a particular personal relationship, or sewing a rug. The criteria we have adopted for declaring a concern are (a) the subject volunteering that he or she is actively pursuing, etc., the object or event, or expressing a strong wish to be (usually requiring evidence of operant thinking about it); (b) the subject describing one or more occasions on which he or she has pursued (etc.) the object or event and further questioning reveals that the pursuit has not yet ended; or (c) the subject expressing positive or negative affect about the object or event and further questioning reveals at least an intention to act on it. These criteria owe much to the system developed by McClelland *et al.* (1953) for scoring "need achievement" in Thematic Apperception Technique stories.

The chief procedure for obtaining the information to which these criteria are applied is a series of structured interviews that delves systematically into every major area of the subject's life. The interviews are supplemented by a "Goals Checklist" of about 220 categories of activities, including vocational, recreational, interpersonal, philosophic, etc., whose primary function is to detect concerns that may have slipped by the interviewer.

Since concerns differ in the degree to which they influence cognitive activities, our procedures are also directed at assessing characteristics of subjects' incentive relationships that might affect the relative influence of their concerns. Content analysts working with interview protocols rate each identifiable concern on each of twelve variables: eight to assess different aspects of the value of the incentive to the subject, two directed at the perceived probability that the pursuit will succeed, and two others to indicate how far off consummation is expected to be in the future. These variables correspond to ideas with long histories in the motivational literature: expectancy X Value theories (e.g., Atkinson, 1964) and Miller's (1944) approach and avoidance gradients. Median interrater reliabilities for the twelve variables range from 0.28 to 0.80, with a median of 0.64, considering only those concerns which both of the two raters independently identified in the interview protocols (based on four pairs of raters working with concerns drawn from seven interviews, yielding 84 correlations). Of those concerns identified by any one rater, an average of 67% were also independently identified by the other rater in the pair.

In addition to completing periodic interviews and Goals Checklists, those subjects whom we have observed longitudinally complete a structured diary, the Daily Personal Log, with which we attempt to keep track of daily activities that might bear on major concerns. It also incorporates four of Wessman and Ricks' (1966) Personal Feeling Scales: anchored self-rating scales for Fullness vs. Emptiness of Life, Elation vs. Depression, Harmony vs. Anger, and Energy vs. Fatigue.

COGNITIVE MEASURES

An important assumption of our research is that cognitive processing operates on input of some kind—that at each step it operates on information generated by an earlier step in the system. This information may consist of sensory events, perceptual transformations of them, or ideational constructions. Whatever form it may take, we shall refer to instances of it as "cues."

There appeared to be three aspects of cognitive functioning which are obviously important for adaptive responding and which could be measured experimentally: attention, retention, and thought content. They are in some respects distinct, in that it is quite possible to attend to something without retaining or thinking about it, and retention in no way compels thought.

However, it seems unlikely that one could think about something not retained or to retain something never attended. In that case, whatever controls attention also controls retention, and whatever controls retention controls thought.

Our chief vehicle for obtaining information on these cognitive variables is a series of individual "thought-sampling" sessions. Each subject listens dichotically through headphones to two channels of a tape recording playing two different 15-min excerpts of descriptive or fictional writing. Each tape contains 12 25-sec "embedding sites" at irregular intervals, synchronized for the two channels. At each embedding site the script for the narration is altered to relate unobtrusively to one of the subject's concerns on one channel and to something comparable to which the subject appears uncommitted (a "nonconcern") on the opposite channel.

To illustrate the embedding, the excerpt that follows is taken from Samuel Beckett's *Stories and Texts for Nothing* (1968), except that the italicized portions are our own compositions to allude to the subject's concern about entering a profession in which she could help people in need:

> "Someone said, perhaps the same, What possessed you to come? I could have stayed in my den, snug and dry, I couldn't . . . *I felt a need, a need to help to help those who listen.* My den, I'll describe it no, I can't. It's simple, I can do nothing any more, that's what you think, *but maybe, yes, maybe I can aid some one else, another coming.* I say to the body, Up with you now, and I can feel it struggling, struggling no more, struggling again, till it gives up." (Tone. Tape stops.)

Nonconcerns are selected from a 517-item "Concerns Pool," the items of which had been rated by eight judges on the four dimensions of category width, commonness, emotional grippingness, and social desirability. (Interrater agreement is expressed in intraclass correlations of 0.35, 0.50, 0.19, and 0.42, respectively, $p<0.0005$ in all cases.) Nonconcerns are selected to resemble the concerns with which they are paired as closely as possible on these four dimensions as well as in general content. Any one tape contains material written to allude to four pairs of concerns and nonconcerns, each pair represented at three embedding sites. Blind ratings by an independent judge established that embeddings written to allude to concerns approximately equaled embeddings for nonconcerns on the three rating dimensions of blending into the context of the embedding, recallability of the content, and noticeability of the writing style.

In order to measure attentiveness to a particular channel of the tape, subjects indicate continuously with a toggle switch to which ear they are listening, selecting among three possible positions of the switch to indicate right ear, left ear, or neither/both. The position of the switch is recorded continuously with an ink-writing recorder. The tape stops 10 sec after the end of each embedded portion, and the subjects report their last thoughts and the last tape contents they can recall, using a prearranged reporting system. These responses, then, provide measures of recall and thought content.

In order to prepare for this rather complicated procedure, subjects undertake an eight-step training program to sensitize them to their inner experience, learn the thought-reporting procedure, become habituated to the setting, and routinize the switching. Thought reporting takes the form of a "Thought Sampling Questionnaire" modified from a dream-reporting procedure used by Allan Rechtschaffen. The first question asks subjects to describe their last thought segments, beginning with the most recent segment and working backwards. Subsequent questions concern the duration, imaginal qualities, time orientation, and directedness of the most recent segment, as well as last recollections of the taped narrative. During experimental sessions, the questions are presented only by cue cards attached to the wall, subjects automatically running through their answers at each interruption of the tape.

To assess the relatedness of thoughts to stimuli, judges blindly rate the resemblance of each thought segment to each of the two embedded passages that just preceded it, using a scale of six categories: (1) containing identical or unmistakably similar words and meaning, (2) identical words but different meaning, (3) same thematic content but different language, (4) metaphoric parallels between thought and passage or similarity in thematic content requiring little inference to detect, (5) relationships requiring extended trains of inference or tenuous assumptions to establish, and (6) total lack of relationship. The latter two categories are scored as nonresemblance of thought segments to embedded passages; the first four are accepted as evidence of a relationship. The reliability of this rating procedure is reflected in agreement 92% of the time between two independent judges as to whether or not there is a relationship (based on a sample of 120 passages) and 73% of the time when all six categories are considered separate outcomes (based on a sample of 163 reported thought segments).

In order to probe subjects' cognitions about the experiment, they are given postsession questionnaires after each session and they are called in periodically to fill out an Inventory of Reactions to the Experiment. Both instruments were designed to provide subjects with many opportunities to reveal their hypotheses about the purpose and design of the research. We employed this strategy, rather than funnel questionnaires that contain progressively more direct questions, in order to keep close track of subjects' conceptions without needlessly communicating the fact of embedding to subjects whom we hoped to retain for repeated sessions over several months. The thought-sampling procedure, unlike most experimental procedures, also contains its own internal probe, since subjects report spontaneous thoughts about the experiment along with other thoughts.

EFFECTS OF CONCERNS ON COGNITIVE ACTIVITIES

We have incorporated all of the procedures described above into a longitudinal study that follows each subject for up to nine months of data-gathering. The following analyses of cognitive activity employ data from 16 experimentally naive college students, ten male and six female, for a total of 57 thought-sampling sessions, ranging from one to six sessions per subject.

Attention

Subjects' attention, as signaled by the toggle switch, clearly favored the concern-related over the nonconcern-related passages (Table 1). Subjects switched toward concern-related passages sooner, spent more time listening to them, and were more inclined to remain listening to them for the entire length of the passage. The data suggest that "preattentive processes" (Neisser, 1967) are sensitive to concern-related cues.

Recall

Retention similarly favors concern-related material. Subjects recalled portions of approximately twice as many concern-related as of nonconcern-related passages (Table 1). Furthermore, concern-relatedness affected recall more strongly than attention ($p<0.01$). The passages subjects recalled had attracted no more attention shifts than other passages, but subjects had spent more time listening to them ($p<0.05$) and more often listened to the whole passages ($p = 0.003$). This suggests that recall constitutes a second-round processing step admission to which is more stringent, involving a clearer confirmation of concern-relatedness, than the cruder, less discriminating preattentive processes.

Thought Content

About a third of the embedded passage pairs were reflected in subjects' immediately subsequent thought content. Consistent with the results for recall, thought content was related to approximately twice as many concern-related passages as nonconcern-related passages (Table 2).

Subjects recalled most of the passages they thought about, and they thought about a majority of the passages they recalled (Table 3). Of course, these statements apply only to reported thoughts, which just preceded the thought-

Table 1. Attention to and Recall of Concern-Related and Nonconcern-Related Embedded Passages[a]

Dependent variable	Type of embedded passage		Significance	
	Concern	Nonconcern	Binomial[b]	t-Test[c]
Attention (number per session)				
Shifts toward passage	3.06	2.59	0.07	$p < 0.025$
First shift to passage	1.67	1.29	0.05	$p < 0.025$
Passages attended from start to end	1.86	1.35	0.02	$p < 0.01$
Attention (mean secs.)				
Latency of first shift from nonattention	6.03	8.09	–	$p < 0.005$
Time spent listening per session	76.35	57.01	–	$p < 0.005$
Recall				
Number of passages recalled per session	3.10	1.47	< 0.00001	$p < 0.0001$

[a]The t-tests are for correlated data, session taken as the unit of analysis, the variance estimate pooled across subjects. $D_f = 35$. The binomial tests and the per-session statistics reported here are based on 51 sessions. Six other sessions, run during or after two subjects expressed suspicions that material had been embedded to relate to them individually, are excluded

[b]Assumes independence of data from different embedded passages per session.

[c]Makes conservative assumption of intrasession dependencies and aggregates all data within sessions.

Table 2. Frequency of Thoughts Related to Concern- and Nonconcern-Related Passages (N = 42 Sessions)

Type of thought-passage relationship	Type of embedded passage		Significance
	Concern	Nonconcern	(Binomial tests)
Dependent variable: Thought segments per session[a]			
Same thematic content and language	1.48	0.62	0.0059
Same thematic content, different language	0.88	0.38	0.0098
Same language, different thematic content	0.71	0.38	0.0122
Thematic similarities or metaphoric parallels	0.57	0.36	0.1405
Tenuous relationship	0.76	0.48	0.2025
Dependent variable: Passages per session related to thoughts[b]	2.71	1.26	<0.0001

[a]For the significance tests only, to guard against interdependence, one thought segment was sampled randomly from the responses to each passage pair. A chi-square comparing the classes of relationships was not significant.

[b]Excludes tenuous relationships.

Table 3. Frequencies of Recall and Thought Related to Embedded Passages, per Session[a]

Type of thought-passage relationship	Recall of passage: Recalled	Recall of passage: Not recalled
Relation of thought segments to passages[b]		
Same thematic content and language	1.67	0.40
Same thematic content, different language	0.69	0.57
Same language, different thematic content	0.71	0.38
Thematic similarities or metaphoric parallels	0.21	0.71
Passages related to thought content (all of above categories)[c]	2.12	1.74
Passages unrelated or tenuously related to thought content[c]	2.10	5.47

[a]Data are taken from 42 sessions.
[b]Thought segment taken as the unit of analysis. Chi-square = 28.15 $p < 0.001$. For the test only one thought segment was sampled randomly from responses to each passage pair.
[c]Passage taken as the unit of analysis. The Fisher Exact Test, performed on the raw frequencies, yields $p < 0.00001$. An index that aggregates the relevant information for each session, and hence does not assume intrasession independence of observations, yielded $t(28) = 2.50$, $p < 0.01$ one-tailed. Considering only embedding sites at which at least one passage was recalled, the recalled passages were far more likely than the unrecalled passages to be judged related to thought segments, $t(27) = 5.84$, $p < 0.0005$.

sampling interruptions. Since subjects probably forgot some thoughts that occurred early in the embedding and would normally have begun to think about some passages after the thought-sampling interruption, the strong relation between recall and thought content in these data probably underestimates the true degree of correspondence. Accordingly, it seems reasonable to conclude that retaining information and working it over explicitly in thought are closely related processing steps, and both are highly selective in favor of material related to current concerns.[1]

It is true, of course, that the concern-related passages were not in all cases attended to, recalled, and thought about, and that nonconcern-related passages sometimes were. Some of the deviations from a perfect fit, expecially the processing of nonconcern-related cues, would be damaging to the theory if it could be assumed that the passages written to be related to each subject's concerns were always so related, that the language of the embedding always clearly articulated with the subject's own semantic network, and that the passages written to be nonconcern-related successfully avoided all language that

[1] Holding recall constant, thought content still is more closely related to concern-related than to nonconcern-related passages. For each session, we subtracted the proportion of recalled nonconcern-related passages to which at least one thought segment was related from the proportion of recalled concern-related passages to which at least one thought segment was related. A t-test over sessions to determine whether the mean difference (0.13) was greater than zero yielded $t\ (17) = 2.62$, $p<0.01$, one-tailed.

could be interpreted by the subject as concern-related. Obviously, however, these assumptions are bound to be false. During pilot testing of procedures for these experiments, we worked for a time with single-channel recordings, and were amazed at subjects' ability to relate even very remotely relevant cues to their concerns. For instance, one subject, hearing an incidental reference to a gray blob, was reminded of her close friend who loved elephants. The dichotic listening system, by pairing ostensibly relevant cues with ostensibly irrelevant ones, reduces the effect of unintended reminders of concerns but does not eliminate them. Furthermore, our system for assessing concerns cannot prevent short-term or obscure concerns, such as doing the laundry, from creeping into the picture and sensitizing the subject to language that the passage writer had no reason to believe was related to a subject's concerns. The method, therefore, is far from completely precise, and the data must be interpreted accordingly. Granting these limitations, the relationships are nevertheless statistically strong enough to warrant conclusions and to serve as the focus of parametric studies aimed at their further dissection.

Subjects' Perceptions of the Experiment

A majority of the subjects began their participation during freshman year, and eight of them began some weeks before their freshman orientation program. It is therefore not surprising that one of the subjects' more popular hypotheses (six individuals) was that we were engaged in an intensive study of the psychological transition from high school to college or from freshman to sophomore year. Other subjects thought we were investigating the way in which people cope with their lives (five individuals) or had miscellaneous idiosyncratic and incorrect notions. Two of the sixteen subjects, however, came to suspect that the stimulus tapes had been altered to refer to their particular lives. (The case of a third subject is doubtful, due to ambiguous language in two thought samples without corroboration in questionnaire responses.) Potentially, such awareness could greatly affect the nature of subjects' cognitive processes during thought-sampling sessions. However, these suspicions surfaced unambiguously in only one of the thought samples (our subjects were not, incidentally, reluctant to report other kinds of negative thoughts), and they did not seem to the subjects to constitute a major factor in their thoughts or behavior during the sessions. The data suggested that during those sessions in which the two subjects developed their suspicions, they listened more persistently to passages of all kinds. Apart from this, the two subjects' records revealed experimental effects virtually identical with those of other subjects, and comparisons between the two subjects' records before and after their realizations revealed no consistent differences. Insofar as trends toward differences existed, they were in the direction of awareness weakening the experimental effects. These analyses therefore suggest that the phenomena reported here cannot be attributed to awareness and are relatively robust with respect to suspicion of embedding.

Conclusion

The data provide the model with strong support. People are sensitized to attend to cues that may be related to their current concerns. Some of the cues they attend to are discarded, and others, which are more frequently and clearly concern-related, they retain and think about. This pattern of results suggests that whatever properties of concerns give them their directing influence over information processing are crucial to successful adaptation.

Given these results, our search for further insights has taken two directions. One is to ask what properties of people's relationships to their incentives are most effective in influencing their cognition and inner experience. The other is to ask what factors, internal or external, affect those properties. Both kinds of inquiry are beginning to produce useful data.

INCENTIVE-RELATIONSHIP PROPERTIES AND THOUGHT CONTENT

As a preliminary approach to the first of the two questions, we asked a group of 45 student volunteers drawn from several undergraduate classes to take just the Concern Dimensions Questionnaire which asked subjects to list up to seven "things" that they had thought about most during the previous 36 hr, in rank order of the amount of time they had spent thinking about them, and then up to seven more "things" that are important to the subjects but that they had thought about little or not at all. Each of these listed "things" the subjects rated on the same twelve scales used by content analysts to rate concerns identified in interview protocols. The basic hypotheses were that the amount of time that people spend thinking about something ought to be predicted best by two kinds of interactions: incentive value times expectancy of success, and incentive value times nearness to consummation. Since the instructions for the Concern Dimensions Questionnaire tend to elicit ratings mostly of important incentives, and therefore reduce greatly the range of incentive value, it was also regarded as likely that expectancy and nearness would constitute linear (i.e., main-effect) predictors, and that the linear predictiveness of incentive value alone would be underestimated in these data. In addition to these hypotheses, we also tested the triple interaction of incentive value, expectancy, and nearness, as well as the individual variables squared.

As indicated earlier, each of these classes of variables was represented in the data by two or more variables each. For incentive value, these were affective positivity (strength of appetitive wish), affective negativity (strength of avoidant wish), importance, desire (how much "fired up"), previous investment in the incentive, degree of anticipated emotional investment, anticipated future cost of pursuing the incentive, and the extent of loss that would be experienced in not attaining it. Expectancy was measured by subjective probability of success and feeling that the enterprise was in jeopardy. Nearness in time was measured by the estimated nearness to consummation and the amount of time left in which something could still be done about attaining the incentive. As might be expected, the variables in each cluster were at least moderately intercorrelated. Factor analysis of the ratings (principal components, varimax rotation) revealed four factors, one bipolar defined by affective positivity and negativity, and three unipolar factors that corresponded to the remaining incentive value variables, the expectancy variables, and the nearness variables. Since inspection cast some doubt on the practical independence of the positivity factor from the incentive value factor, and since there were no separate hypotheses for it, the two variables, positivity and negativity, were treated along with the other incentive-value variables.

We analyzed these data in series of stepwise multiple regression analyses in which each incentive ("thing") listed and rated served as the unit of analysis. Each rating was adjusted by subtracting the subject's individual mean for that variable and adding five to eliminate negative values. The criterion variable, of course, was the rank assigned to each incentive to indicate how much the subject had thought about it in the past 36 hr. The predictors were potentially all of the twelve rating variables considered as main effects, products of scores on each incentive value variable times scores on each possible expectancy variable, products of scores on each incentive value variable times scores on each nearness variable, all possible triple interactions that combine one variable from each of the three groups, and variables constructed out of the squares of each single variable. The stepwise process started by choosing one predictor of each type on the basis of the factorial purity and interjudge reliability of its component ratings and then stepwise adding other predictors of the same type. This permitted tests of specific hypotheses without undue proliferation of potential predictors in particular analyses. Table 4 presents simple correlations between the criterion variable and each predictor (except triple interaction terms) about which there had been a hypothesis.

The most powerful single-term predictors are clearly the interactions of incentive value times nearness (or time available). That is, subjects reported thinking most about those incentives that were at the same time most valuable and most pressing. However, the second most powerful simple predictors are the time-related variables uncomplicated by multiplication with value. When one controls for the two simple correlations of time spent thinking about something

Table 4. Correlations of Self-Rated Time Spent Thinking About Incentives with Value Ratings and with Products of Value Times Other Ratings[a]

		Correlations with value *X*			
Value variables	Simple correlations	Probability of success	Jeopardy	Nearness	Time available
Positivity	0.14	0.15	0.12	0.30	0.35
Negativity	0.01	0.12	–0.02	0.23	0.27
Importance	0.20	0.13	0.20	0.35	0.41
Previous investment	0.18	0.19	0.12	0.34	0.38
Emotional investment	0.21	0.21	0.15	0.37	0.41
Cost	0.11	0.15	0.06	0.29	0.34
Loss	0.11	0.14	0.07	0.30	0.35
Desire	0.22	0.21	0.15	0.36	0.43

[a]Signs of all correlations, which were with a ranked variable, have been reversed where necessary so as to reflect the direction of relationship between the variables as conceptualized. Time spent thinking about an incentive correlated 0.14 with Probability of Success, –0.04 with Jeopardy, 0.28 with Nearness, and –0.35 with Time Available. Missing data reduced some correlations to 533 cases, from 569. With 487 degrees of freedom, reflecting 533 incentives listed by 45 subjects, correlations pooled across subjects, the correlation required for significance at the 0.05 level is 0.09 and at the 0.01 level is 0.12, two-tailed.

with (a) a particular value variable (for instance, amount of emotional investment anticipated) and with (b) a particular time-pressure variable (for instance, amount of time available to act), the partial correlation of time spent thinking with the interaction of value and time available becomes nonsignificant. The failure of the interaction to remain significant may be attributable to the restricted range of value ratings in these data.

The data confirm the importance of temporal approach and avoidance gradients in determining incentive effects on mental activity. However, since Time Available is a somewhat better predictor than Nearness, and, as we shall see, Time Available and Nearness both enter stepwise into a multiple regression equation, time pressure appears to be a factor independent of an approach gradient effect.

The correlations of time spent thinking about something with the various products of value times probability of success are considerably smaller, but here two interactions of value variables (Desire and Positivity) with Jeopardy remain significant when value and Jeopardy are separately partialed out.

Using the "stepwise-stepwise" approach described above, the most efficient equation appears to be that tabulated in Table 5, yielding a multiple correlation of 0.492 ($p<0.001$). The equation contains four value variables (one of which is the square of another: Emotional Investment), two time variables, and one variable consisting of the products of a value variable and a probability-of-

Table 5. Multiple Regression Prediction of Time Spent Thinking About Incentive

Variable	Beta	Significance
Time available for acting	0.299	< 0.001
Nearness of consummation in time	0.166	< 0.001
Square of anticipated emotional investment	0.782	< 0.01
Anticipated emotional investment	–0.563	0.05
Desire	0.187	< 0.001
Sense of loss if unsuccessful	–0.136	0.01
Positivity X jeopardy	0.099	< 0.025

success variable. It is interesting to note that two of the value variables, Loss and Emotional Investment, act as suppressors for the squared Emotional Investment variable and for Desire. Evidently, the effect of incentive value accelerates with higher values.

MOOD AND MENTAL ACTIVITY

The factors that govern the influence of concerns on mental activity vary as a function of numerous situational and internal variables. We have just begun to assess these, and the one we have focused on first is mood, especially elation-depression. Mood itself depends on situational events: with sustained frustration, organisms eventually become depressed. However, mood can be altered artificially through drugs, by having subjects recall sad events, or by showing them depressing films. There is some evidence that altering mood in this way produces some of the same effects as spontaneous mood states do (Gouaux, 1971; Strickland *et al.,* 1974; Weiss *et al.,* 1974), which suggests that mood mediates at least some of the effects of frustrating situations on behavior and therefore indicates the worthwhileness of examining mood effects in their own right.

A large clinical literature describes extensive changes in mental activity during depression. The theory of current concerns readily accounts for these and suggests specific reasons why they should occur (Klinger, 1975). The key factor appears to be a decline in the incentive value of incentives other than the one about whose loss the person is depressed. In severe depression, people become pervasively apathetic to virtually the whole array of incentives around which they had previously organized their lives. Therefore, we would predict that during depressed moods the effects of current concerns on attention, recall, and thought content should also weaken.

Our thought-sampling experiments are beginning to generate some confirmation of these hypotheses. There are two different ways to look at these data: differences between subjects and differences within the same subjects over time.

In order to sample occasions across subjects, we chose each subject's first thought-sampling session for which we had complete data. The number of analyzable sessions obtained in this way ranges from 10 to 14 for various analyses. This number, small as it is, has nevertheless generated several suggestive results, some sufficiently strong to achieve acceptable levels of statistical significance. (Tests involving attention and recall are tested one-tailed, other tests two-tailed.)

During each thought report, subjects rate their thoughts according to how directed and how spontaneous each thought segment felt. Subjects who rated themselves as relatively elated reported their thoughts to be less directed ($r(12) = -0.55$, $p<0.05$) and more spontaneous ($r(12) = 0.49$, $p<.10$) than did relatively depressed subjects.

One way to think about these data is to suppose that the preattentive processes of the more elated subjects are strongly attuned to concern-related cues, which then elicit more spontaneous respondent thoughts related to them. This interpretation is supported by a further finding: subjects are most responsive to concern-related passages, in the sense of shifting attention toward them rather than toward nonconcern-related passages, when they are most elated, and least responsive when they are most depressed ($r(12) = 0.55$, $p<0.025$). Correlations of other indexes with mood parallel this one but do not achieve significance: number of times the subject's first attention shift was toward a concern-related rather than a nonconcern-related passage ($r(12) = 0.40$), and number of concern-related rather than nonconcern-related passages recalled ($r(12) = 0.43$) or thought about ($r(11) = 0.44$), $p<0.10$ in each case. By contrast, correlations of mood with indiscriminate responsiveness, such as total number of attention shifts or total number of passages recalled, were close to zero.

Relationships within subjects between mood and the various indexes of cognitive activity failed to confirm the relationships found with sessions from different subjects. We have no theoretical explanation, but it seems likely that the very weak relationships obtained may be attributable to the unexpectedly small variation in individual subjects' mood self-ratings over sessions.

DISCUSSION

The results permit a number of conclusions. Current concerns guide attention, recall, and thought content toward relevant cues. People are especially likely to think about incentives that are valuable, particularly if these incentives are in jeopardy, and even more so if there is little time left to act on them. Finally (and somewhat more tentatively), people become less responsive to concern-related cues when they are relatively depressed. The results therefore sketch out some relationships between motivational states and cognitive events, and they do so in such a way as to suggest some of the factors that sharpen or enfeeble people's capacity to cope.

In order for people to respond adaptively, they must respond in some appropriate relation to external events, which requires that they notice and accommodate relevant features of those events. The relevant features, however, are not always necessarily the obvious ones, and a considerable amount of important processing goes on outside the arena of concerted action, during periods of preparation or just respondent, undirected thought. Creative problem solving seems to involve respondent processing of cues that have not before consciously been considered relevant to solving the problem (Klinger, 1971; Koestler, 1964). Children who are most creative seem better than others at using cues in their immediate environment (Ward, 1969). Appropriate responsiveness to the right cues lies at the heart of successful adaptation, and the research reported here identifies some of the variables that control it: current concerns that conform to the challenges posed by objective circumstances; correct perceptions of incentive values, accessibility, and the margin of time available for acting; and depression.

Relationships between cognition and motivation are a two-way street. For over a decade, theory and research in this area (e.g., Heckhausen and Weiner, 1972; Zimbardo, 1966) have produced extremely useful insights concerning the ways in which motivation depends on the way people construe themselves and their situations. The theory and data presented here deal with the opposite arc of the cognition-motivation circle. They deal with the dependency of cognitive processes on motivation—with the question of what guides the construing process before it has had a chance to manufacture the cognitions that in turn affect motivation.

Implications for Theory and Method

The results confirm the operation of what must be a continuing brain process that is specific to the particular incentives to which people have become committed. How else can we explain people's demonstrated responsiveness to cues that are related to their incentive pursuits only semantically, and then in a situation that does not permit their doing anything about the incentive, that does not present the incentive itself, and that cannot reasonably be said to foster relevant states of drive? Yet people notice those cues, retain them at least long enough to report them later, and think about them. It is as if people go along continuously in readiness to respond to information related to their current concerns, to respond as desultorily or as effectively as the situation permits. Furthermore, if we suppose that internal, ideational cues play the same kind of role as the external cues we manipulated, the effects on cognitive activity suggest not only a model of stimulus effects on mental activity but also a model of the flow of thought from one segment to the next.

Moreover, these results present the demonstrated relationships in quantitative form. The methods used to obtain them permit reasonably close inspection of moment-to-moment mental activity, and they generate quantitative indexes of that activity that reliably reflect experimental manipulations. These results and methods therefore provide a kind of leverage on the study of mental events that has not heretofore been available. The variables thus created promise to be sensitive enough to serve as vehicles for observing the effects of motivation, emotion, and properties of life situations. Specifically, it should be possible to assess with greater precision the effects of becoming committed to incentives, or of having to disengage from them, of involvement with different kinds of incentives, and of mood states created by nonincentive factors.

The kind of research approach described here arose out of a certain view of motivation, one created to account for mental events as these had been investigated by earlier methods. However, as this research unfolds it should in turn suggest a further reformulation of the relevant motivational theory. It may be of interest to illustrate this kind of development with two new kinds of motivational variables, both of which depend on the kind of joint quantification of motivation and mental activity described here: Concern Mass and Concerns Entropy. In the simplest case, imagine that it is possible to construct a regression equation which can predict a person's responsiveness to cues related to a concern, using as predictors such variables as incentive value, time pressure, and jeopardy. This predicted influence may be called an "Index of Concern Influence." Concern Mass may then be described as the combination (let us say sum) of the predicted influences (of the Index of Concern Influence values) exerted by all of a person's concerns at any moment in time. Such an index should be capable of distinguishing an incentive-rich from an incentive-poor situation. One might differentiate this index with respect to the relative weight of appetitive and aversive concerns. Concerns Entropy may be described by the conventional entropy formula applied to the ratios (p_i) of each Index of Concern Influence value to the person's Concern Mass:

$$CE = -(p_1 \ln p_1 + p_2 \ln p_2 + \ldots + p_n \ln p_n)$$

The index of Concerns Entropy should be capable of distinguishing cases of an individual preoccupied with one or a very few dominant concerns from an individual carrying many concerns of approximately equal influence. These two indexes, Concern Mass and Concerns Entropy, can then become analytic tools for assessing the effects of differing life situations on motivation, cognition, inner experience, and adaptive capacity generally.

SUMMARY

Psychological adaptation is a matter of responding appropriately to cues that bear on survival. The research reported here examines the effects of "current concerns"—motivational states of having become committed to particular incentives—on responsiveness to cues. Sixteen college students' current concerns were assessed on repeated occasions through interviews and questionnaires. Subsequently, they listened dichotically to narratives modified intermittently to relate to their concerns and nonconcerns simultaneously on opposite channels. They continuously indicated their attention to a particular channel with a toggle switch and gave periodic verbal reports of their most recent thoughts and recollections of the taped material. They significantly more often attended to, recalled, and reported thoughts related to concern-related rather than nonconcern-related passages. The results suggest that preattentive processes perform a relatively crude scan for cues related to current concerns, followed by a more refined selection of concern-related information for recall and thought processing. Therefore, the directing influence of concerns over information processing and the conditions that moderate their influence are crucial to successful adaptation. Subjects were more specifically attentive to concern-related cues and reported that their thoughts depended less on deliberate self-direction if they were relatively elated rather than relatively depressed. Forty-five other subjects reported thinking most about those incentives that they regarded as affectively most valuable, which they had the least time left to act on, and which were jointly most valuable and jeopardized.

ACKNOWLEDGMENTS

It is hard to determine where author status should leave off and credit begin. The following people made clearly definable conceptual and methodological contributions to the research: Roxanne M. Anderson, Rachel Froiland Quenemoen, Deborah A. Smith, and Susan Stumm. We thank the following for extensive technical and observational contributions: John F. Andrews, Jane M. Delage, Paul F. Heyl, Mary K. Martin, George A. Peterson, and Stephen C. Peterson. Wei-Ching Chang provided valuable statistical consultation. We thank the following for their general assistance: Cheryl Barta, Charles E. Cornillie, Laura A. Falteisek, David Farmer, Bayne E. Holley, Susan E. Jackson, Madeline Maxeiner, Jean A. Nolting, Anthony Palmer, Kathleen M. Reiman, Gail Rixen, Gloria J. Rixen, Yvonne Storck, and Charlotte Syverson. The research was supported by Grant No. 1-R01-MH24804 from the National Institute of Mental Health. Essential pilot studies not reported here were made possible by grants-in-aid from the University of Minnesota Graduate School.

REFERENCES

Atkinson, J. W. (1964). *An Introduction to Motivation.* Princeton, N. J.: Van Nostrand.

Beckett, S. (1968). *Stories and Texts for Nothing.* New York: Grove.

Gouaux, C. (1971). Induced affective states and interpersonal attraction. *J. Pers. Soc. Psychol. 20,* 37–43.

Heckhausen, H., and Weiner, B. (1972). The emergence of a cognitive psychology of motivation. In *New Horizons in Psychology II* (P. C. Dodwell, Ed.). Baltimore: Penguin, pp. 126–147.

Klinger, E. (1971). *Structure and Functions of Fantasy.* New York: Wiley.

Klinger, E. (1975). Consequences of commitment to and disengagement from incentives. *Psychol. Rev. 82,* 1–25.

Koestler, A. (1964). *The Act of Creation.* New York: Macmillan.

McClelland, D. C., Atkinson, J. W., Clark, R. A., and Lowell, E. L. (1953). *The Achievement Motive.* New York: Appleton-Century-Crofts.

Miller, N. E. (1944). Experimental studies of conflict. In *Personality and the Behavior Disorders* (Vol. 1) (J. McV. Hunt, Ed.). New York: Ronald, pp. 431–465.

Neisser, U. (1967). *Cognitive Psychology.* New York: Appleton-Century-Crofts.

Strickland, B. R., Hale, W. D., and Anderson, L. K. (1974). Effect of induced mood states on activity and self reported affect. Paper delivered at the annual convention of the American Psychological Association, New Orleans.

Ward, W. C. (1969). Creativity and environmental cues in nursery school children. *Dev. Psychol 1,* 543–547.

Wessman, A. E., and Ricks, D. F. (1966). *Mood and Personality.* New York: Holt, Rinehart & Winston.

Weiss, J. M., Glazer, H. I., and Pohorecky, L. A. (1974). Neurotransmitters and helplessness: A chemical bridge to depression? *Psychol. Today 8*(7), 58–62.

Zimbardo, P. G. (1966). The cognitive control of motivation. *Trans. NY Acad. Sci.* Ser. II, *28*(7), 902–922.

Stereotyped Behavior and Stress

ELLIOT S. VALENSTEIN

It does not require a very extensive exposure to the relevant literature to appreciate the apparent impossibility of defining stress very rigorously (Mason, 1975*a*). Stress cannot be defined in terms of the physical properties of the eliciting stimulus as cognitive and emotional factors can change the impact of any stimulus on the same person or animal (Lazarus, 1967; Mason, 1975*b*). It is equally difficult to define stress as a specific set of physiological responses. Lacey (1967), for instance, has repeatedly pointed out that each "stressor" tends to elicit its own automatic response pattern, while at the same time there is a tendency for individuals to have unique physiological response profiles. Mason (1975*b*), who measures gonadal, thyroidal, growth hormone, adrenal medullary and insulin activity as well as pituitary-adrenal cortical changes, also emphasizes the specific adaptive pattern of hormone response depending upon the type of stress. Selye's (1974) definition of stress as the nonspecific response of the body to any demand (whether pleasant or unpleasant) is likely to be too broad (and also too restricting in its focus on only a few responses) to continue to be useful in the future. Perhaps, agreement can be reached only on a very general statement such as an organism is subjected to stress when its adaptive mechanisms are taxed beyond their normal range of functioning either because of the intensity or the duration of the response required. In spite of the difficulty in defining stress, there may still be value in bracketing a field that concentrates on the consequences of straining adaptive mechanisms. Certainly, Selye (1956) has proven the value of such a focus in his highly productive career.

Traditionally, those who have studied the physiological responses to stress

ELLIOT S. VALENSTEIN · The University of Michigan, Ann Arbor, Michigan.

have emphasized the primary role of the pituitary-adrenal cortical axis. The brain has been drawn into the picture mainly in its role as a "perceiver" of psychological stress as well as a regulator of pituitary-adrenal response. More recently the role of the brain has been extended by the demonstration that a part of the ACTH polypeptide chain may directly activate brain circuits that play an important role in evoking fear (DeWied, 1974). Under the influence of these polypeptides, the capacity of stimuli to induce fear is prolonged and animals cope by continuing to make avoidance responses even when there is no longer anything to avoid. In this report, I would like to speculate about several phenomena in the brain-behavior field that seem to be revealing another type of "coping response"—one that seems to be regulated by specific neural pathways in the brain.

The first of these phenomena concerns the well-established demonstration that electrical stimulation of certain brain regions can induce animals to engage in a great variety of biologically significant behaviors such as eating, drinking, sex, and aggression (Valenstein, 1973*b*). The most common interpretation of these demonstrations is that the electrical stimulus activates the specific neural circuits regulating biological drives such as hunger, thirst, and other "appetites." More recently, however, studies in my own laboratory have suggested an alternative explanation—one more closely related to the concept of a stress-induced "coping response." Briefly, we demonstrated that rats displaying eating or drinking in response to stimulation do not behave as if hungry or thirsty when the testing conditions are varied (Valenstein, 1969; Valenstein *et al.*, 1970). Stimulated animals might eat one type of food, but when offered another food—they commonly do not eat even though the stimulus parameters are identical (Valenstein *et al.*, 1968*b*). Such animals may not even eat the same food when it is changed in texture as when food pellets are ground and offered as a dry or wet mash. It has also been shown that the taste preference of an animal when it is thirsty is very different from when it is drinking in response to brain stimulation (Valenstein *et al.*, 1968). Particularly significant was the observation that when water was removed from the drinking bottle, brain stimulation continued to evoke lapping, seemingly without any extinction (Valenstein *et al.*, 1968). *Such observations strongly suggested that the execution of a stereotyped response was more important for maintaining the stimulation-evoked behavior than any substance that was ingested.* In an earlier theoretical paper, Glickman and Schiff (1967) had also suggested that the execution of a response might be reinforcing by itself.

These observations and others raised doubts about the relevance of biological drives for the behavior evoked by brain stimulation and encouraged us to reexamine some of the pioneering studies in this field. This seemed particularly important because the early studies were undertaken prior to the establishment of a strong bias in favor of the biological drive interpretation (Valenstein,

1973*b*). The earlier reports strengthened our conviction that the response might be more important than the substance ingested. Hess (1957), for example, described the eating elicited by brain stimulation as follows:

> "Stimulation here produces bulimia. If the animal has previously taken neither milk nor meat, it now devours or drinks greedily. As a matter of fact, *the animal may even take into its mouth or gnaw on objects that are unsuitable as food, such as forceps, keys, or sticks*" [p. 25, italics added].

Similarly, Greer's (1955) description of the drinking that could be induced by stimulating the diencephalon of a rat strongly suggested that the maintenance of the evoked response was not dependent on the availability of water. Greer first noted that his rat would vigorously lick the glass wall every time brain stimulation was administered. When the stimulation was delivered intermittently during the night, the behavior gradually changed. Greer described this event as follows:

> "It was now found that stimulation would result in violent drinking activity. The non-specific licking response had been lost. As soon as the current was turned on, the animal would jump for the water bottle and continue to drink avidly until the switch was turned off. If the water bottle was removed and the current then turned on, the rat would go back to its 'licking' behavior of the previous day, but would immediately transfer it to drinking behavior when the water bottle was replaced" [Greer, 1955, pp. 60–61].

Our conclusion, that stimulation was probably not evoking natural motivational states was supported by other observations as well. Animals induced to eat by stimulation and therefore presumed to be hungry would often switch to drinking or some other behavior (Valenstein *et al.*, 1968*a*). These results did not seem consistent with the hypothesis that a specific drive state had been induced. Although there were several alternative ways of interpreting the data suggested by Wise (1968) and Roberts (1969) their objections were answered by subsequent experiments and arguments (see Valenstein, 1969, 1971, 1973*b*, 1975*a,b*, 1976*b*). The conclusion that brain stimulation did not evoke specific drive states and that the behavioral changes it produces can be modified by changing the environmental contingencies now seems compelling. This conclusion applies equally to observations of humans experiencing brain stimulation as well as to animals (Valenstein, 1973*a*, 1976*a*).

It is not meant to imply that stimulation at any brain site can evoke any behavior if the environmental contingencies are arranged appropriately. This does not seem to be the case for several reasons. One of the main factors that may determine the types of behaviors which can be evoked is the activation (or facilitation) of fragmentary motor responses such as chewing and licking. While these motor responses do not duplicate natural motivational states, they may play an important role in determining the behavior that is displayed during stimulation. Berntson and Hughes (1974), for example, elicited eating in the cat by medullary reticular formation stimulation, but the behavior of their animals did not support the view that a hunger drive was evoked. These investigators described the behavior in the following way:

> "With stimulation of medullary 'eating' points, cats would eat not only cat chow, but nonnutritive items, such as small pieces of bedding. The cats, however, would not lap up liquified tuna fish, a normally highly preferred food. These results suggest that stimulation did not induce a generalized hunger 'drive.' Rather, stimulation may have selectively facilitated sensorimotor mechanisms for specific components of eating behavior, such as chewing and swallowing, but not lapping" [Berntson and Hughes, 1974, pp. 260–261].

Earlier, Berntson and his colleagues reported almost identical behavior elicited in cats by cerebellar stimulation and noted that "hungry cats which were lapping highly preferred liquified tuna before stimulation, promptly left it and began eating dry cat chow when stimulated" (Berntson *et al.,* 1973). It is obvious that the facilitation of motor responses—a factor, which may be involved to a greater or lesser extent in any given instance—will also influence the evoked behaviors that are likely to substitute for one another.

In addition to the sensitization of particular responses, the positive or aversive motivational or emotional states evoked by the brain stimulation will also serve to channel behavior. There are very likely to be clusters of behaviors that are more naturally linked to either "positive" or "negative" emotional states. Stimuli that irritate, cause pain, or in any way are aversive facilitate different behaviors than stimuli that induce pleasant or positive motivational states. A more detailed analysis of the ways that the induced motor and motivational effects of brain stimulation influence the behaviors that are expressed has been presented elsewhere (Valenstein, 1976*b*).

Even though the sensitization of motor responses and the evocation of positive and aversive emotional states set some limits to the behavioral possibilities that are likely to be observed during brain stimulation there is much plasticity that remains. If an animal is prevented from engaging in one behavior during stimulation it is likely to adopt a different response pattern. It becomes important to ask, therefore, why animals seem to be compelled to become involved in some behavior during stimulation. A possible clue to the answer to this question may be provided by studies of what appear to be two closely related phenomena.

Recently, Antelman and his colleagues (Antelman and Szechtman, 1975; Antelman *et al.,* in press) have observed that pinching the tail of rats reliably induces eating, gnawing, licking, and other behaviors in almost all rats tested. Typically, when pressure is applied to its tail, a rat sniffs and explores for a few seconds before approaching a food pellet and eating it. In many respects the tail-pinch behavior is strikingly similar to that evoked by brain stimulation. For example, eating gradually becomes more reliably evoked by tail-pinch over successive trials as does eating in response to brain stimulation (Valenstein, 1971). If food is not available during tail-pinch, rats will switch to other behaviors such as gnawing or licking. Rats can be made to overeat and become obese by tail-pinching (Rowland and Antelman, 1976) as Steinbaum and Miller (1965) observed with brain stimulation. As was shown for brain stimulation (Devor *et al.,* 1970), after the animals consume a considerable amount of food in response to tail-pinch, the intensity of this stimulus needs to be increased in order to induce the animal to eat. If all the goal objects are removed, rats become either hyperactive or engage in vigorous escape maneuvers when pressure is applied to their tails. This is very similar to observations of brain-stimulated animals made in my laboratory. Furthermore, because animals may eat only certain foods and not others during tail-pinch, this behavior seems to have some of the "finicky" characteristics my colleagues and I have described with brain stimulation (Valenstein *et al.,* 1968*b*; Valenstein *et al.,* 1968).

It may be important to note that even though the behavior induced by brain stimulation and tail-pinching are very similar in many important ways, they are not identical. For example, there is a general preference for wet food during tail-pinch and while this may be considered a type of finickiness the food preferences of brain-stimulated animals are much more individualized. This latter fact probably reflects the facilitation of different motor responses and emotional states by brain stimulation. While tail-pinch as it is normally applied may not be intensely aversive or painful it is clearly annoying. Brain stimulation on the other hand is capable of evoking a very great range of qualitatively different emotional states extending from pleasure to pain. It also would seem to be true that most rats receiving tail-pinch prefer eating over other behaviors, while the first behavior displayed by brain-stimulated animals has much greater variability. Nevertheless, it is true that many behaviors can also be seen during tail-pinch if the environment provides the appropriate stimulus incentives. In some preliminary studies, we have observed animals carrying food pellets (but not eating them) and engaging in coprophagia in response to tail-pinch in a manner similar to that of some animals responding to brain stimulation. Tail-pinch as well as tail-shock has also been observed to induce copulatory behavior (Barfield and Sachs, 1968; Caggiula and Eibergen, 1969; Caggiula, personal communication), aggression (Caggiula, 1972), and maternal behavior (Sherman *et al.,* unpublished observations).

The many similarities to the tail-pinch phenomenon places the brain-stimulation evoked behaviors in a new perspective. In addition to any special properties of brain stimulation such as might be related to the motor and motivational responses it can evoke, there may be a core of commonality with tail-pinch that involves similar mechanisms. Evidence is beginning to accumulate which suggests that tail-pinch and brain-stimulation evoked behaviors depend upon the facilitation of the nigrostriatal dopamine system. Antelman and his colleagues (Antelman and Szechtman, 1975; Antelman *et al.*, 1976) have demonstrated that both dopamine antagonists (haloperidol, spiroperidol, or pimozide) and the application of 6-OHDA, the catecholamine toxic agent, to the nigrostriatal system eliminates, or significantly reduces, tail-pinch behavior. In contrast, alpha and beta norepinephrine receptor blockers (Phentolamine, sotolol) and norepinephrine synthesis inhibitors (FLA-63) had either no effect on tail-pinch behavior, or in the case of the latter, a potentiating influence.

The neuropharmacological basis of the behavior induced by brain stimulation has not been studied extensively, but at least one report suggests that a dopaminergic system might also underlie this behavior. Phillips and Fibiger (1973) have reported that intraventricular injections of 6-OHDA, eliminated brain-stimulation induced feeding even after normal eating had returned to preoperative levels. Parenthetically, this latter observation reinforces the conclusion that normal eating and brain-stimulation induced eating involve different mechanisms. Although intraventricular injections of 6-OHDA depletes both norepinephrine as well as dopamine, Phillips and Fibiger present additional evidence arguing for the importance of a dopaminergic system for the deficits in brain-stimulation induced eating produced by 6-OHDA. Recently, Phillips and Fibiger (Phillips, personal communication) have significantly attenuated "stimulus-bound" behavior by administering the dopamine antagonist, haloperidol. Antelman, Black, and Fisher (unpublished observations cited in Antelman *et al.*, 1976) also have observed that brain-stimulation induced behavior is similar to tail-pinch behavior in its dependence on dopamine.

If facilitation of a dopaminergic system is indeed critical for both brain-stimulation and tail-pinch induced behaviors, it becomes important to speculate about the influence this facilitation might have on behavior. Over the past several years, a considerable body of evidence implicating dopamine in sensory and motor functions has accumulated. Ungerstedt (1974) has recently reviewed the experimental data which demonstrates the importance of the nigrostriatal dopamine system for sensory and motor processes. Following destruction of the nigrostriatal system, animals exhibit a "sensory neglect" syndrome and are unresponsive to either distal or tactual stimuli. The facilitation of dopaminergic pathways increases responses to the same stimuli (see reviews by Ungerstedt, 1974, and Snyder, 1974*a*). In man, dopamine release during amphetamine psychosis can sometimes produce a hypersensitivity to skin sensations and a delusion of skin parasites. Associated with these sensations is a repetitive scratching or "grooming" response. The observation of tail preening and vigorous grooming sometimes produced by brain stimulation and tail-pinch in rats may be a related phenomenon (Valenstein *et al.*, 1970; Valenstein, personal observations; Antelman, personal communication). Snyder (1974*b*) among others has argued that hyperactivity in dopaminergic pathways during schizophrenia may allow even normally insignificant stimuli to have an intrusive and disruptive influence on thought processes. The perceptual distortions and hallucinations commonly experienced by schizophrenics are believed by many to be a basic part of the schizophrenic disorder. Shakow (1971), for instance, has noted that schizophrenics confuse figure and background and the observation of abnormal eye movements in such patients by several investigators (e.g., Holtzman *et al.*, 1973) may reflect their hyperresponsiveness to stimuli. The effectiveness of neuroleptics (antipsychotics) in counteracting the perceptual distortions and mental confusion of schizophrenics is highly correlated with their capacity to antagonize dopamine transmission (see, for example, the recent report by Seeman and Lee, 1975).

The nigrostriatal dopamine system has also been demonstrated to play a significant role in regulating motor responses. The experimental and clinical evidence that this system is critically involved in Parkinson's disease is very extensive and quite generally considered to be totally convincing. Parkinsonian patients, who display among other symptoms an inertia in initiating movements, have been shown to have dopamine deficiencies in the caudate nucleus and they often respond to L-dopa therapy. Moreover, Ungerstedt (1974) has been able to induce lateralized turning responses by unilateral destruction of the nigrostriatal system while the administration of dopamine mimetics and antagonists in such a preparation has provided further insight into the action of this system in regulating motor responses. It has also been shown that the spatial preference of individual intact rats and the direction of turning induced by amphetamine is correlated with asymmetrical levels of dopamine in the caudate nucleus (Glick *et al.*, 1974; Zimmerberg *et al.*, 1974).

All of this evidence has led to the suggestion that the nigrostriatal dopamine pathway regulates the process by which motor commands become engaged with the neural circuitry that translates commands into actions (Matthysse, 1974). This neural circuit appears to exert its influence at the interface between motor commands and their execution. Hyperactivity may result from an overly responsive nigrostriatal system, while inhibition and a general dampening of movement as in Parkinson's disease is associated with deficiencies in this circuit. Paralleling the regulatory influence of this circuit on the motor system is its role in sensory processing. The degree to which stimuli impose themselves on attentional and perhaps emotional mechanisms may be regulated by nigrostriatal system. The general neglect of stimuli that follows destruction of this system has already been mentioned (see also Marshall *et al.*, 1971, 1974). Conversely, as already mentioned, overactivity in this circuit may result in even normally trivial stimuli commanding attention.

Further insight into the influence of the nigrostriatal dopamine system on behavior may be revealed by another phenomenon that appears to have much in common with the behaviors observed during tail-pinch and brain stimulation.

Randrup and Munkvad (1970) have described the increases in sniffing, licking, and gnawing produced by amphetamines when administered to the rat. During the more intense amphetamine induced stereotypy, behaviors such as grooming, eating, and forward locomotion are absent, but these behaviors may be present before and after the stage of most extreme stereotypy. Although both catecholamines are released by amphetamines, dopamine rather than norepinephrine has been implicated because DOPA injections can produce this behavior and inhibiting the synthesis of norepinephrine by diethyldithiocarbamate (DDC) does not block amphetamine-induced stereotypies. Moreover, neither alpha nor beta norepinehprine receptor blockers prevent the appearance of stereotypies.

Stereotyped behaviors have been induced by amphetamines in many species. The stereotyped behaviors adopted, however, are not always the same. Monkeys receiving amphetamine injections may engage in a continuous locomotion over the same route while repeating stereotyped responses in specific parts of the cage. High doses of amphetamines in humans produce a perseveration of single behaviors (see review by Randrup and Munkvad, 1970) and amphetamine psychosis has been mistaken for the stereotyped posturing frequently displayed by schizophrenics. Schizophrenics also display elaborate, ritualistic behavior sequences. For instance, Hanley *et al.* (1972) have described a patient who invariably moved his shoes about "in a repetitive precise manner before putting them on . . . [and then] . . . went through a laborious and seemingly prescribed manner of tying his shoelaces, which he would then undo and start tying over again."

In common with brain-stimulation and tail-pinch induced behaviors is the fact that amphetamine stereotypies usually develop gradually with particular components and manner of expression becoming strengthened over time. Also, Randrup and Munkvad (1970) have observed that "the kind of activity that is stimulated depends on the species, the individual, the environment, previous learning, amphetamine dose and probably other factors." Ellinwood *et al.* (1972) have noted that amphetamine stereotypies in cats and monkeys also evolve gradually into single stereotyped patterns.

Admittedly, the evidence is far from conclusive, but there seems to be sufficient reasons for seriously considering the possibility that the behaviors induced by brain stimulation, tail-pinch, and amphetamines may result in part from a facilitation of dopaminergic pathways. The result of this facilitation may produce a state of high arousal and a hyperresponsiveness to stimuli—a distressing state that can be coped with by resorting to stereotyped behavior. The stereotyped behavior that is expressed may reflect the relative level of spontaneous (but subliminal) activity in motor circuits characteristic of the animal being observed. The particular behaviors seen most commonly probably reflects responses that are likely to be adaptive for a given species, but individual life experiences may also play a role. Engaging in stereotyped behavior is probably adaptive because it helps to limit the variety of stimuli to which the organism is exposed. The amount of stimulus input required to overwhelm and therefore to require the recruitment of such adaptive responses probably varies with mental state and capacity to process incoming information. Thus, stereotypics may be seen during exposure to stress as well as in older people who attempt to limit the complexity, intensity, and novelty of environmental inputs by adopting increasingly routinized activities. This may be a commonly adopted pattern to cope with the stress produced by "perceptual flooding." Schizophrenics also tend to limit their exposure to the novel or strange and Shakow (1971) has included "neophobia" as one of the psychological features of this disorder.

Although dopaminergic transmission has been given a central role in this discussion, it would be unrealistic to rule out a role for most, if not all, neurotransmitters. Norepinephrine may also induce stereotypies. Ayhan and Randrup (1972) have concluded that norepinephrine and dopamine may have a different influence as Ellinwood *et al.* (1972) have emphasized that a balance between norepinephrine and dopamine may determine the relative amount of hyperactivity and stereotypy that occurs. The latter investigators have observed that cats pretreated with antabuse, which inhibits dopamine-beta-hydroxylase and therefore norepinephrine synthesis, exhibit less hyperactivity and more intense stereotyped response patterns. Moreover, although Antelman *et al.* (1976) have argued that tail-pinch behavior depends upon a facilitation of the nigrostriatal dopamine system, the fact that tail-pinch does not increase dopamine turnover suggested to these investigators that neurotransmitters acting antagonistically to dopamine must be involved. They present an argument for the involvement of norepinephrine and gamma-aminobutyric acid (GABA). GABA has also been implicated in schizophrenia according to at least one suggested hypothesis (Roberts, 1972). It is also generally acknowledged that acetylcholine has an antagonistic effect on the nigrostriatal dopamine system. Munkvad *et al.* (1968) have shown, for example, that anticholinergics enhance amphetamine-induced stereotypy while cholinergic drugs weakly antagonize it. The importance of cholinergic activity in the caudate nucleus for Parkinson's disease has also long been recognized (McLennan and York, 1966). Norepinephrine and dopamine have also been implicated in learning by a number of investigators (see Olds, pp. 47–75, and Grossman, pp. 11–46). It is conceivable, therefore, that these neurotransmitters play a critical role in the reinforcement process that leads to the repetition of the same behaviors during stimulation. The fact that many neurotransmitters seem to contribute to the development of stereotyped "coping" responses should not blur the central role of the nigrostriatal dopaminergic system.

Lastly, the way specific behaviors get selected during brain stimulation and why they appear with increasing regularity over time is a question of considerable importance. We have expended considerable effort in trying to link behaviors to reinforcing brain simulation with little success. The pairing of reinforcing brain stimulation with the act of eating when the animals were deprived of food was not successful in producing "simulus-bound" eating (Valenstein and Cox, 1970). It is likely that these attempts were unsuccessful because eating in response to hunger has little to do with the physiological mechanisms underlying eating induced by brain stimulation. The eating and other behaviors induced by brain stimulation and tail-pinch are probably more closely related to the "compulsive" and "neurotic" eating seen in response to stress than it is to that eating which is responsive to nutritional needs.

REFERENCES

Antelman, S. M., and Szechtman, H. (1975). Tail pinch induces eating in sated rats which appears to depend on nigrostriatal dopamine. *Science 189,* 731–733.

Antelman, S. M., Szechtman, H., Chin, P., and Fisher, A. E. (1976). Tail-pinch-induced eating, gnawing and licking behavior in rats: Dependence on the nigrostriatal dopamine system. *Brain Res.* (in press).

Ayhan, I. H., and Randrup, A. (1972). Role of brain noreadrenaline in morphine-induced stereotyped behavior. *Psychopharmacologica 27,* 203–212.

Barfield, R. J., and Sachs, B. D. (1968). Sexual behavior: Stimulation by painful electric shock to the skin in male rats. *Science 161,* 392–394.

Berntson, G. G., and Hughes, H. C. (1974). Medullary mechanisms for eating and grooming behaviors in the cat. *Exp. Neurol. 44,* 255–265.

Berntson, G. G., Potolicchio, S. J., Jr., and Miller, N. E. (1973). Evidence for higher functions of the cerebellum: Eating and grooming elicited by cerebellar stimulation in cats. *Proc. Natl. Acad. Sci. USA 70,* 2497–2499.

Caggiula, A. R. (1972). Shock-elicited copulation and aggression in male rats. *J. Comp. Physiol. Psychol. 80,* 393–397.

Caggiula, A. R., and Eibergen, R. (1969). Copulation of virgin male rats evoked by painful peripheral stimulation. *J. Comp. Physiol. Psychol. 69,* 414–419.

Devor, M. G., Wise, R. A., Milgram, N. W., and Hoebel, B. G. (1970). Physiological control of hypothalamically elicited feeding and drinking. *J. Comp. Physiol. Psychol. 73,* 226–232.

DeWied, D. (1974). Pituitary-adrenal system hormones and behavior. In *The Neurosciences: Third Study Program* (F. O. Schmitt and F. G. Worden, Eds.). Cambridge, Mass.: MIT Press, pp. 653–666.

Ellinwood, E. H., Sudilovsky, A., and Nelson, L. (1972). Behavioral analysis of chronic amphetamine intoxication. *Biol. Psychiatry 4,* 215–230.

Glick, S. D., Jerussi, T. P., Waters, D. H., and Green, J. P. (1974). Amphetamine-induced changes in striatal dopamine and acetylcholine levels and relationship to rotation (circling behavior) in rats. *Biochem. Pharmacol. 23,* 3223–3225.

Glickman, S. E., and Schiff, B. B. (1967). A biological theory of reinforcement. *Psychol Rev. 74,* 81–109.

Greer, M. A. (1955). Suggested evidence of a primary "drinking center" in the hypothalamus of the rat. *Proc. Soc. Exp. Biol. Med. 89,* 59–62.

Hanley, J., Rickles, W. R., Crandall, P. H., and Walter, R. D. (1972). Automatic recognition of EEG correlates of behavior in a chronic schizophrenic patient. *Am. J. Psychiatry 128,* 1524–1528.

Hess, W. R. (1957). *The Functional Organization of the Diencephalon.* New York: Grune and Stratton.

Holtzman, P., Proctor, L. R., and Hughes, D. W. (1973). Eye tracking patterns in schizophrenia. *Science 181,* 179–181.

Lacey, J. I. (1967). Somatic response patterning and stress: Some revisions of activation theory. In *Psychological Stress: Issues in Research* (M. H. Appley and R. Trumbull, Eds.). New York: Appleton-Century-Crofts, pp. 14–37.

Lazarus, R. S. (1967). Cognitive and personality factors underlying threat and coping. In *Psychological Stress: Issues in Research* (M. H. Appley and R. Trumbull, Eds.). New York: Appleton-Century-Crofts, pp. 151–169.

Marshall, J. F., and Teitelbaum, P. (1974). Further analysis of sensory inattention following lateral hypothalamic damage in rats. *J. Comp. Physiol. Psychol. 86,* 375–395.

Marshall, J. F., Turner, B. H., and Teitelbaum, P. (1971). Sensory neglect produced by lateral hypothalamic damage. *Science 174,* 523–525.

Mason, J. W. (1975). A historical view of the stress field. Part II. *J. Human Stress, 1,* 22–36.

Mason, J. W. (1975). Emotion as reflected in patterns of endocrine integration. In *Emotions–The Parameters and Measurement* (L. Levi, Ed.). New York: Raven Press.

Matthysse, S. (1974). Schizophrenia: Relationship to dopamine transmission, motor control, and feature extraction. In *The Neurosciences: Third Study Program* (F. O. Schmitt and F. G. Worden, Eds.). Cambridge, Mass.: MIT Press, pp. 733–737.

McLennan, H., and York, D. H. (1966). Cholinergic mechanisms in the caudate nucleus. *J. Physiol. 187,* 163–175.

Munkvad, I., Pakkenberg, H., and Randrup, A. (1968). Aminergic systems in basal ganglia associated with stereotyped hyperactive behavior and catalepsy. *Brain Behav. Evol. 1,* 89–100.

Phillips, A. G., and Fibiger, H. C. (1973). Deficits in stimulation-induced feeding after intraventricular administration of 6-Hydroxydopamine in rats. *Behav. Biol. 9,* 749–754.

Randrup, A., and Munkvad, I. (1970). Biochemical, anatomical and psychological investigations of stereotyped behavior induced by amphetamines. In *Amphetamines and Related Compounds* (E. Costa and S. Garratini, Eds.) New York: Raven Press, pp. 695–713.

Roberts, E. (1972). An hypothesis suggesting that there is a deficit in the GABA system in schizophrenia. In *Prospects for Research in Schizophrenia* (S. Kety and S. Matthysse, Eds.). Neuroscience Research Program Bulletin, Vol. 10 (No. 4), November pp. 468–482.

Roberts, W. W. (1969). Are hypothalamic motivational mechanisms functionally and anatomically specific? *Brain, Behav. Evol. 2,* 317–342.

Rowland, N. E., and Antelman, S. M. (1976). Stress-induced hyperphagia and obesity in rats: A possible model for understanding human obesity. *Science 191,* 310–312.

Seeman, P., and Lee, T. (1975). Antipsychotic drugs: Direct correlation between clinical potency and presynaptic action on dopamine neurons. *Science 188,* 1217–1219.

Selye, H. (1956). *The Stress of Life.* New York: McGraw-Hill.

Selye, H. (1974). *Stress Without Distress.* New York: Lippincott.

Shakow, D. (1971). Some observations on the psychology (and some fewer, on the biology) of schizophrenia. *J. Nerv. Ment. Dis. 153,* 300–316.

Snyder, S. H. (1974*a*). Catecholamines as mediators of drug effects in schizophrenia. In *The Neurosciences: Third Study Program* (F. O. Schmitt and F. G. Worden, Eds.). Cambridge: MIT Press, pp. 721–732.

Snyder, S. H. (1974*b*). *Madness and the Brain.* New York: McGraw-Hill.

Steinbaum, E. A., and Miller, N. E. (1965). Obesity from eating elicited by daily stimulation of hypothalamus. *Am. J. Physiol. 208,* 1–5.

Ungerstedt, U. (1974). Brain dopamine neurons and behavior. In *The Neurosciences: Third Study Program* (F. O. Schmitt and F. G. Worden, Eds.). Cambridge: MIT Press, pp. 695–703.

Valenstein, E. S. (1969). Behavior elicited by hypothalamic stimulation. A prepotency hypothesis. *Brain, Behav. Evol. 2,* 295–316.

Valenstein, E. S. (1971). Channeling of responses elicited by hypothalamic stimulation. *J. Psychiat. Res. 8,* 335–344.

Valenstein, E. S. (1973*a*). *Brain Control: A Critical Examination of Brain Stimulation and Psychosurgery.* New York: John Wiley.

Valenstein, E. S. (1973*b*). Invited comment: Electrical stimulation and hypothalamic function: Historical perspective. In *The Neuropsychology of Thirst: New Findings and Advances in Concepts* (A. N. Epstein, H. R. Kissileff, and E. Stellar, Eds.). Washington, D.C.: Winston & Sons, pp. 155–161.

Valenstein, E. S. (1975*a*). Brain stimulation and behavior control. In *Nebraska Symposium on Motivation* (J. K. Cole, Ed.). Lincoln, Nebraska: University of Nebraska Press.

Valenstein, E. S. (1975*b*). Persistent problems in the physical control of the brain. James Arthur Lecture on the Evolution of the Human Brain. Published by the American Museum of Natural History, New York.

Valenstein, E. S. (1976*a*). Brain stimulation and the origin of violent behavior. In *Issues in Brain/Behavior Control* (W. L. Smith and A. Kling, Eds.). Holliswood, New York: Spectrum Publications, Inc.

Valenstein, E. S. (1976*b*). The interpretation of behavior evoked by brain stimulation. In *Brain-Stimulation Reward* (A. Wauquier and E. T. Rolls, Eds.). Amsterdam: North Holland Publ. Co.

Valenstein, E. S., and Cox, V. C. (1970). The influence of hunger, thirst and previous experience in the test chamber on stimulus-bound eating and drinking. *J. Comp. Physiol. Psychol. 70,* 189–199.

Valenstein, E. S., Cox, V. C., and Kakolewski, J. W. (1968*a*). Modification of motivated behavior elicited by electrical stimulation of the hypothalamus. *Science 159,* 1119–1121.

Valenstein, E. S., Cox, V. C., and Kakolewski, J. W. (1968*b*). The motivation underlying eating elicited by lateral hypothalamic stimulation. *Physiol. Behav. 3,* 969–971.

Valenstein, E. S., Cox, V. C., and Kakolewski, J. W. (1970). Reexamination of the role of the hypothalamus in motivated behavior. *Psychol. Rev. 77,* 16–31.

Valenstein, E. S., Kakolewski, J. W., and Cox, V. C. (1968). A comparison of stimulus-bound drinking and drinking induced by water deprivation. *Comm. Behav. Biol. 2,* 227–233.

Wise, R. A. (1968). Hypothalamic motivational systems: Fixed or plastic neural circuits. *Science* 162, 377–379.

Zimmerberg, B., Glick, S. D., and Jerussi, T. P. (1974). Neurochemical correlate of a spatial preference in rats. *Science 185,* 623–625.

WORKSHOP I. Neurophysiological Mechanisms of Adaptive Behavior

Edited by ELIOT STELLAR

Quite typical of a rapidly advancing field of study, progress in the area of neurophysiological mechanisms of adaptive behavior has been marked by successive advances in both concept and technique. The major modern advance in concept was Papez's idea (1937) that the rhinencephalon was not primarily concerned with olfactory functions in man, but rather with the control and expression of emotional behavior. Particularly he was thinking of the circuit involving the hippocampus, fornix, hypothalamus (mammillary bodies), mammillo-thalamic tract, anterior thalamus, and cingulate cortex.

His conception was remarkable, for subsequent empirical studies showed that indeed these structures were involved in not only emotional behavior but also drive and motivation. Kluver and Bucy (1939), for example, showed that lesions of the tip of the temporal lobe plus underlying amygdala and hippocampus produced "taming" in wild monkeys, exaggerated oral tendencies, and indiscriminate hypersexuality among other symptoms. About the same time Hetherington and Ranson (1942) showed that lesions of the ventromedial hypothalamus led to hyperphagia and obesity, and even earlier W. R. Hess (1957) showed that electrically stimulating various sites in the hypothalamus

ELIOT STELLAR · Institute of Neurological Sciences, University of Pennsylvania. (Workshop moderated by Eliot Stellar.)

and thalamus of waking cats resulted in remarkable, natural-looking, emotional, and motivational responses such as rage, flight, eating, sleeping, etc.

A second important conceptual advance was Richter's (1942–43) concept of "self-regulatory behavior." These behaviors were adaptive behaviors that contributed to homeostasis or the equilibrium of the internal environment. A good example is salt hunger in the adrenalectomized rat which had dangerously low sodium levels because it is obliged to lose salt at the kidney. Hunger, thirst, and thermoregulatory behavior also fit this paradigm. Not only do they contribute to the maintenance of the internal environment, and thus are adaptive behaviors, but they also are the drives leading to motivated behavior and hedonic experience in man. In fact, it turns out that a whole range of motivated and emotional behaviors (rage, fear, sex, maternal as well as hunger, thirst, and thermoregulatory behavior) have the same basic neurological mechanism in the rhinencephalon, are clearly motivated and adaptive behaviors, and in man, have the subjective qualities of pleasurable or aversive experience (Stellar, 1954, 1974).

The success of building these striking empirical facts on the conceptions of Papez, Richter, and others depended upon certain technological advances that came into widespread use only in recent years. First among these was the use of the stereotaxic method for accurately reaching structures deep within the brain. Developed initially right after the turn of the century by Horsley and Clarke (1908) for use in human neurosurgical patients, it was adapted by Ranson and his coworkers to animal work. Now investigators could make lesions in the hypothalamus, for example, implant electrodes there for stimulation or pippettes for chemical, thermal, and osmotic manipulation.

A second technical advance was the capability of maintaining the chronic animal preparation with implanted electrodes and pippettes as well as lesions in good condition to respond behaviorally. W. R. Hess was a pioneer in this field, but so were Olds and Grossman.

The third advance was the development of behavioral techniques for the sensitive measurement of changes in motivated behavior and emotional response. Particularly we are indebted to Skinner (1959), but also in a most important way to the ingenious use of instrumental conditioning by Miller (1956).

In fact, the striking thing about our authors and discussant is the fact that they are all pioneers in both the concepts and techniques that led to our modern understanding of the neurophysiological mechanism of adaptive behavior. Indeed they are still pioneering with new leads as made clear in Valenstein's discussion, for they all show a lively interest in neurotransmitter mechanisms involved in the recently discovered catecholamine pathways, extending all the way from the brainstem to forebrain structures. Quite clearly these are mechanisms concerned with emotional and motivated behavior, and particularly in the disease states that make up the psychopathology of human adaptation, and apparently in the drug therapies designed for their amelioration or even cure.

After these scholars' presentations, and Valenstein's discussion, a workshop was held. The discussion was opened by Professor Mendels, a psychiatrist from the University of Pennsylvania, particularly interested in depression and the pharmacological and biochemical variables affecting patients. Thus he cautioned against the temptation to believe that a single catecholamine mechanism could be identified as the main variable in either depression or schizophrenia since this disease has multiple etiological factors.

Secondly Dr. Mendels pointed out the important difference between the acute and the chronic effects of many of the drugs used to treat these diseases. It has been thought, for example, that drugs that deplete amines might induce depression and drugs that restore or increase the amount of amines will relieve depression.

A third point is that simple reduction in the absolute amount of amine does not cause depression in normal subjects. For example, administration of drugs like alphamethylparatyrosine which depletes norepinephrine or parachlorophenoalanine which depletes serotonin do not induce anything at all like clinical depression when given to normal subjects in high enough doses to bring about a reduction in these amines on the order of 40, 50, 60, and 70%. This is not to deny the relationship between brain amines and affective disorders, but rather to point out that the mechanism may be very complicated.

In response to Dr. Mendels' presentation, Dr. Valenstein raised the question of whether the effect of drugs on depression might be due to changes in the cell surface receptors for the various catecholamines. He pointed out that Edelman has reported that there can be rapid increases in insulin receptors of postsynaptic neurons when insulin secretion is reduced. While there appears to be no comparable data on changes in catecholamine receptors, it was agreed that the insulin example provided an excellent model for future research.

At this point, the discussion was then turned over to the panelists and Dr. Grossman began with a statement agreeing with Valenstein that dopamine is not the magic amine. He went on to make the very important point from his own work and that of others that the dopaminergic pathway and the noradrenergic pathway (or one of the noradrenergic pathways at any rate) were necessary in order to integrate complex learned behavior successfully.

Secondly, Dr. Grossman pointed out that a range of arousal deficits are produced by chemical lesion of the nigrostriatal projections. For example 6-hydroxydopamine injections into the substantia nigra, the ventricle, the lateral hypothalamus, or the globus pallidus produce very severe arousal problems, sensory neglect, and very severe motor problems such as akinesia and rigidity of limbs. The remarkable thing is that fine knife cuts of the lateral hypothalamus that interrupt ascending dopaminergic fibers and deplete 95% of striatal dopamine do not produce the arousal deficits, sensory neglect, or lasting motor problems, but only the complex learning deficits. Since hypothalamic and forebrain noradrenalin are normal or close to it, then any effect that the knife cuts had on noradrenergic fibers in the lateral hypothalamus must be minimal. Therefore, it seems that chemicals like 6-hydroxydopamine must produce their devastating deficits by depleting both the dopaminergic and the adrenergic pathways.

The second panelist, Dr. Miller, began his discussion by agreeing with the cautions expressed about oversimplified conclusions on the mechanism of action of drugs, but pointing out that there is enough converging evidence to indicate that there is something important in this area. For example, he cited the work of Weiss and his colleagues at Rockefeller. They found that if they subjected rats to periods of unpredictable, uncontrollable shocks that not only was their behavior in avoidance learning depressed, but their noradrenaline levels were depleted. The same kind of depression of avoidance learning can also be produced by tetrabenazine, a reserpine-like drug, which has a short-acting depleting effect on the amines. Furthermore, if the rats are subjected to the stress of shocks over a period of 15 days, they become "toughened up" so that, as might be expected from enzyme induction, they are less depleted of noradrenaline after being subjected to the sixteenth stress than they are after being subjected to the first stress and are also more resistant to the depression of avoidance learning. Even prettier is the converse effect. If the rats are subjected to tetrabenazine for 15 days, they are not only toughened up so that they resist the tetrabenazine better, they are also toughened up so they resist electric shock better.

Dr. Miller then turned to a facet of his own work and pointed out that the effects of lateral hypothalamic lesions are certainly not limited to effects on food and water ingestion. To support this conclusion, he cited the work of Fonberg on dogs in Konorski's laboratory at the Nencki Institute in Warsaw. Here it was shown that lateral hypothalamic lesions not only produced deficits in eating and drinking, but reduced their motivation for all social events so that they no longer came up to the front of the cage and jumped to be let out.

Referring to Valenstein's work in the shift from elicitation of eating to elicitation of drinking during the course of stimulating the same point in the lateral hypothalamus over a period of days, Dr. Miller pointed out that in his laboratory stimulating these same points in the absence of food or water led to a very agitated animal. As soon as food was presented, however, the animals calmed down. Dr. Miller believes that the same thing would happen if water were presented or if an object that could be gnawed were presented and then argues that the animals may be learning to reduce the general excitement caused by brain stimulation by simply engaging in a consumatory act. Since the same sort of drive reduction or rewarding effect of food can occur in natural hunger, he believes that the channeling of drives may be a very general proposition.

In response to this discussion, Dr. Valenstein agreed that food or water seems to help the animal cope with the highly aroused state produced by electrical stimulation of the lateral hypothalamus. However, he doubts that the act of eating the food in that situation has a great deal to do with hunger and the act of eating when hungry. He draws this conclusion because he has been unable to increase the frequency or intensity of eating evoked by brain stimulation by the device of pairing active eating under normal hunger with electrical stimulation of the lateral hypothalamus. He points out that hard pressure on the rat's tail will induce eating 100% of the time and that he believes that pairing tail pressure and brain stimulation might be more successful in increasing the evocation of eating by brain stimulation alone because they are more similar in that they are inducing eating by arousing the animal more generally.

In regard to the first part of Dr. Miller's discussion, Dr. Stellar raised the question of how similar the failure to learn avoidance behavior was under conditions of the unpredictable, uncontrollable shock of Weiss on the one hand and the lateral hypothalamic knife cuts of Grossman on the other. Both procedures influence the aminergic systems as evidently do reserpine or tetrabenzene, electrolytic lesions of the lateral hypothalamus used by Teitelbaum, and 6-hydroxydopamine injections used by Ungerstedt. And all of these procedures cause deficits in avoidance learning. Dr. Grossman felt that all of these procedures implicated the striatal pathway more than any other.

The third panelist, Dr. Olds, discussed the special properties of the catecholamines for our thinking about neuronal mechanisms of drives, aversive responses, and rewards. First of all, he pointed out that the catecholamine pathways have a small focus (in their cells of origin in the brainstem and their tight bundling in the hypothalamus) and yet a wide broadcast to the forebrain striatum and cortex, cerebellum, and hypothalamus. This is a favorable arrangement for arousal and for reinforcement.

Second, Dr. Olds pointed out that there is an interesting relation between the catecholamines in that noradrenalin and dopamine have a common precursor which is tyrosine and one of them is the precursor of the other. So it goes tyrosine, adrenalin, dopamine in such a way that more of one could mean less of the others and so forth. The main way of blocking the catecholamines is by blocking production with AMPT (alphmethylparatyrosine). If you block them between dopamine and noradrenalin, you get quite a sick animal, but if you block both, you don't.

The third point is that the same second messenger is used by serotonin, dopamine, noradrenalin, and the peptide hormones. This is cyclic AMP. This implies a relationship in which a peptide could be a drive and a catecholamine be a reward.

Fourth, Olds discussed reuptake. In the case of the neurotransmitter acetylcholine, its message is terminated by degradation of the transmitter. Catecholamine transmitters are also degraded by enzyme action, but in addition utilize the mechanism of reuptake to remove the transmitter from the synaptic space. One surprising feature of the reuptake mechanism is that dopamine and noradrenalin neurons might compete for the same catecholamine transmitter by taking it up equally well. Thus the reuptake mechanism could be involved in the interactions of the dopamine and adrenalin systems and serves to regulate the overall supply of available neurotransmitter substances. Interestingly enough, acetylcholine degradation is so fast as to occur in "neurophysiological time" (fractions of a millisecond), whereas catecholamine mechanisms occur in the much longer behavioral time (fractions of a second).

Fifth, when catecholamines are applied directly to the brain, they have an inhibitory effect on neurons and this might be part of the mechanism whereby catecholamines can play a reinforcing function in behavior.

Sixth, catecholamines play an important role in controlling the hormone system, the adeno-hypophysis. Noradrenaline, for example, simultaneously inhibits the ACTH system while exciting the gonadotropic hormone system.

There is even the possibility that the catecholamines have much of their behavioral effect by way of their coupling to the peptide hormone mechanism of the brain. Catecholamines have, for example, inhibitory effects when directly applied (by electrophoresis) to peptide emitting neurons (the oxytosin ones of the parafascicular nucleus at least). It could therefore be that the relation of catecholamines to peptide matches and possibly even mediates the relation of rewards to drives; and this might be the main mechanism of the rewarding effect.

Dr. Corson supported the notion expressed by Dr. Mendels that one must differentiate between acute and chronic effects of psychotropic drugs. He pointed out further that it is also important to differentiate between acute and chronic effects of psychologic stressors. Dr. Corson's studies on Pavlovian conditioning in dogs (with aversive reinforcement) demonstrated that it is *longitudinal experiments* and not single acute experiments that made it possible to delineate significant and reproducible constitutional differences in physiologic reactions of the animals to the same psychologically stressful situations.

Dr. Corson further emphasized that in investigating neurophysiological mechanisms underlying adaptive behavior to psychologic stressors, it would be advisable to (1) study individuals with different constitutional makeup (i.e., genetic plus developmental factors); (2) study comparative reactions to avoidable and unavoidable stressors; (3) investigate *patterns* of neuroendocrine, endocrine, and psychophysiologic reactions rather than recording only one variable. Dr. Corson supported Drs. Grossman and Mendels in cautioning against simplistic notions of relating behavior to a single neurotransmitter.

Dr. Corson referred to the well-controlled studies on neurochemistry and behavior by Engel (1972) suggesting that, while dopamine is essential for elementary motor functions, more complex integrated behavioral acts involve also norepinephrine. In referring to the extensive and highly sophisticated studies by Mason (1974) on *specific patterns* of psychoendocrine responses to different stressful situations, Dr. Corson suggested that multidisciplinary investigations involving the measurement *also of physiologic and behavioral parameters* would make it possible to elucidate patterns of psychobiologic adaptation designed to maintain physiologic homeostasis.

In investigating psychophysiologic responses of several breeds of dogs to repeated exposure to unavoidable stressful situations, Dr. Corson delineated two major types of animals: (a) Low-adaptation dogs (e.g., many wirehair fox terriers, cocker spaniels, and border collies) characterized by a persistent generalized quintet of reactions: tachycardia, polypnea, copious salivation, vasopressin-induced antidiuresis, and high-energy metabolism. These dogs behaved physiologically as though they were engaged in a "fight or flight" activity leading to increased thermogenesis. Thus, the release of vasopressin serves a compensatory function of providing body water for thermoregulatory homeostasis. (b) Dogs exhibiting high adaptation to repeated exposure to unavoidable stressful situations (e.g., most beagles and other hounds). These dogs adapt rapidly, and after several experimental sessions do not exhibit the above-described quintet of psychophysiologic reactions.

No significant differences in plasma thyroxine (T_4) levels were found between high- and low-adaptation dogs. In contrast, low-adaptation dogs were characterized by high urinary excretion of epinephrine and also of norepinephrine when placed in the psychologically aversive environment. High-adaptation dogs had a much smaller output of catecholamines in the kennels as compared to the low-adaptation dogs, and showed much smaller catecholamine responses to the aversive room. In fact, in high-adaptation dogs the urinary catecholamine excretion in the aversive room was of the same order of magnitude as the urinary catecholamine excretion of the low-adaptation dogs in the control room. On the basis of these observations, Dr. Corson suggested that catecholamines are most likely implicated in the increased energy metabolism of low-adaptation dogs (Corson *et al.,* 1973, 1974; and unpublished data). In this connection, Dr. Corson called attention to clinical reports on psychogenic fever in humans (Friedman, 1950).

As might be expected, anxiolytic drugs (minor tranquilizers, e.g., meprobamate, Valium) inhibited the generalized psychovisceral turmoil in the low-adaptation dogs. Paradoxically, low doses of amphetamine also inhibited these reactions, presumably by enhancing inhibitory influences on limbic structures (Corson *et al.,* 1975; Eysenck, 1963).

Dr. Corson indicated that his studies on psychogenic release of vasopressin would tend to support Dr. Olds' suggestion regarding a possible coupling of peptide hormones to catecholamines and possibly to other neurotransmitters and hormones. A number of studies suggested that vasopressin may be involved in the release of ACTH and of growth hormone. Dr. Corson cited recent studies by Zimmerman *et al.* (1973) demonstrating that vasopressin and neurophysin do reach the anterior pituitary via the portal circulation. Thus, Dr. Corson concluded, psychogenic vasopressin release in low-adaptation dogs may have far-reaching influences on a wide spectrum of endocrines, and thus affect patterns of psychophysiologic and behavioral adaptive reactions.

Dr. Klinger pointed out that his analysis is one at a very different level of description from that of the other panelists, but he sees points of possible convergence somewhere in the distance. He feels that stress and untoward emotional reactions result from breaks in expectancy. As a result he believes that the neural models are similar in properties to those Sokolov described with respect to his theory of habituation, to Karl Pribram, in relationship to attention, or to James Olds' integrated motor skills which are most appropriate to this eventual convergence of thinking.

Dr. Olds expressed interest in the Sokolov model, and Dr. Miller drew the parallel with dogs that were reared in isolation and then suddenly exposed to a regular environment without preparation. Without appropriate models, so to speak, or expectations or habits, they were terrified.

In another vein, Dr. Miller referred to a pathway through the central gray that seems to inhibit the firing of neurons that fire when the animal is subjected to pain. This pathway can be electrically stimulated and produce the inhibition or it can be stimulated by minute amounts of morphine placed in the same locus in the brain. Dr. Stellar pointed out that this is a pathway that can be reached by stimulation of the lateral hypothalamus at placements that yield good self-stimulation. Recent work at Pennsylvania suggests that this pathway may be involved more generally in the suppression of responses to a wider range of aversive stimuli than pain, for the same stimulation of the lateral hypothalamus seems to inhibit the startle response to strong auditory stimuli. Furthermore, there may be a facilitatory pathway as well as an inhibitory one, for Flynn has shown that lateral hypothalamic stimulation in the cat facilitates the biting response in predatory attack. The stronger the electrical stimulation of the lateral hypothalamus the larger is the receptive field around the mouth that elicits the biting response upon tactile stimulation.

The discussion ended with an emphasis by Dr. Miller and Dr. Wolf on the importance of the control of stress by inhibitory mechanisms, both neurological and neurohumoral. Clearly we have a long way to go to bridge the gap between neurophysiological mechanisms of adaptive behavior in animals and the psychopathology of human adaptation, yet remarkable progress is being made. At the neurological level, we now appreciate the significance of the rhinencephalic structures, particularly the catecholamine parhways, in the genesis of drive, emotional response, and reinforcement. In the normal individual the relationship between drive and reinforcement is well-regulated, arousal and emotion are kept within adaptive bounds, and reward and aversive events serve to reinforce new adaptive behavior in successful learning. In psychopathology something goes wrong with the regulation either because the arousing stimulation is too strong or too stressful or because the responding neural system is too sensitive or too refractory, possibly due to depletion of catecholamines. In such cases, extremes of emotional response characterize the individual, and learning and adaptation fail. Obviously this is a gross oversimplification of the import of this section on neurophysiological mechanisms of adaptive behavior. The more important question is how do we get from neurophysiological mechanisms to the psychopathology of human adaptation as seen clinically? This is the task of the rest of the symposium.

REFERENCES

Corson, S. A., Corson, E. O'Leary., Kirilcuk, V., Kirilcuk, J., Knopp, W., and Arnold, L. E. (1973). Differential effects of amphetamines on clinically relevant dog models of hyperkinesis and stereotypy: relevance to Huntington's Chorea. In *Advances in Neurology, Vol. 1, Huntington's Chorea, 1872–1972* (A Barbeau, T. N. Chase, and G. W. Paulson, Eds.). New York: Raven Press, pp. 681–697.

Corson, S. A., Corson, E. O'Leary., Kirilcuk, V., Kirilcuk, J., and Vanecko, S. (1974). Urinary epinephrine and plasma thyroxine in dogs with high and low adaptation to psychologic stress. *Fed. Proc. 33*(3): 463.

Corson, S. A., Corson, E. O'Leary., Arnold, L. E., and Knopp, W. (1975). Interaction of psychopharmacologic and psychosocial therapy in behavior modification of animal models of violence and hyperkinesis. In *Psychopathology and Human Adaptation* (G. Serban, Ed.). New York: Plenum Press.

Engel, J. (1972). Neurochemistry and Behavior. A Correlative Study with Special Reference to Central Catecholamines. Göteborg, Sweden: University of Göteborg.

Eysenck, H. J. (1963). Biological basis of personality. *Nature 199* (4898). 1031–1034.

Friedman, M. (1950). Hyperthermia as a manifestation of stress. In *Life Stress and Bodily Disease. Res. Publ. Assoc. Res. Nerv. Ment. Dis., Vol. 29.* (H. G. Wolff, S. G. Wolf, Jr., and C. C. Hare, Eds.). Baltimore: Williams and Wilkins, pp. 433–444. Reprinted New York: Hafner Publishing Company.

Hess, W. R. (1957). *The Functional Organization of the Hypothalamus.* New York: Grune and Stratton.

Hetherington, A. W., and Ranson, S. W. (1942). The spontaneous activity and food intake of rats with hypothalamic lesions. *Am. J. Physiol. 136,* 609–617.

Horsley, V., and Clarke, R. H. (1908). The structure and function of the cerebellum examined by a new method. *Brain 31,* 45–125.

Klüver, H., and Bucy, P. C. (1938). An analysis of certain effects of bilateral temporal lobectomy in rhesus monkeys, with special reference to "psychic blindness." *J. Psychol. 5,* 33–54.

Mason, J. W. (1974). Specificity in the organization of neuroendocrine response profiles. In *Frontiers in Neurology and Neuroscience Research 1974.* First International Symposium of the Neuroscience Institute. (P. Seeman and G. M. Brown, Eds.). Toronto: University of Toronto Press, pp. 68–80.

Miller, N. E. (1956). Effects of drugs on motivation, the value of using a variety of measures. *Ann. NY Acad. Sci. 65,* 318–333.

Papez, J. W. (1937). A proposed mechanism of emotion. *Arch. Neurol. Psychiatry 38,* 725–743.

Richter, C. P. (1942–1943). Total self-regulatory functions in animals and human beings. *Harvey Lect. 38,* 63–103.

Skinner, B. F. (1959). *The Cumulative Record.* New York: Appleton-Century-Crofts.

Stellar, E. (1954). The physiology of motivation. *Psychol. Rev. 61,* 5–22.

Stellar, E. (1974). Brain mechanisms in hunger and other hedonic experiences. *Proc. Am. Philosoph. Soc. 118,* 276–282.

Zimmerman, E. A., Carmel, P. W., Husain, M. K., Ferin, M., Tannenbaum, M. Frantz, A. G., and Robinson, A. G. (1973). Vasopressin and neurophysin: high concentrations in monkey hypophyseal portal blood. *Science 182*(4115): 925–927.

Psychopathology of Adaptive Learning: Motivation, Anxiety, and Stress

Stress without Distress

HANS SELYE

Stress is "the nonspecific response of the body to any demand made upon it," that is, the rate at which we live at any one moment. All living beings are constantly under stress and anything, pleasant or unpleasant, that speeds up the intensity of life, causes a temporary increase in stress, the wear and tear exerted upon the body. A painful blow and a passionate kiss can be equally stressful.

The financier worrying about the stock exchange, the laborer or the baseball player straining his every muscle to the limit, the journalist trying to meet a deadline, the patient fighting a fever, all are under stress. But so is the baseball fan who merely watches an interesting game, and the gambler who suddenly realizes that he has lost his last cent or that he has won a million dollars.

Contrary to widespread belief, stress is not simply nervous tension nor the result of damage. Above all, stress is not something to be necessarily avoided. It is associated with the expression of all our innate drives. Stress ensues as long as a demand is made on any part of the body. Indeed, complete freedom from stress is death!

THE SYNDROME OF "JUST BEING SICK"

It was in 1925 that I first suspected the existence of what I later called stress and the General Adaptation Syndrome (G.A.S.). When studying medicine at the University of Prague, during one of the initial lectures in internal medicine, we

HANS SELYE · Institut de médecine et de chirurgie expérimentales, Université de Montréal, Québec, Canada.

were shown several patients in the earliest stages of various infectious diseases. The professor carefully pointed out all the specific signs and symptoms characteristic of each disease but what struck me most was that each of these patients felt and looked ill, had a coated tongue, and complained of more or less diffuse aches and pains in the joints and of intestinal disturbances with loss of appetite and loss of weight. The patients we had just seen also had a common syndrome, but the professor attached very little significance to the signs that were common to all these diseases because they were "nonspecific" and hence "of no use" to the physician in making a correct diagnosis or prescribing the appropriate treatment.

Instead of concentrating exclusively on specific manifestations of disease, would it not be even more important to learn something about the mechanism of being sick and the means of treating this "general syndrome of sickness" which appeared to me as being superimposed upon all individual diseases? I could not understand why our professor did not pay any attention to it. However, as a student, I accepted the fact that "this is so," just as physicians had done ever since the dawn of medical history.

Not until ten years later did these same questions confront me again. At that time I was working in the biochemistry department of McGill University in Montreal, trying to find a new ovarian hormone in extracts of cattle ovaries. All the extracts, no matter how prepared, produced the same syndrome characterized by (1) enlargement of the adrenal cortex, (2) gastrointestinal ulcers, and (3) involution of the thymus and lymph nodes. Although at first I ascribed all these changes to some new ovarian hormone in my extract, it soon became apparent that extracts of other organs—in fact, toxic substances of all kinds—produced the same changes. It was only then that I suddenly remembered my classroom impression of the "syndrome of just being sick." In a flash, I realized that what I had produced with my impure extracts and toxic drugs was an experimental replica of this condition.

This simple hunch of a connection between the almost forgotten, purely speculative, clinical concept of student days and the reproducible, objectively measurable changes in the animal experiments at hand was the basis for the development of the entire stress concept.

TRIPHASIC RESPONSE

During this reaction, all the organs of the body showed involutional or degenerative changes; only the adrenal cortex actually seemed to flourish on stress. I suspected this adrenal response to play a useful part in a nonspecific adaptive reaction, which I visualized as a "call to arms" of the body's defense forces and therefore designated as the "*alarm reaction*."

Later, we noted that this alarm reaction was not the entire response. Upon continued exposure to a stressor capable of eliciting the initial reaction, a stage of adaptation or resistance ensued, since no organism can be maintained continuously in a state of alarm. If the stressor is so severe that continued exposure becomes incompatible with life, the animal dies within a few hours or days during the alarm reaction. If it does survive, the initial response is necessarily followed by a *stage of resistance* during which most of the initial symptoms diminish or vanish. After still more prolonged exposure to the stressor, this acquired adaptation is lost and the animal enters into a third phase, the *stage of exhaustion,* since the "adaptation energy" or adaptability of an organism is finite.

These three stages are analogous to the three stages of man's life: childhood (with its characteristic low resistance and excessive responses to any kind of stimulus), adulthood (during which adaptation to most commonly encountered agents has occurred and resistance is increased), and, finally, senility (characterized by irreversible loss of adaptability and eventual exhaustion) ending with death.

Subsequent investigations revealed that a decisive part of this defense mechanism is the excitation of the hypothalamus and, in particular, its most caudal portion, the median eminence (ME), to discharge a chemical messenger, the corticotrophin-releasing factor or CRF. It was a great merit of my former graduate student, Guillemin (in conjunction with Schally and Saffran), to find quite independently, after leaving our institute, that CRF is produced in the hypothalamus and increases the secretion of ACTH, of which we had already known that it stimulates the adrenal cortex to produce those hormones for which I introduced the term corticoids. Most important among the corticoids are the glucocorticoids, such as cortisone, which cause thymus atrophy and influence glucose metabolism, especially by stimulating glycogen formation.

Ever since Cannon described his classic emergency response to threatening stimuli, it was known that, at least in acute emergencies, the adrenal medulla and the sympathetic nerves also play a decisive role through the discharge of catecholamines. These go to various parts of the body and help to adjust metabolism and the cardiovascular system for conditions of "fight or flight." They also play an important role in the formation of peptic ulcers which constitute a part of the alarm reaction triad and for which I therefore suggested the term "stress ulcers" (Fig. 1).

DISEASES OF ADAPTATION

Various derangements in the secretion of these hormones can lead to maladies which I called "diseases of adaptation," because they are not directly

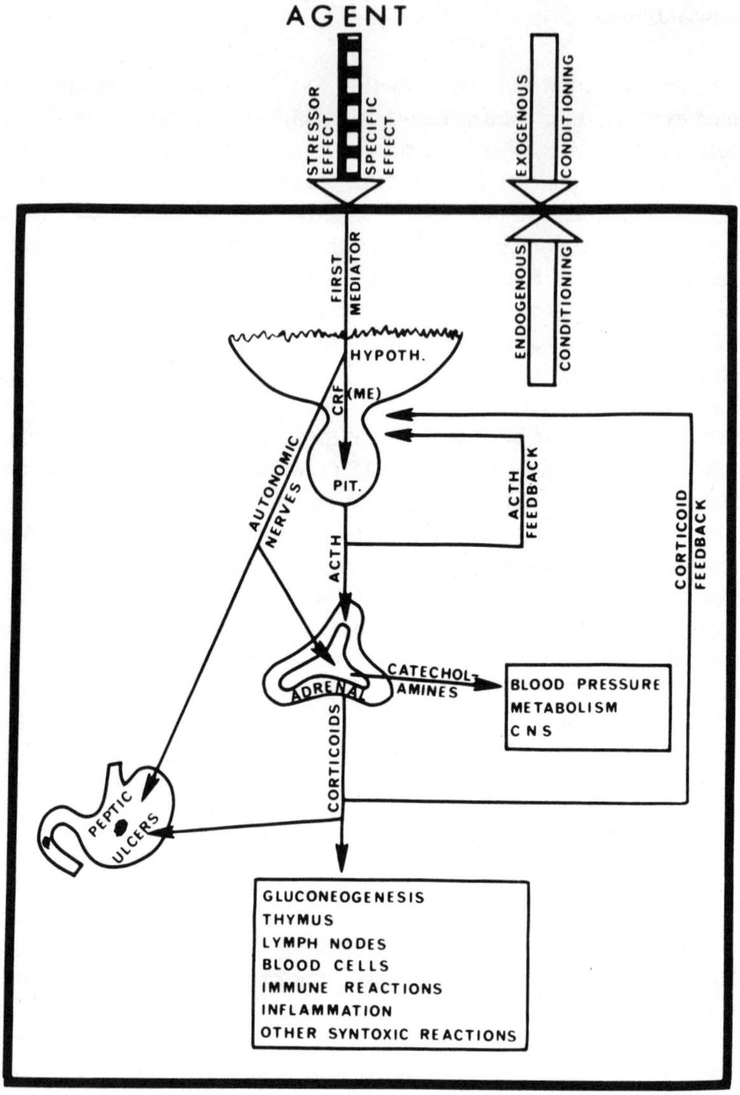

Fig. 1. Principal pathways mediating the response to a stressor agent and the conditioning factors which modify its effect. As soon as any *agent* acts upon the body (thick outer frame of the diagram) the resulting effect will depend upon three factors (broad vertical arrows pointing to the upper horizontal border of the frame). All agents possess both nonspecific *stressor* effects (solid part of arrow) and *specific* properties (interrupted part of arrow). The latter are variable and characteristic of each individual agent; they will not be discussed here more than to state that they are inseparably attached to the stressor effect and invariably modify it. The other two heavy vertical arrows, pointing toward the upper border of the frame, respectively represent exogenous and endogenous *conditioning factors* which largely determine the reactivity of the body. It is clear that since all stressors have some specific effects, they cannot elicit exactly the same response in all organs; furthermore, even the same agent will act differently in different individuals, depending upon the internal and external conditioning factors which determine their reactivity. (From *Stress in Health and Disease,* courtesy of Butterworths, Reading, Mass.)

caused by any particular pathogen but indirectly by a faulty adaptive response to the stressor effect of some pathogen. In this sense, many ailments, such as various emotional disturbances, schizophrenia, migraine headaches, insomnia, hypertension, peptic ulcers, certain types of asthma, as well as cardiovascular and renal diseases appear essentially to be initiated or enhanced by the body itself because of its faulty adaptive reactions to only potentially injurious agents.

Apparently, conditioning (particularly hereditary predisposition, diet, environmental factors) determines which organ or system is weakest and breaks down most readily under the influence of systemic stress, that is, a nonspecific demand made upon the organism as a whole.

Among the derailments of the G.A.S. which may cause disease, the following were the first to be recognized:

1. An absolute excess or deficiency in the amount of corticoids and somatotrophic or growth hormone (STH) produced during stress.
2. A disproportion in the relative secretion, during stress, of ACTH and glucocorticoids on the one hand, and of STH and mineralocorticoids on the other.
3. Production by stress of metabolic derangements, which abnormally alter the target-organ's response to STH, ACTH or corticoids (through conditioning).
4. Finally, we must not forget that although the hypothalamus–hypophysis–adrenal mechanism plays a prominent role in the G.A.S., other organs which participate in the latter (e.g., the nervous system, liver, kidney) may also respond abnormally and become the cause of disease during adaptation to stress. Indeed, stress occurs even in unicellular organisms when they are faced with demands for adaptation.

CATATOXIC AND SYNTOXIC RESPONSES

Numerous studies on stress have shown that the internal stability of the human body (homeostasis) is maintained by two types of reactions: syntoxic (from "syn," meaning together, as in syndicate) and catatoxic (from "cata," meaning down, against). The former defend the organism by ordering the cells to put up and coexist peacefully with the intruders. In many cases, the aggressors (indirect pathogens) are harmless and damage is caused only by our own uncalled-for defense reactions. Hence, it is wise to suppress this hostility. But when the agents are inherently harmful (direct pathogens), the body must protect itself by destroying them. In essence, these scientific observations show that there are two roads to survival: tolerance and fight. The former, of course, is generally more advantageous. Flight is possible only if the aggressor (drug, antigen, microbe) has not entered the body.

Tolerance is the simplest, but by no means the best form of adaptation. It is useful because it prevents friction and conflicts, but it does just that and no more; it cannot protect, for instance, against overcrowding, shortages of essential materials, etc. Therefore, disciplined, mutually advantageous collaboration and teamwork are far more suitable. During evolution from unicellular organisms to man, the billions of individual cells in our body have learned to work as a team for the benefit of the whole person. Breakdown of such cooperation is best illustrated by the growth of a cancer, which lives off the other parts of the body until it kills its host. It destroys itself by its uncontrolled egoistic development.

Teamwork between various animal species manifests itself usually in the form of an ecological group which becomes an individual entity. In human society, this cooperation is evident in the family, tribe, corporation, federation, and even in the United Nations. The strength of the team depends upon collaboration between its individual units.

This is what we should try to achieve through the philosophy of "altruistic egoism," as outlined later.

MAN MUST WORK

Can the scientific study of stress help us to formulate a precise program of conduct? Can it teach us the wisdom to live a rich and meaningful life which satisfies our needs for self-expression and yet is not marred or cut short by the stresses of senseless struggles?

I think we have to begin by clearly realizing that work is a biological necessity. Just as our muscles become flabby and degenerate if not used, so our brain slips into chaos and confusion unless we constantly use it for some work that seems worthwhile to us. The average person thinks he works for economic security or social status, but when, at the end of a most successful business career, he has finally achieved this, there remains nothing to fight for–no hope for progress, only the boredom of assured monotony.

Do not listen to the tempting slogans of those who keep repeating, "There is more to life than just work," or, "You should work to live, not live to work." This sounds pretty convincing, but is it really? Of course, these statements are true in themselves, yet your principal aim should be not to avoid work but to find the kind of occupation which, for you, is play. The best way to avoid harmful stress is to select an environment (wife, boss, friends) which is in line with your innate preferences–to find an activity which you like and respect. Only thus can you eliminate the need for frustrating constant readaptation that is the major cause of distress.

The Western world is being wrecked right now by the unsatiable demand for less work and more pay. Less work to get more time for what? More pay to do what? Few people give much thought to what they will do with their free time and extra money after they have reached a comfortable minimum income. Of course, there is such a thing as a minimum living standard; but in practice, the urgency of the clamor for improvement does not depend so much upon the number of working hours or the salaries earned, as upon the degree of dissatisfaction with life. We could do much—and at little cost—by fighting this dissatisfaction.

THE NEED FOR SELF-EXPRESSION

> *"No wind blows in favor of the ship that has no port of destination."*
> (Montaigne).

After a pilot has left the ground in a plane he cannot stop his motor before he gets back to earth again. He must complete his mission back to earth. Yet there is very much he can do, through voluntary choice of conduct, to get as far as possible with a given airplane and fuel supply under given climatic conditions. For instance, he can fly at a speed and on a course best suited to his machine under the prevailing weather conditions. The two great limiting factors over which, once in flight, he has no control are the fuel supply and the wear and tear that the weakest vital part of his plane can tolerate.

When a human being is born he cannot stop either before he has completed his mission on earth. Yet he too can do much, through voluntary choice of conduct, to get as far as possible with a given bodily structure and supply of adaptation energy, under given social conditions. For instance, he can live and express his personality at a tempo and in a manner best suited to his inherited talents, under the prevailing social conditions. The two great limiting factors—which are genetically fixed when a man is born—are his supply of adaptation energy and the wear and tear that the weakest vital part of his body can tolerate.

So, actually, we can accomplish a great deal by living wisely in accordance with natural laws. We can determine our optimum speed of living, by trying various speeds and finding out which one is most agreeable. We can determine our course by the same empirical method, keeping in mind, however, that occasional deviations have a virtue of their own: they equalize the wear and tear throughout the body, and thereby give overworked parts time to cool down.

Man certainly does not get the feeling of happiness, of having completed his mission on earth, just by staying alive very long. On the contrary, a long life without the feeling of fulfillment is very tedious. And yet, when (and if) they analyze their lives, most people get the feeling of merely muddling through, of drifting aimlessly, from one day to another. Just staying alive, no matter how comfortably and securely, is no adequate outlet for man's vital adaptation energy. Comfort and security make it easier for us to enjoy the great things in life, but they are not, in themselves, great and enjoyable aims.

ALTRUISTIC EGOISM

A world in which each creature refuses to protect itself is unimaginable, but so is a world in which uncontrolled egoism, with total disregard for the interests of others, is the leading principle of behavior. To my mind, the only philosophy which necessarily transforms all aggressive egoistic impulses into altruism, without curtailing any of their self-protecting values, is altruistic egoism. It has amply proved its value throughout evolution from the simplest multicellular organism to man. In lower animals, whenever it developed, it was of considerable help but, being unpremeditated, could arise only through the power of its survival value. Wherever it developed—be it even by accident—it created new strengths which increased resistance.

Single cells combined into multicellular organisms and these into larger groups on the basis of this principle, although they were not aware of it. Similarly, individual people have formed the cooperative "mutual insurance" groups of the family, tribes, and nations within which altruistic egoism is the key to success. It is the only way to preserve teamwork, whose value is ever increasing in modern society.

Each person must find a way to relieve his pent-up energy without creating conflicts with his fellow men. Such an approach not only ensures peace of mind but also earns the goodwill, respect, and even love of our fellow men, the highest degree of security, the most noble "status symbol" to which man can aspire.

This philosophy of hoarding a wealth of respect and friendship is merely one reflection of the deep-rooted instinct of man and animals to collect. It is as characteristic of ants, bees, squirrels, and beavers as of the capitalist who collects money to put away in the bank. The same impulse drives entire human societies to amass a system of roads, telephones, cities, fortifications, that strike them as useful means of accumulating the ingredients of future security and comfort.

In man, this urge first manifests itself when children start to amass match boxes, shells, or stickers; it continues when adults gather stamps or coins. The natural drive for collecting is certainly not an artificial, indoctrinated tradition. By collecting certain things, you acquire status and security in your community. The guideline of earning love merely attempts to direct the hoarding instinct toward what I consider the most permanent and valuable commodity that man can collect: a huge capital of goodwill which protects him against personal attacks by his fellow men.

"EARN THY NEIGHBOR'S LOVE"

"Love thy neighbor as thyself," one of the oldest guidelines for purpose and conduct, was propounded (even before Christ) to please God and thereby offer security to man. Since our philosophy is based on natural laws, it is perhaps not surprising that, for centuries, throughout the world, so many of its elements have turned up again and again–in the most diverse religions and political doctrines–though usually supported by mysticism and blind trust in an infallible authority rather than by science. The people in whose cultures one or the other of these elements appeared were quite unrelated and often did not even know of each other's existence. Their creeds had only one thing in common: they were acceptable to the human brain and reflected the natural evolution of its functional mechanism.

This is perhaps why we felt we should update the guideline by rephrasing it to "*Earn* thy neighbor's love." In this form, it is biologically sound and cannot conflict with any religion or philosophy; in fact, ardent believers in any one of these can use our code to complement their own. In it, they will find scientific support not only for one of the most deep-rooted and generally accepted religious precepts of the brotherhood of man but also for that of atheistic communism, with its avowed goal: "From each according to his capabilities, to each according to his needs," a slogan which otherwise might only encourage laziness. The laws of Nature, which we have used to construct our doctrine, apply to everybody, irrespective of his formalized and labeled creed.

Viewed from the pinnacle of the eternal general laws governing Nature, we are all surprisingly alike. Nature is the fountainhead of all our problems and solutions; the closer we keep to her the better we realize that, despite the apparently enormous divergences in interpretation and explanation, her laws have always prevailed and can never become obsolete. The realization of this truth is most likely to convince us that, in a sense, not only all men but all living beings are brothers. To avoid the stress of conflict, frustration, and hate, to achieve peace and happiness, we should devote more attention to a better understanding of the natural basis of motivation and behavior. No one will be disappointed if, in daily life, he learns to follow the guiding light of "*Earn* thy neighbor's love."

So the whole translation of the laws governing resistance of cells and organs to a code of behavior comes down to three precepts:

1. *Find your own natural stress level.* People differ with regard to the amount and kind of work they consider worth doing to meet the exigencies of daily life and assure their future security and happiness. In this respect, all of us are influenced by hereditary predispositions and the expectations of our society. Only through planned self-analysis can we establish what we really want; too many people suffer all their lives because they are too conservative to risk a radical change and break with traditions.

2. *Altruistic egoism.* The selfish hoarding of the goodwill, respect, esteem, support, and love of our neighbor is the most efficient way to give vent to our pent-up energy and create enjoyable, beautiful, or useful things.

3. *EARN thy neighbor's love.* This motto, unlike love on command, is compatible with man's natural structure and, although it is based on altruistic egoism, could hardly be attacked as unethical. Who would blame him who wants to assure his own homeostasis and happiness only by accumulating the treasure of other people's benevolence toward him? Yet, this makes him virtually unassailable, for nobody wants to attack and destroy those upon whom he depends.

I myself have tried to follow this philosophy as best I could, and it has made my life a happy one. Let me frankly admit that, in looking back, I realize that I have not always succeeded to perfection, but my failures were due to my own shortcomings, not to those of the philosophy itself.

As I have said in *Stress without Distress,* the builder of the best racing car is not necessarily its best driver.

ACKNOWLEDGMENT

This contribution is based largely on the author's various lectures, articles, and books, especially his latest publications, *Stress without Distress* (J. B. Lippincott) and *Stress in Health and Disease* (Butterworths).

Selectivity of Corticosteroid and Catecholamine Responses to Various Natural Stimuli

JOHN W. MASON, JOHN T. MAHER,
L. HOWARD HARTLEY,
EDWARD H. MOUGEY, MARK J. PERLOW,
and LEEROY G. JONES

For about the past 15 years, we have been involved in an experimental approach to the study of the neuroendocrine motor system of the brain which views the many *interdependent* neuroendocrine systems on as broad a scope as possible. Figure 1 presents a schematic view of the assemblage of the principal neuroendocrine systems known at the present time. When we began our studies during the 1950's, interest in neuroendocrine regulation was heavily concentrated on just two systems, the pituitary–adrenal cortical and sympathetic–adrenal medullary systems. The view was steadily emerging in endocrinology, however, that the hormones of the various glands are highly interdependent in their metabolic effects throughout the body, being variously aligned in patterns of antagonistic, synergistic, or additive relationships with each other. No single hormone acts entirely alone in isolation. Houssay, among others, has emphasized that there is

---

JOHN W. MASON, EDWARD H. MOUGEY, and MARK J. PERLOW • Division of Neuropsychiatry, Walter Reed Army Institute of Research, Walter Reed Army Medical Center, Washington, D. C. JOHN T. MAHER, L. HOWARD HARTLEY, and LEEROY G. JONES • Army Research Institute of Environmental Medicine, Natick, Massachusetts.

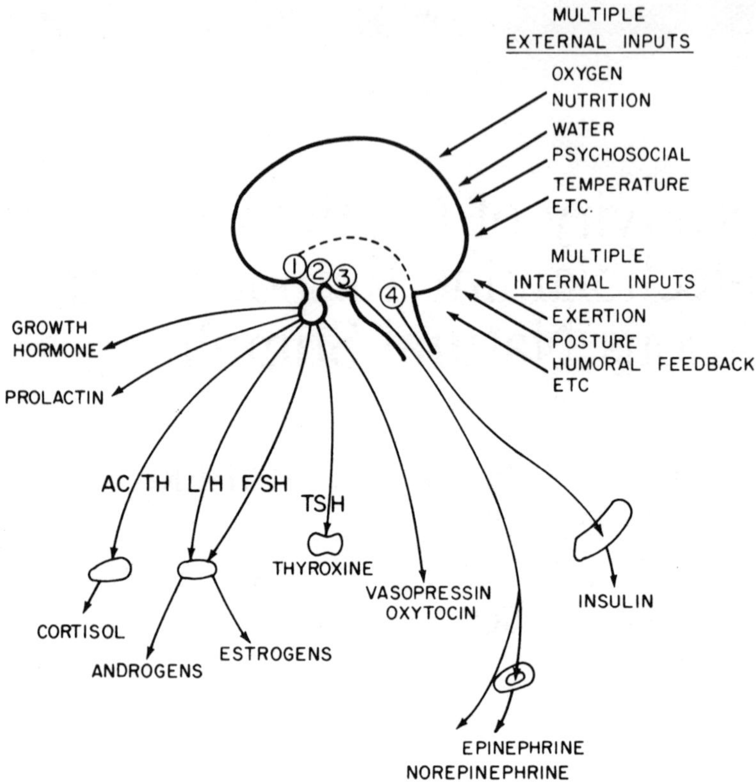

Fig. 1. The assemblage of interdependent neuroendocrine systems.

always a balance between opposing and cooperating hormones acting on any particular metabolic process (Houssay, 1957). *Our guiding working hypothesis has been, therefore, that if hormones are so closely interdependent in their actions at the metabolic level, then it seems logical that the integrative mechanisms controlling endocrine secretion are organized in a manner closely in keeping with such interdependencies and that the secretion of the many hormones probably is accordingly coordinated on an overall basis* (Mason, 1968*c*).

In testing this hypothesis, then, we have generally been conducting experiments in which levels of many different hormones are measured *concurrently,* in relation to various, specific natural input stimuli from both external and internal sources, as shown in Fig. 1, with the aim of achieving as broad a view as possible of multiple hormonal responses in terms of "response *patterns,*" or "response *profiles*" or "shifts in *overall hormonal balance*." While most of our efforts have been focused upon neuroendocrine responses to psychological stimuli, we have

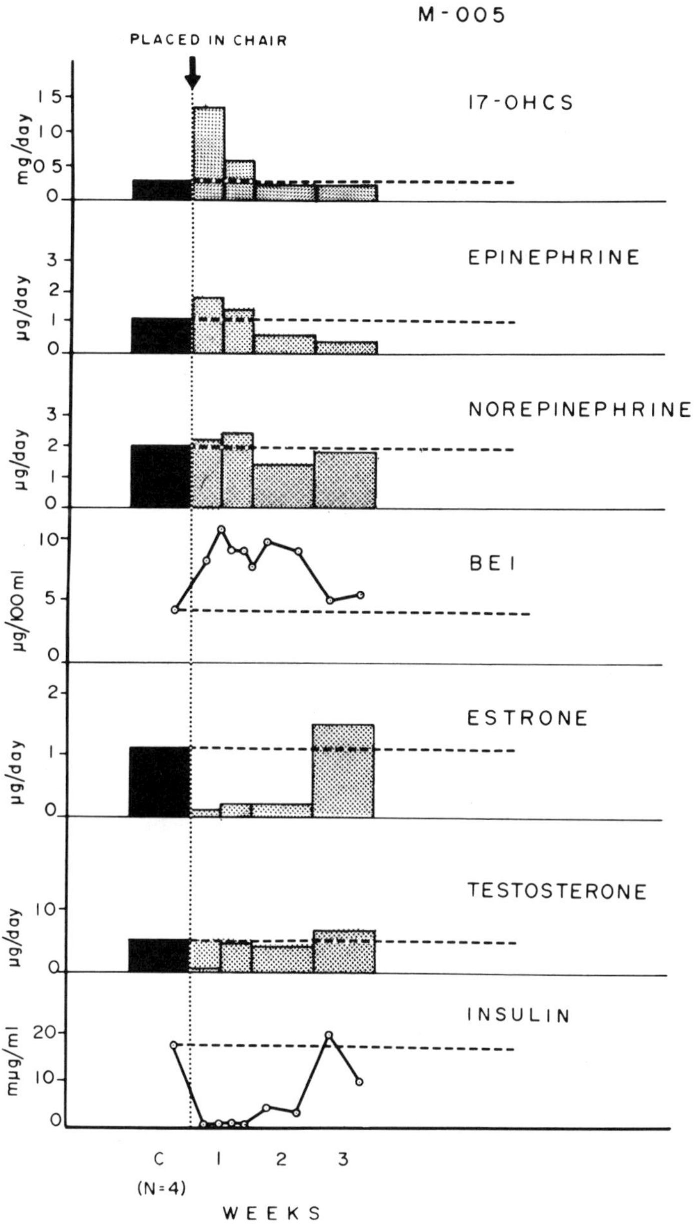

Fig. 2. Pattern of hormonal responses to chair restraint in a monkey.

also studied a number of other stimuli, including exercise, fasting, heat, and cold.

In order to convey some feeling for the type of hormonal response "pattern" or "profile" data which has been the central concern in our program, Fig. 2 presents an example of an experiment on the profile of psychoendocrine responses to the first experience of capture and acute chair restraint in a male rhesus monkey. After a four-week period of baseline measurements, while the monkey is housed in a cage, the monkey is captured and transferred to a primate restraining chair in the same quiet experimental chamber. It is evident that a broad range of hormonal responses follow onset of chair restraint, with urinary 17-hydroxycorticosteroid (17-OHCS), epinephrine, norepinephrine, and plasma butanol extractable iodine (thyroid hormone) levels rising, while urinary estrone, testosterone, and plasma insulin levels are sharply lowered. This overall pattern of change is very similar to that we have previously observed in conditioned avoidance sessions in the monkey and appears to be generally oriented in a way that might be expected to promote mobilization of energy resources, as has been discussed at length elsewhere (Mason, 1968*b*, 1974).

As our work has progressed, we have made continuing efforts to improve and refine our hormone assay methodology so as to approach a higher level of confidence that our measurements reflect actual endocrine secretory activity. We have also been concerned with developing improved control measures for isolating our independent variables as cleanly as possible. Both of these points are perhaps illustrated in a comparison of Fig. 3 with Fig. 2. Figure 3 presents a recent experiment in which we have moved toward the measurement of as many hormones in plasma as possible to supplement our earlier studies based largely upon less direct urinary hormone measurements. It is the same type of acute chair restraint experiment described in connection with Fig. 2, except that plasma hormone measurements are taken at repeated short intervals during the initial hours after capture and restraint of the monkey. In addition to about a three-fold increase in cortisol level at 4 hr, marked elevations also occur in plasma prolactin and growth hormone levels, peaking at about 20 min and 40 min, respectively. Plasma thyrotropin (TSH) levels also rise rather early, but thyroxine levels move very slowly upward over a 48-hr period. After a transient rise, plasma testosterone levels decline markedly at the 24- and 48-hr points and plasma insulin levels show a substantial and sustained decline. All of the above plasma hormonal measurements, incidentally, were made by radioimmunoassay except that for total thyroxine concentration which was determined by a competitive protein-binding procedure.

In general, then, this study adds some new hormones to our response profile (growth hormone, prolactin), refines our assessment of other endocrine systems (thyroid, testes) and by the demonstration of marked hormonal changes during

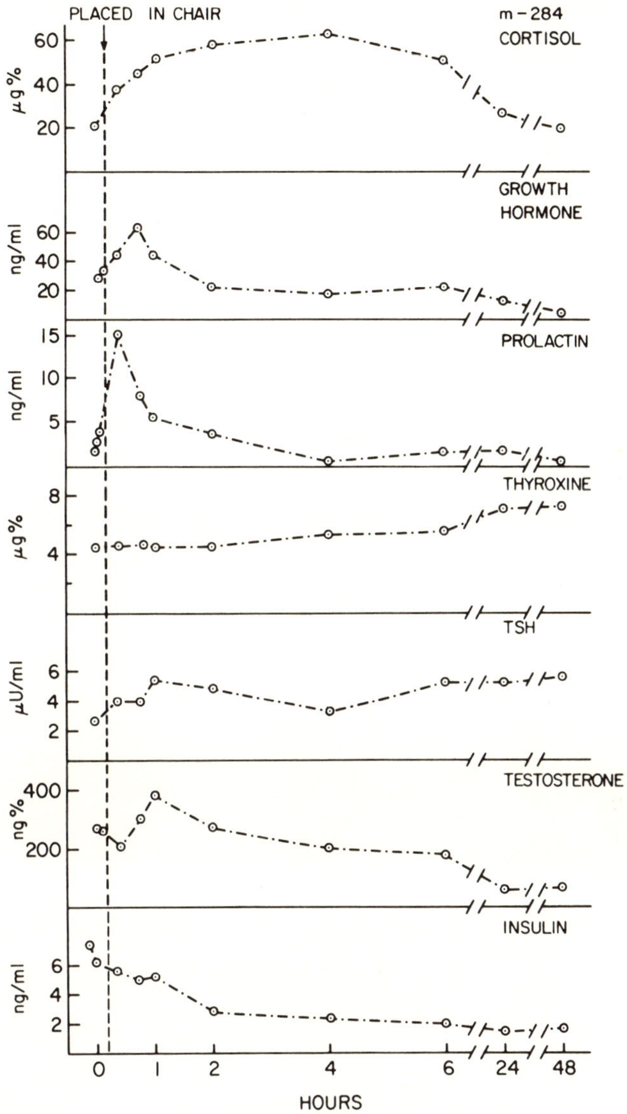

Fig. 3. Pattern of plasma hormonal responses to acute capture and chair restraint in a monkey.

the early hours of restraint, reduces greatly the likelihood that other independent variables involved in more chronic restraint, such as reduced food intake, disruptions in sleep, postural changes, etc., can be implicated as major determinants of the hormonal response profile during acute restraint. This experiment is just one of a great many from many different laboratories, incidentally, which reflect the extraordinary sensitivity of neuroendocrine systems to psychological stimuli and which form the foundation of the burgeoning field of psychoendocrinology (Mason, 1968*a,* 1975*a,* 1975*d,* 1975*e*).

While our main conceptual approach has been along these lines of the study of multihormonal response profiles to a wide range of psychological and other natural stimuli, during the course of this work over many years we have made a number of observations which appear to have implications for the evaluation of "stress" theory, particularly the "general adaptation syndrome" or "nonspecificity" concepts introduced by Selye (Selye, 1936, 1950, 1956; Selye and Heuser, 1956). Very briefly, the "nonspecificity" concept formulated by Selye holds that such diverse agents as cold, heat, x-rays, exercise, trauma, fasting, emotional stimuli, and many others all elicit certain common or nonspecific reactions in the body, including stimulation of the pituitary-adrenal cortical system. Such responses have been described recently as "totally nonspecific and common to all types of exposure" (Selye, 1973) or as resulting from "any demand" made upon the body (Selye, 1974). The main purpose of the remaining discussion, then, will be to separate out from our profile work a series of observations on the pituitary–adrenal cortical and the sympathetic adrenal medullary systems which have raised some compelling questions concerning the validity of the "nonspecificity" concept.

One of the early experiments to raise such questions involved the study of fasting in monkeys. In a crude, initial effort, we simply stopped feeding two monkeys out of a group of eight monkeys housed in the same room. During the 3-day fasting period, it was noticed that the deprived monkeys often vocalized in apparent protest when the caretaker bypassed them in the handing out of food pellets to the other nonfasting monkeys. It was not until we observed marked 17-OHCS elevations in these two monkeys on the very first day of food deprivation, however, as shown in Fig. 4, that it dawned upon us that there might be psychological as well as nutritional factors operating as independent variables in these experiments (Miller and Mason, unpublished observations). In this particular laboratory setting, the fasting animals were being exposed to all the sights, sounds, and smells associated with the routine feeding procedure, as well as the social variables related to the presence of nonfasting monkeys. Accordingly, we devised the following experimental precautions or measures to minimize possible psychoendocrine reactions during these fasting experiments.

1. Minimize novelty and uncertainty. A lengthy period (> 2 wk) was allowed for adaptation to chair restraint. No location changes were made. The animals

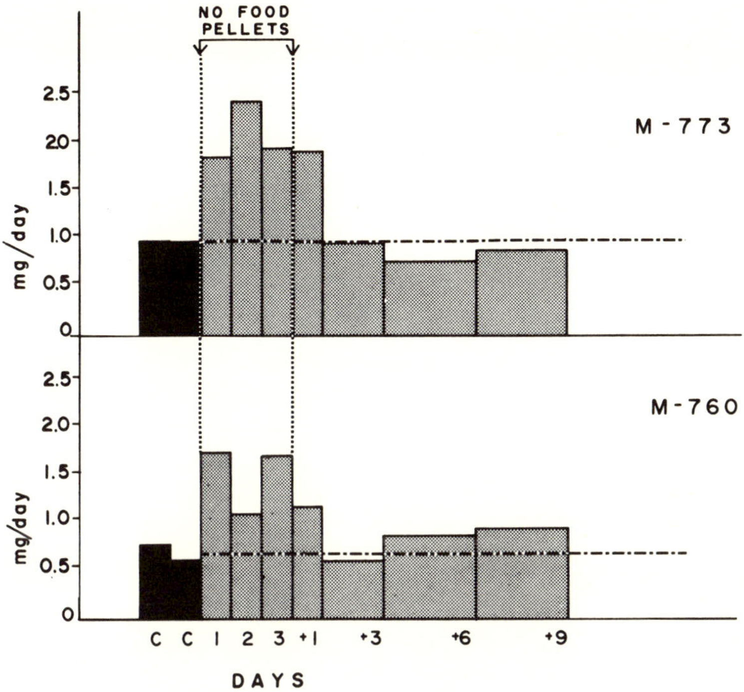

Fig. 4. Effect of "fasting" on urinary 17-OHCS Levels in the monkey when psychological variables were not considered.

were adapted to urine and plasma collections during a precontrol period. The animals were handled only by familiar persons.

2. Minimize extraneous psychosocial stimuli. Animals were kept in a private, sound-resistant booth housed in cubicle containing four booths. Cubicles were entered only for feeding, cleaning, sample collection, or other experimental procedures.

3. Minimize discomfort. "PLACEBO FOOD" (non-nutritive fruit-flavored cellulose pellets) were given to animals to reduce discomfort of empty gastrointestinal tract, as well as to prevent any change in social variables associated with routine feeding.

When fasting was studied in monkeys with the above precautions, quite a different picture emerged, as shown in Fig. 5. When four monkeys were simply placed in a quiet, private cubicle without the presence of nonfasting monkeys, very little change, perhaps a slight rise, in mean 17-OHCS levels occurred during the 3 days of fasting, and certainly dramatically less change than was observed in the initial two monkeys, mentioned earlier, who were in a less well-controlled social environment, designated as "Busy Lab" in Fig. 5. When the additional step

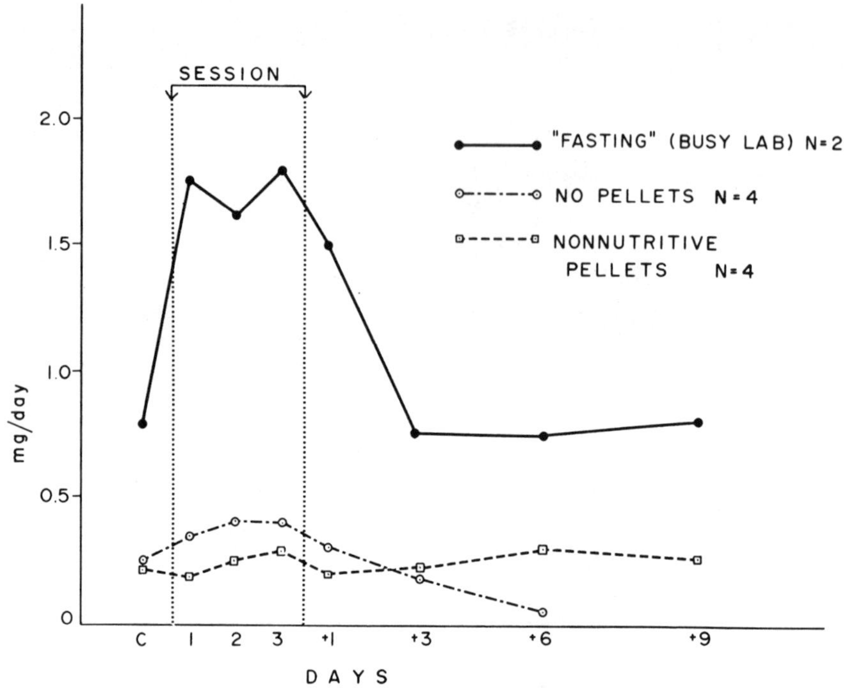

Fig. 5. Urinary 17-OHCS responses to fasting in the monkey under three different experimental conditions.

was taken of providing the nonnutritive pellets, or "placebo food," for the monkeys to eat, no significant changes in 17-OHCS levels were associated with the fasting sessions. As our series of observations was extended to 11 experiments in eight monkeys, this finding remained consistent, as shown in Fig. 6 which presents mean values and standard errors. It is of interest, however, that while corticosteroid levels are not significantly changed in association with the fasting sessions, there is an appreciable rise in epinephrine excretion ($p < 0.001$) and a marked decrease in norepinephrine excretion ($p < 0.001$) under these same conditions.[1] In a preliminary report elsewhere, we have also indicated that characteristic changes occur in levels of testosterone, thyroid hormone, and insulin when a broader response profile is viewed (Mason, 1974). *The pituitary–*

[1] *Statistical Procedure.* In this and all other instances in which multiple experimental values are compared with a control value, statistical evaluation was carried out by a preliminary analysis of variance with repeated measures followed by Dunnett's "t" test which corrects for multiple comparisons. In cases where only two values are compared in an experiment, the two-tailed Student's paired "t" test is used.

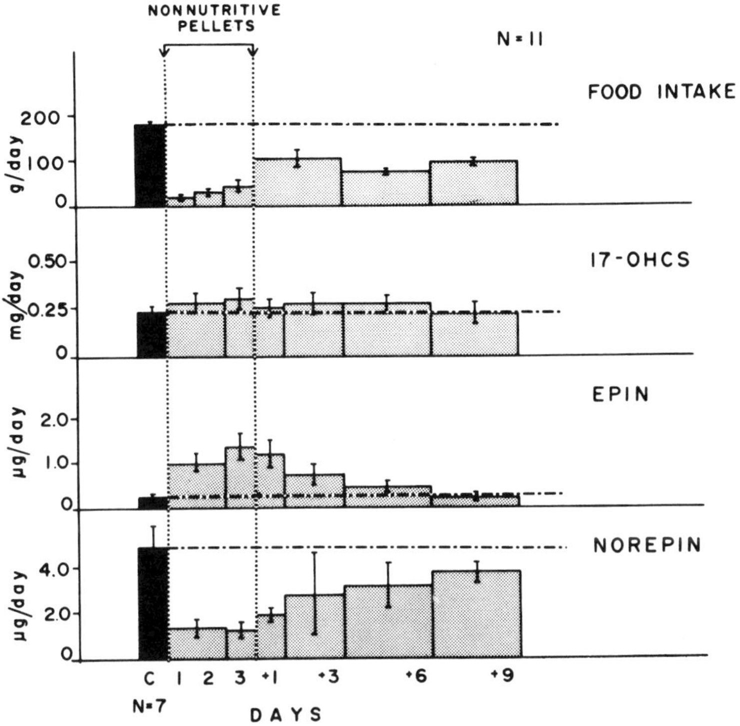

Fig. 6. Effect of fasting (nonnutritive pellets given) on corticosteroid and catecholamine levels in the monkey.

adrenal cortical system, however, does not appear to be stimulated in "non-specific" fashion by fasting, under these conditions when psychoendocrine factors are taken into account.

Our early studies of muscular exercise also provided another body of evidence suggesting that psychological variables could frequently be interfering or extraneous variables in studies of physical or physiological demands upon the organism. In some pilot experiments in which chair-restrained monkeys were required to lift heavy weights in order to obtain their daily food pellets, for example, we did sometimes observe elevated urinary 17-OHCS levels in relation to 3-day periods of heavy muscular work, as shown in Fig. 7. In session No. 6 of lifting heavy weights for food, this monkey showed roughly a two-fold increase in 17-OHCS excretion over the 3-day period, with a workload ranging up to over 4000 kg meters per day. In session No. 7 on the same monkey, however, a still greater 17-OHCS response was observed on an occasion when the monkey apparently decided that he would rather not eat for 3 days than put forth the

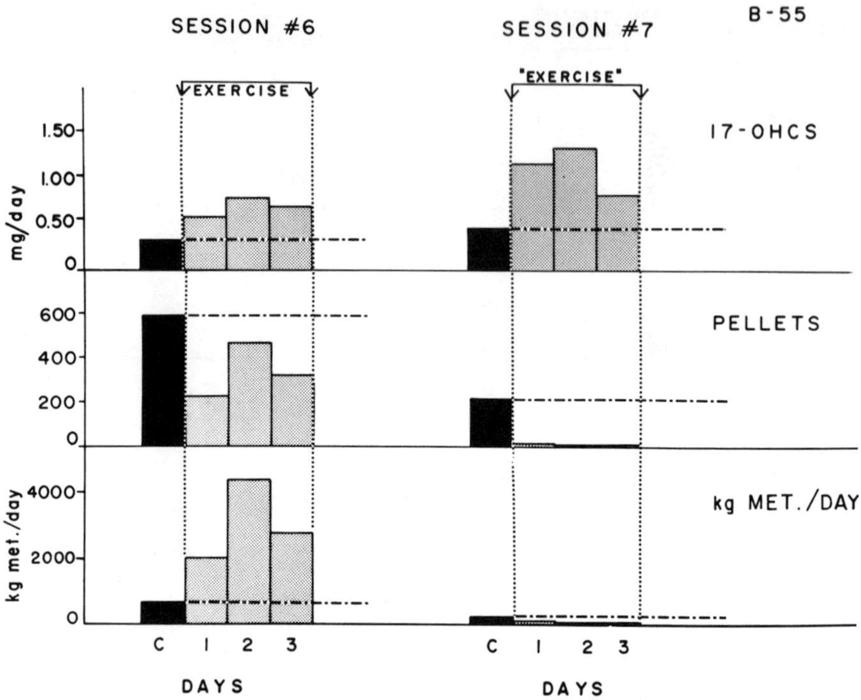

Fig. 7. Corticosteroid responses in monkey on program of lifting heavy weights for food reward.

heavy muscular effort required to obtain food pellets. These experiments suggested caution, then, in ascribing the 17-OHCS response in session No. 6 to muscular work, per se, since the findings in session No. 7 indicate that the weight-lifting task was apparently rather aversive to the animal and probably generally attended by some degree of psychological reaction (Miller and Mason, 1965).

In an effort to devise a more naturalistic, and hopefully less aversive, exercise task for the monkey, we next built a tall cage in which the monkey must climb up and down a ladder repeatedly in order to obtain his daily food. Figure 8 illustrates that a monkey usually readily performed considerable work under these conditions, with this particular monkey climbing about 1500 meters or more per day, in repeated sessions, as depicted by the lightly shaded bars. In spite of climbing in the vicinity of about a mile per day, however, relatively little change was observed in urinary 17-OHCS levels, shown by solid black bars, particularly after repeated experience with the procedure, as in sessions 3 and 4

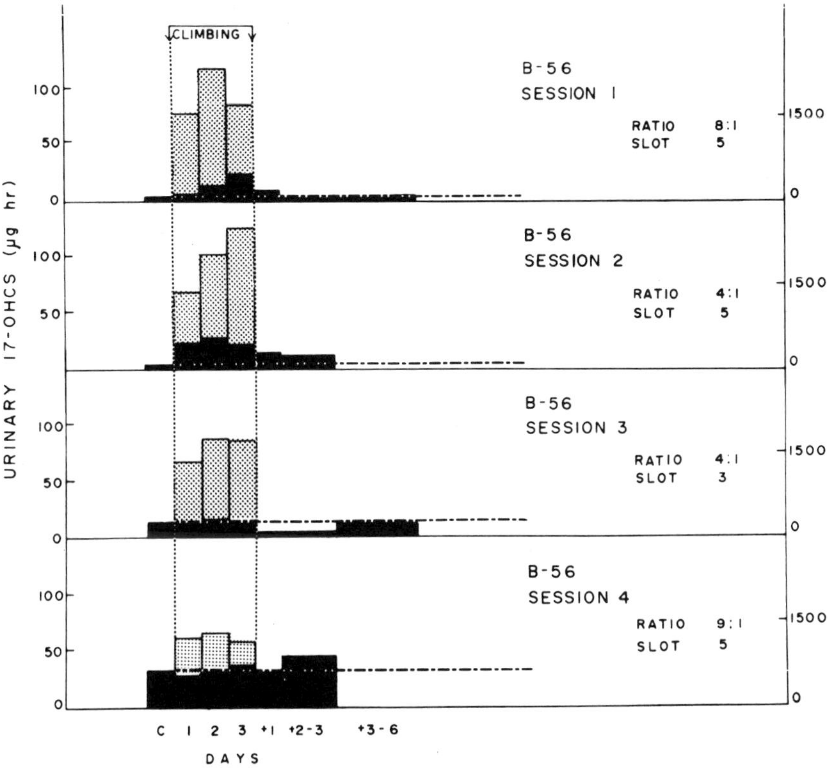

Fig. 8. Corticosteroid levels in monkey exercising by climbing for food reward.

(Miller and Mason, 1965). While these experiments raised further doubts about the potency of muscular work per se as a stimulus to the pituitary–adrenal cortical system, this experimental model clearly had limitations, particularly with regard to the difficulty in obtaining blood samples from the free-ranging monkey in a very large cage.

At this point we were able to progress to human studies which provided greatly superior opportunities to minimize and evaluate psychological variables as interfering factors in exercise experiments. On the basis of much information about relevant psychological parameters from earlier psychoendocrine studies, the following measures were taken to minimize psychological reactions during muscular exercise experiments involving normal young men.

1. Minimize novelty and uncertainty. Subjects were familiarized with the laboratory setting and procedures in a sham, pre-experimental experience a week earlier. Subjects were informed in advance of exactly what was to be done in

each experiment. I-V catheter was installed at least 2 hr before control observations started to allow time for recovery from possible attendant psychoendocrine reactions.

2. Minimize extraneous psychological stimuli. Experiments were performed in a quiet, stable laboratory with only essential research personnel present. Measures were taken to minimize any competitive overtones regarding exercise performance.

3. Minimize discomfort. Extremely heavy work levels were avoided and only graded levels of mild to moderate exercise were studied. Indwelling I-V catheter for blood sampling was used to avoid repeated venipuncture.

The exercise workload in all of these experiments was carefully quantitated on the basis of the maximal oxygen uptake (max. Vo_2) level determined for each individual in preliminary sessions. Two different work levels, 40% and 70% of the maximal oxygen uptake level, were studied in 3-hr sessions in two groups of eight subjects exercising on a bicycle ergometer.

The lack of urinary 17-OHCS response to exercise is very consistent in our data, and quite in contrast to urinary catecholamine changes. Figure 9 shows the data for urinary corticosteroid, epinephrine, and norepinephrine excretion during 3-hr exercise sessions at the 40% max. Vo_2 level in comparison to values on a control day with no exercise. The successive 3-hr 17-OHCS levels do not differ on the exercise day as compared to the control day, showing only the well-known diurnal trend downwards. By contrast, both epinephrine ($p < 0.01$) and norepinephrine ($p < 0.01$) excretion is markedly increased in association with the exercise sessions. It should be noted, incidentally, that urinary 17-OHCS levels do not show any evidence of delayed increase in either the 1400–1700 hr (recovery) or 1700–0800-hr (overnight) samples following the exercise sessions.

Figure 10 shows that a very similar picture is seen in the urinary excretion of these hormones even at the 70% max. Vo_2 level of sustained exercise, which represented muscular work to the point of exhaustion in seven out of eight subjects. In this experiment, involving a second group of eight subjects, there again is no significant urinary 17-OHCS change, but marked increases in urinary epinephrine ($p < 0.001$) and norepinephrine ($p < 0.01$) levels occur in association with muscular work. A pattern of additional hormonal changes in this study involving other endocrine systems has been reported elsewhere (Hartley *et al.*, 1972*a*).

Plasma cortisol levels were also studied in these experiments. As shown in Fig. 11, which focuses just on the acute phase of response during the first hour of exercise, there was a large, graded mean increase in both oxygen uptake and heart rate at these two work levels as evidence that there was appreciable muscular exertion in both cases. Yet at the 40% level, there is no change in plasma cortisol levels after 1 hr of exertion, values remaining very steady at 7 μg%. At the 70%, or exhausting work level, there is a preexercise, apparently

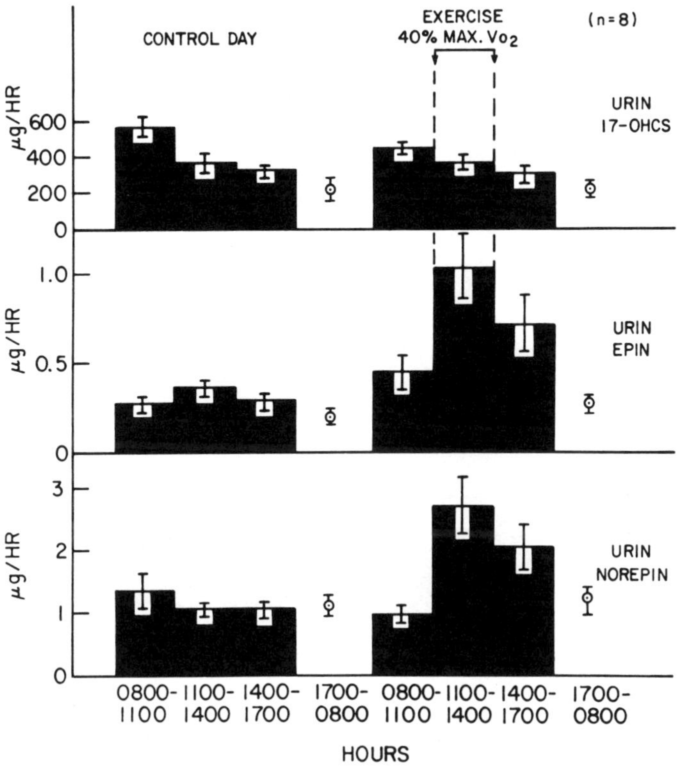

Fig. 9. Urinary 17-OHCS and catecholamine excretion during 3-hr exercise session (40% max. Vo_2 work level) in human subjects.

anticipatory, psychoendocrine cortisol response from 5 to 9 μg% ($p < 0.01$) during the 20 min immediately prior to onset of exercise (Mason *et al.,* 1973), followed by a smaller increase ($p < 0.05$) to 11 μg% 50 min after the onset of work.

Of the two work levels, it was clear that the 40% max. Vo_2 level most nearly met our criteria for an experiment in which there was appreciable exertion with apparently minimal attendant psychoendocrine reaction. All subjects completed the full session, none ventured complaints of marked subjective discomfort, and there was no evidence of anticipatory psychoendocrine reactions. As the full course of this experiment was studied, as shown in Fig. 12, it is evident that as the muscular exertion continues there is a mild to moderate increase in plasma cortisol levels after the second ($p < 0.05$) and third ($p < 0.01$) hours, although this change is not reflected in urinary 17-OHCS levels collected during, and after exercise. While the findings of no cortisol change during the first hour of

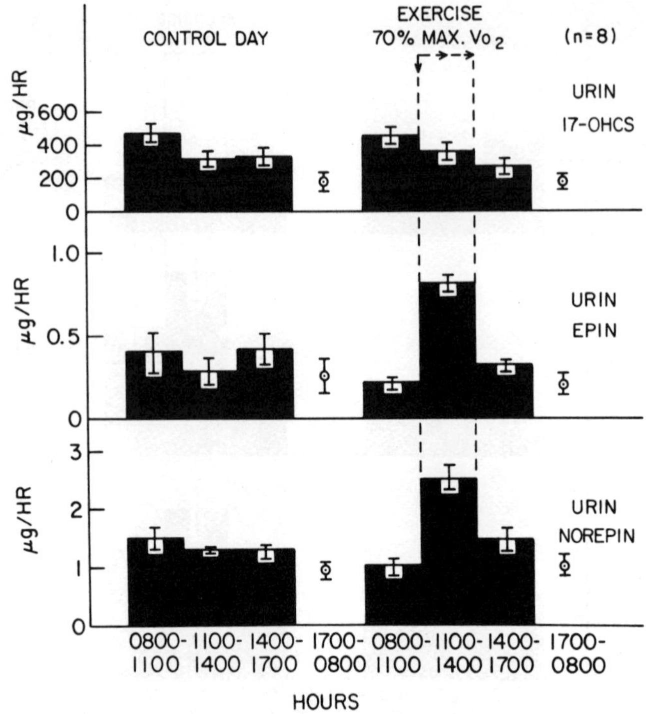

Fig. 10. Urinary 17-OHCS and catecholamine excretion during exercise to exhaustion (70% max. Vo_2 work level) in human subjects.

exertion are consistent with similarly negative findings in related studies of 10-min graded exercise sessions which we have reported previously (Hartley *et al.*, 1972*b*), the occurrence of some elevation as exertion is prolonged beyond 1 hr raises the question as to whether this represents hormonal response to prolonged exercise itself or possibly to mounting psychological reactions as the obligatory work must be continued and lunch delayed.

The present data on exercise, then, suggest that there is no significant stimulation of the pituitary–adrenal cortical system with moderate exercise of short duration, perhaps up to 1 hr, but the possibility remains that there may be some stimulation of the system with prolonged muscular exertion, although it appears that it may be very difficult experimentally to separate the role of psychological versus physiological stimuli to cortisol secretion during relatively severe or prolonged muscular exertion. In this connection there is an interesting study by Frankenhaeuser *et al.* (1969), in which a correlation was found between urinary catecholamine excretion at different workloads and the intensity of concomitant subjective efforts; these results may be interpreted "as

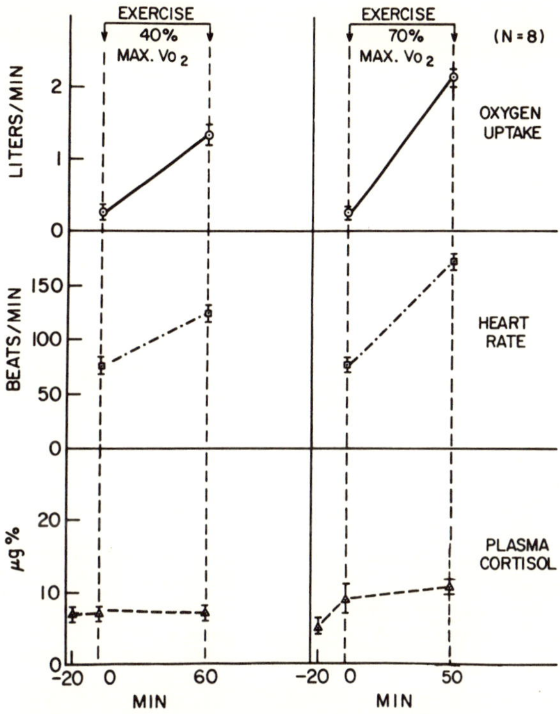

Fig. 11. Plasma cortisol levels during first hour of mild and moderate exercise in human subjects.

indicating that an increase in adrenal-medullary secretion occurs during muscular work when the work grows so heavy that feelings of emotional stress and unpleasantness are evoked." The corticosteroid data in the present study would seem to add further support to the view that subjective reactions at different levels of exercise should be carefully evaluated in future work in this field in an effort to separate hormonal responses to muscular work per se from attendant psychological reactions.

It should also be emphasized that, while our present data include both plasma and urinary corticosteroid measurements, the evaluation of pituitary–adrenal cortical responses to exercise cannot be considered as reasonably conclusive until studies similar to those described above have been conducted with the additional measurement of secretion or production rates of cortisol by isotopic methods. The theoretical possibility still remains that increased cortisol secretion may occur in exercise, but that such changes are not fully reflected in plasma or urinary levels because of altered turnover or metabolic disposition of the secreted cortisol. *The bulk of evidence we have at present, however, does not*

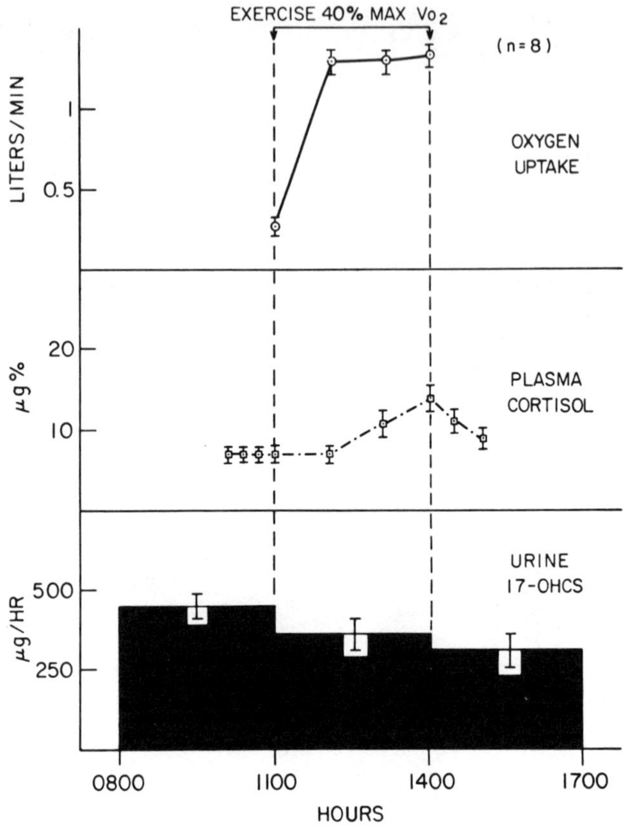

Fig. 12. Plasma and urinary corticosteroid levels during sustained exercise (max. Vo_2 40% work level) in human subjects.

support the conclusion that exercise is a nonspecific stimulus to the pituitary–adrenal system, if psychoendocrine parameters are taken into account.

Perhaps the most compelling data militating against the "nonspecificity" concept which have emerged from our work so far have been in relation to heat exposure. Much as in the case of fasting and exercise, our initial crude efforts to study heat in the monkey raised questions of interfering psychological reactions. As before, we decided that it was necessary to establish some measures for minimizing such psychological reactions. These measures were similar to those described for fasting experiments with regard to minimizing novelty and extraneous psychosocial stimuli. Two additional measures were observed in our heat exposure studies:

1. *Severe* temperature changes were avoided and graded heat levels of mild to moderate intensity were studied.

2. *Sudden* temperature changes were avoided by raising temperature gradually in small increments over a period of hours.

When these precautions were observed, heat exposure in the monkey produced the striking findings shown in Fig. 13. When ambient temperature was elevated from 72°F to 85°F gradually over a 15-hr period and then sustained at 85° for two weeks, urinary 17-OHCS levels in a chair-restrained monkey not only didn't rise, but actually appeared to be suppressed consistently below the control baseline, representing a mean of the 23 days prior to onset of heat exposure, with values rising above this level again only about a week after return to normal ambient temperature. Urinary norepinephrine levels show an even more pronounced lowering during heat exposure, while urinary epinephrine levels show some trend downwards, but not as convincingly (Mason, 1974; Poe *et al.*, unpublished observations).

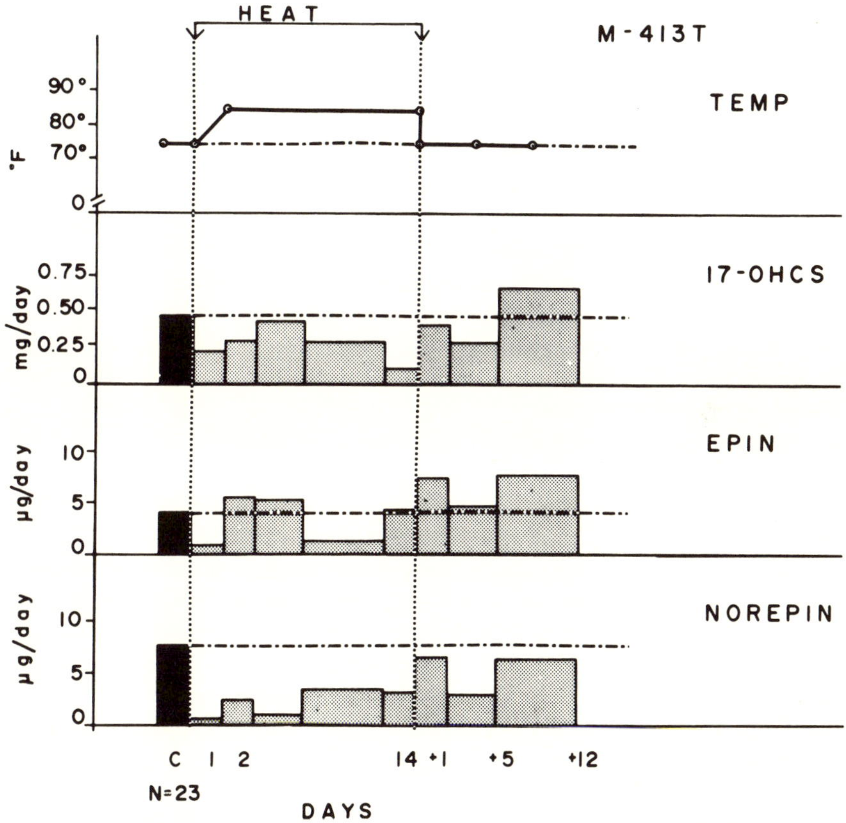

Fig. 13. Suppression of corticosteroid and catecholamine levels by sustained heat exposure in a monkey.

Using the general principles developed in these monkey experiments, we have recently carried out extensive studies of acute heat exposure in normal human subjects. In these experiments, a group of eight normal young men were studied longitudinally before, during, and after 3-hr periods of exposure to 74°F (control condition), 95°F, 100°F, and 105°F ambient temperatures, with weekly intervals between each experiment at the different heat exposure levels. During control and recovery periods, temperatures were always maintained at 74°F and relative humidity was maintained at 50% at all times. During heat exposure sessions, temperatures were gradually elevated to the desired heat level over the first 30 min of each 3-hr session. Subjects were clothed only in shorts. All blood samples were obtained through indwelling intravenous catheters which were installed about 90 min before onset of control hormonal measurements.

Figure 14 presents the mean urinary corticosteroid and catecholamine levels during the control experiment in which the subjects remained in the large experimental chamber at 74°F throughout the day. In this experiment, there is no significant elevation in mean rectal temperature, nor in the three successive 3-hr urinary values for epinephrine, norepinephrine, or 17-OHCS excretion. Figure 15 shows a slight mean elevation ($p < 0.01$) in rectal temperature when the eight subjects were exposed to 95°F ambient temperature for a 3-hr period, but there is no significant elevation in catecholamine or corticosteroid excretion. Figure 16 shows a more substantial elevation ($p < 0.01$) in mean rectal temperature during the session at 100°F ambient temperature, but again there is no associated elevation of catecholamine or corticosteroid levels. Figure 17 presents the final experiment, this time at the rather high level of 105°F 50% RH, in which the rectal temperature shows a substantial elevation ($p < 0.01$) from 98.6°F to 100.2°F. Even with a body temperature rise of this magnitude, however, there is no elevation in corticosteroid or catecholamine levels either during or after heat exposure.

The plasma cortisol results for these four experiments are summarized in Fig. 18. It is clearly evident that there is no elevation in mean plasma cortisol levels in any of the heat exposure sessions, including the one at 105°F ambient temperature in which mean rectal temperature rose 1.6°F. The general conclusions to be drawn from both the urinary and plasma corticosteroid measurements in the study, then, are that *the pituitary–adrenal cortical system is not nonspecifically stimulated by heat exposure which produces mean rectal temperature elevations up to 100.2°F, and which involves the present measures designed to minimize possible interfering psychoendocrine reactions.*

In summary, then, the experimental findings presented above on fasting, moderate exercise, and heat exposure appear to raise increasing questions concerning the validity of Selye's concept of "nonspecificity" in which the pituitary–adrenal cortical system is viewed as responding to any demand or to all types of exposure irrespective of the nature of the evocative stimulus. *The*

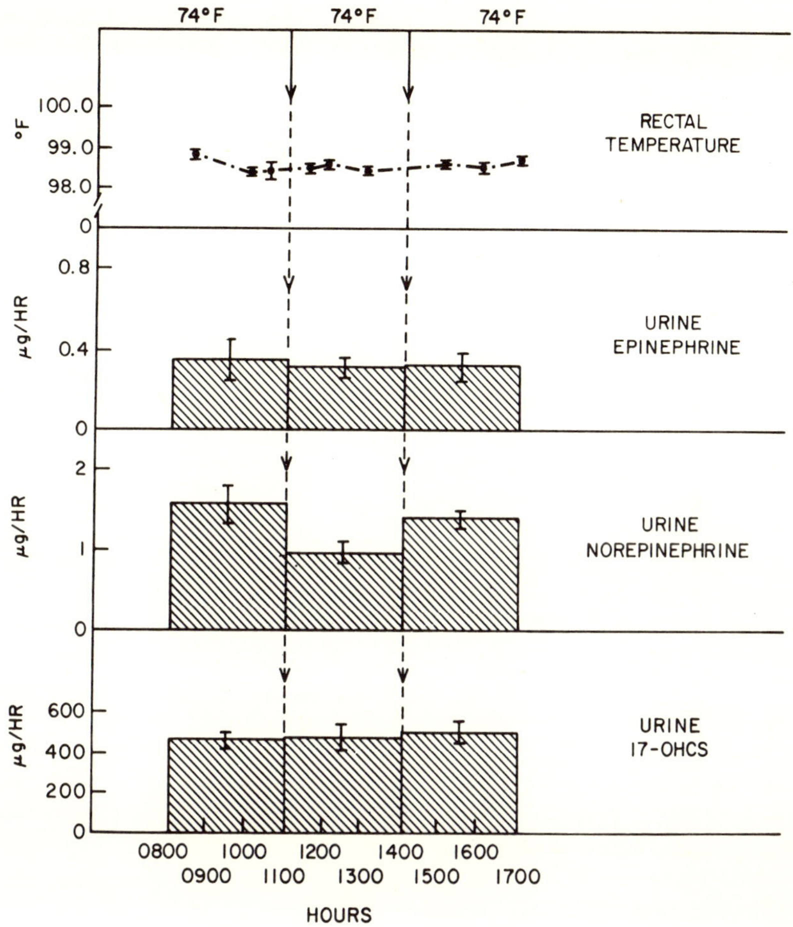

Fig. 14. Corticosteroid and catecholamine levels under control conditions (74°F, 50% RH) in normal human subjects.

present findings suggest that the conditions to which animals or humans are subjected in the laboratory study of physical stimuli or "stressors" commonly involve attendant and substantial psychoendocrine reactions. In turn, there is now an enormous body of psychoendocrine data which has established firmly that psychological or emotional stimuli rank very high among the most potent as well as prevalent natural stimuli to the pituitary–adrenal cortical system. It seems likely, then, that lack of awareness of the potency and prevalence of such psychological variables may have led to the erroneous impression in earlier work that the pituitary–adrenal cortical system responds to a broader and more

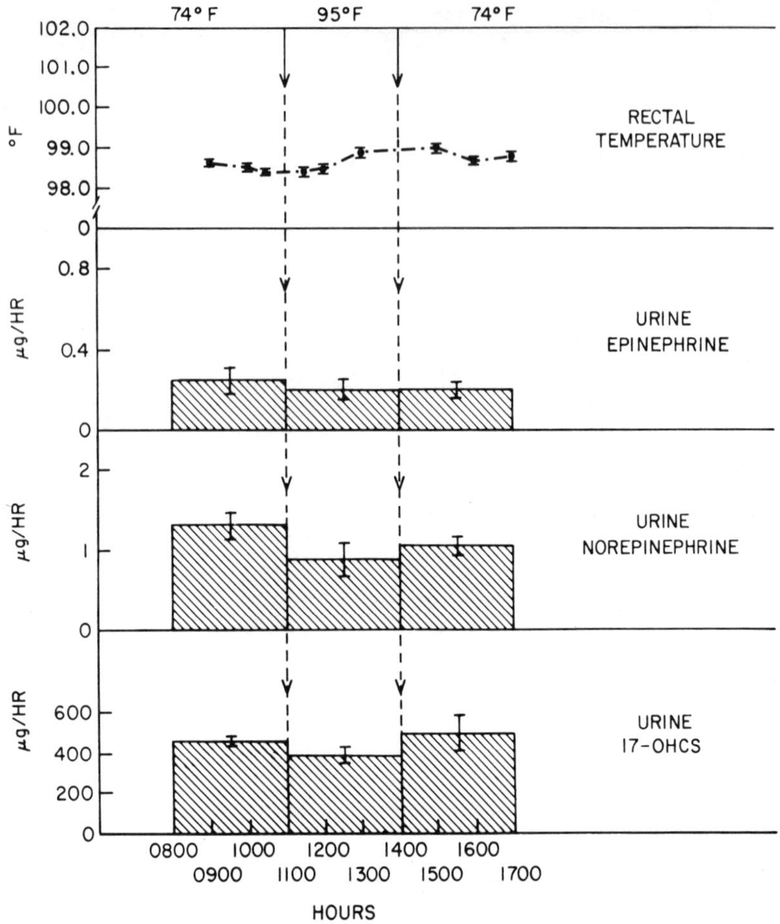

Fig. 15. Corticosteroid and catecholamine levels during acute heat exposure (95°F, 50% RH, for 3 hr) in normal human subjects.

comprehensive range of stimuli than is actually the case. In other words, it now appears likely that the "nonspecific" element in many early "stress" experiments may have been to a considerable extent a reflection of ubiquitous psychological reactions to various experimental conditions rather than a consequence of the afferent messages of diverse physical stimuli acting directly at the lower level of the final common neuroendocrine pathways (Mason, 1971, 1975*b,c*).

It should be emphasized, however, that this is *not* to conclude that the

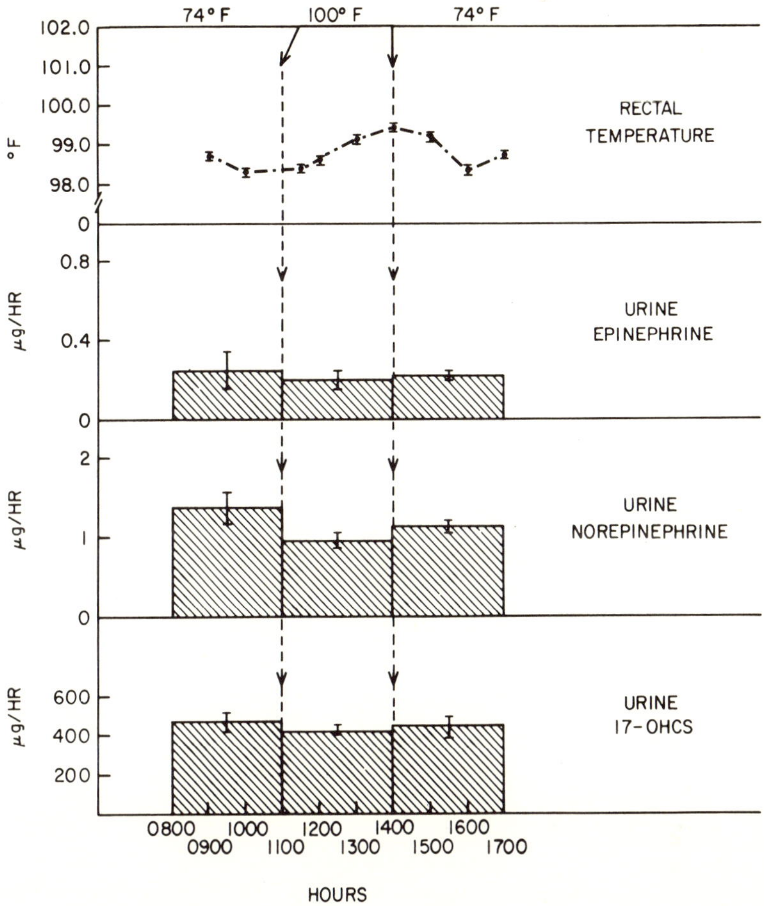

Fig. 16. Corticosteroid and catecholamine levels during acute heat exposure (100°F, 50% RH, for 3 hr) in normal human subjects.

pituitary–adrenal cortical system responds *only* to psychological stimuli, *nor that all* "stressors" act through psychological mechanisms, since present data indicate, for example, that cold, hypoxia, and probably some other stimuli activate this neuroendocrine system directly (Mason, 1974). Rather, the point is simply to emphasize that psychological influences upon the pituitary–adrenal cortical system are potent, often subtle, easily overlooked, and more ubiquitously present as interfering variables in laboratory experiments than has been generally realized. It appears rather reasonable to suggest, therefore, that stress theory with regard to the "nonspecificity" and "general adaptation syndrome"

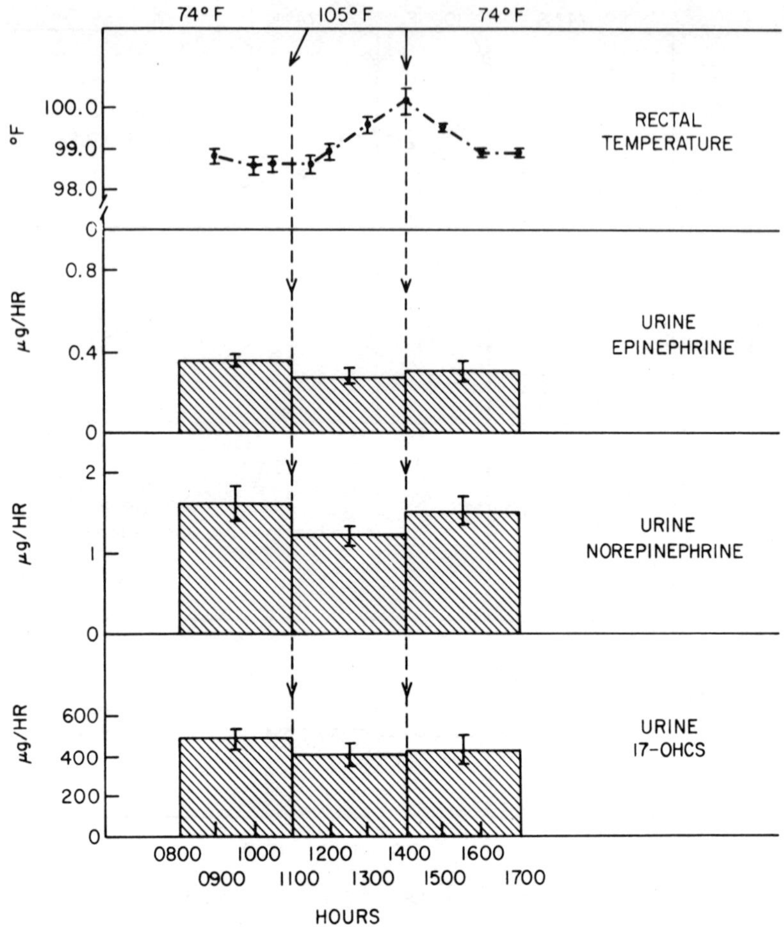

Fig. 17. Corticosteroid and catecholamine levels during acute heat exposure (105°F, 50% RH, for 3 hr) in normal human subjects.

concepts be regarded as still open to question and be subjected to continuing and thoroughgoing experimental reevaluation in the light of these emerging insights.

The possible further theoretical implications of these developments cannot be fully explored here, but they have been discussed at length elsewhere (Mason, 1975*b,c*). In concluding, it may be useful, however, to return to the broader perspective of the research strategy and program from which the present data has emerged. As mentioned earlier, our main interest has been the study of multihormonal patterns of response to various natural stimuli with a view to the

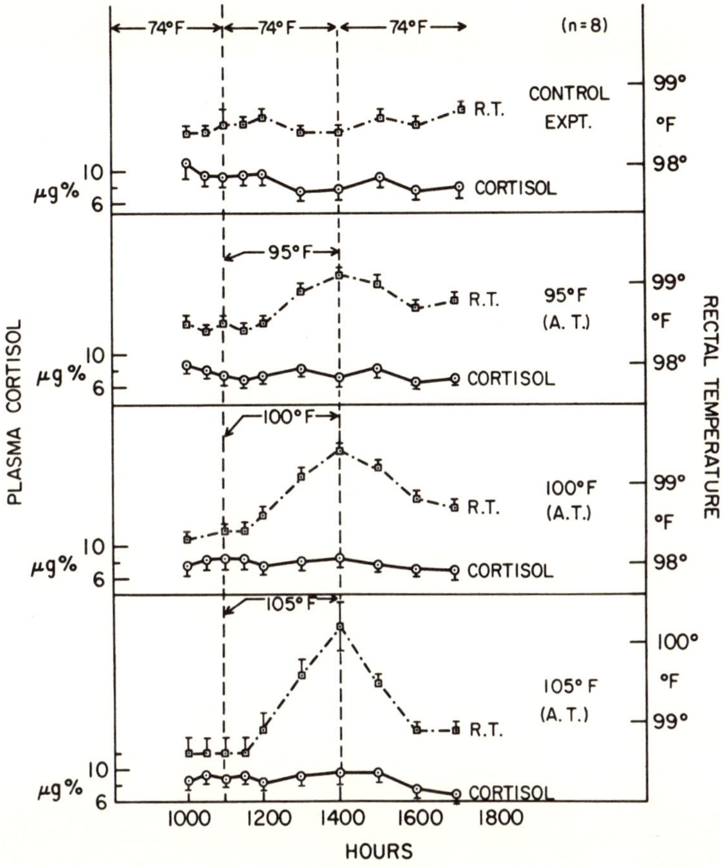

Fig. 18. Lack of plasma cortisol response to graded acute heat exposure in normal human subjects.

"overall" organization of integrative neuroendocrine responses. There have emerged from these studies, which have been reviewed in detail elsewhere (Mason, 1974), some general conclusions which appear relevant to the development of research strategies in future studies of neuroendocrine aspects of "stress:"

1. Most stimuli, or "stressors," studied so far evoke not just a few scattered hormonal responses, but generally elicit a broad scope of multiple, concurrent responses involving many endocrine systems. There is a clear need, then, to extend increasingly our view of hormonal responses in "stress" research beyond the adrenal systems to include such hormones as TSH, thyroxine, testosterone, LH, FSH, estrogens, growth hormone, prolactin, insulin, vasopressin, and so on (Mason, 1974, 1968*a*).

2. The multiple neuroendocrine responses to various stimuli we have studied so far appear to be organized broadly into overall "patterns" or "profiles" which appear to be relatively distinctive, or selective, and rather consistently reproducible as a whole for each particular stimulus in question (Mason, 1974).

3. While all hormones studied so far appear to respond to multiple stimuli, there appears to be no single hormonal response of any neuroendocrine system which occurs nonspecifically as a common element of all the various response profiles studied. It appears, rather, that the level of any given hormone may sometimes rise, sometimes fall, or sometimes not change, depending upon the nature of the stimulus (Mason, 1974).

Finally, it should be emphasized that our work along these lines must still be regarded as largely of a preliminary nature and in need of much further, increasingly rigorous experimental evaluation by the most refined methods which become available. It already seems quite clear, however, that such attempts to break away from the conventional study of single endocrine systems in isolation and to view the highly interdependent neuroendocrine apparatus as nearly as a coordinated whole as possible is very likely to continue to carry us to new, higher plateaus of insight into the principles underlying the organization of the central mechanisms integrating the unit functions of the internal environment.

REFERENCES

Frankenhaeuser, M., Post, B., Nordheden, B., and Sjoeberg, H. (1969). Physiological and subjective reactions to different physical work loads. *Percept. Mot. Skills, 28,* 343–349.

Hartley, L. H., Mason, J. W., Hogan, R. P., Jones, L. G., Kotchen, T. A., Mougey, E. H., Wherry, F. E., Pennington, L. L., and Ricketts, P. T. (1972*a*). Multiple hormonal responses to prolonged exercise in relation to physical training. *J. Appl. Physiol. 33,* 607–610.

Hartley, L. H., Mason, J. W., Hogan, R. P., Jones, L. G., Kotchen, T. A., Mougey, E. H., Wherry, F. E., Pennington, L. L., and Ricketts, P. T. (1972*b*). Multiple hormonal responses to graded exercise in relation to physical training. *J. Appl. Physiol. 33,* 602–606.

Houssay, B. C. (1957). Comments in *Hormonal Regulation of Metabolism.* Springfield, Ill.: Thomas, p. 27.

Mason, J. W. (1968*a*). Organization of psychoendocrine mechanisms. *Psychosom. Med. 30,* 565–808.

Mason, J. W. (1968*b*). Organization of the multiple endocrine responses to avoidance in the monkey. *Psychosom. Med. 30,* 774–790.

Mason, J. W. (1968*c*). "Over-all" hormonal balance as a key to endocrine organization. *Psychosom. Med. 30,* 791–808.

Mason, J. W. (1971). A Re-evaluation of the concept of "non-specificity" in stress theory. *J. Psychiat. Res. 8,* 323–333.

Mason, J. W. (1974). Specificity in the organization of neuroendocrine response profiles. In *Frontiers in Neurology and Neuroscience Research* (P. Seeman and G. Brown, Eds.). Toronto: Univ. of Toronto, pp. 68–70.

Mason, J. W. (1975*a*). Emotion as reflected in patterns of endocrine integration. In *Emotions–Their Parameters and Measurement* (L. Levi, Ed.). New York: Raven Press, pp. 143–181.

Mason, J. W. (1975*b*). A historical view of the stress field, Part I. *J. Human Stress 1*(1); 6–12.

Mason, J. W. (1975*c*). A historical view of the stress field, Part II. *J. Human Stress 1*(2); 22–36.

Mason, J. W. (1975*d*). Clinical psychophysiology: Psychoendocrine mechanisms. In *American Handbook of Psychiatry* (M. Reiser, Ed.).

Mason, J. W. (1975*e*). Psychological stress and endocrine function. In *Topics in Psychoendocrinology* (E. J. Sachar, Ed.). New York: Grune and Stratton.

Mason, J. W., Hartley, L. H., Kotchen, T. A., Mougey, E. H., Ricketts, P. T., and Jones, L. G. (1973). Plasma cortisol and norepinephrine responses in anticipation of muscular exercise. *Psychosom. Med. 35,* 406–414.

Miller, R. E., and Mason, J. W., (1965). Changes in 17-hydroxycorticosteroid excretion related to increased muscular work. *Symposium on Medical Aspects of Stress in the Military Climate, WRAIR.* GPO, Washington, D.C.: GPO, pp. 353–372.

Miller, R. E., and Mason, J. W. Unpublished observations.

Poe, R. O., Ehle, A. L., and Mason, J. W. Unpublished observations.

Selye, H. (1936). A syndrome produced by diverse nocuous agents. *Nature 138,* 32.

Selye, H. (1950). *The Physiology and Pathology of Exposure to Stress.* Montreal: ACTA, Inc.

Selye, H. (1956). *The Stress of Life.* New York: McGraw-Hill Book Co.

Selye, H. (1973). The evolution of the stress concept. *Am. Sci. 61,* 692–699.

Selye, H. (1974). *Stress without Distress.* Philadelphia: J. B. Lippincott Co.

Selye, H., and Heuser, G. (1956). *5th Annual Report of Stress.* New York: MD Publications, Inc.

The Role of Peripheral Catecholamines in Adaptation to Understimulation and Overstimulation

MARIANNE FRANKENHAEUSER

INTRODUCTION

A key problem in the current debate about man in future society concerns the application of technological advances. Will it be possible to direct new applications in such a way that they contribute to realizing social and human goals? So far, technological innovations have been used mainly to increase automation and mechanization, thereby maximizing the effectiveness of human production and communication. Today, there is a growing awareness that improvements gained by automation and mechanization may have negative side effects on the psychological level. The psychosocial environment has changed drastically, conditions characterized by understimulation and overstimulation have been created, and the demands on man's adaptive capacity continue to increase.

These problems call for new efforts to use the tools provided by the biological and the behavioral sciences in searching the environment for potentially harmful factors and in identifying high-risk individuals and groups, so as to enable us to propose preventive measures and remedial action.

MARIANNE FRANKENHAEUSER • Department of Psychology, University of Stockholm, Stockholm, Sweden.

The new concern about psychosocial factors is clearly manifested in the area of working life, a major sociopolitical issue in Sweden today. Among the workers themselves, among industrial managers, trade unions, and policy makers, there is a growing recognition of psychosocial risk factors. This has generated a new readiness to pay attention to knowledge gained by research in social psychology, showing that highly automated jobs, characterized by constriction of freedom and low utilization of personal talent, constitute a threat to health and well-being. Moreover, there is a growing realization that ill effects of psychologically unrewarding work conditions tend to spread to life outside work and, hence, may color the individual's total life situation. The view that the worker would be able to compensate for a dull and boring job by stimulating and enriching activities in his free time, is being replaced by a new insight into the strong links between a job that is circumscribed and repetitious and a leisure which is passive and psychologically unrewarding. In other words: those individuals whose work is restricted and monotonous are less likely to engage in leisure activities requiring planning, cooperation, and effort (Gardell, 1971).

When viewing our experimental, psychobiological research in this context, we were struck by the similarity—on the conceptual level—between our problems and those of the social psychology of working life. These two areas have had little in common, but it seemed that by bridging the gap, one would strengthen the knowledge gained by each of them. Data integrating knowledge from the two areas would carry more weight with those responsible for planning the work environment and working conditions.

Accordingly, we have made the problems of working life part of our experimental stress research program. Using the tools of psychoendocrinology, experimental psychology and social psychology, we are approaching problems of stress and adaptation by combining two research strategies. One strategy involves extracting specific subproblems from the real-life environment and bringing them into the laboratory where they can be examined under controlled conditions. The other takes our laboratory-based, experimental techniques into the field, where they can be applied to the study of people engaged in their daily activities. So far, laboratory experiments have formed the main part of our work, but emphasis is shifting toward the field, as will become apparent in the course of this paper.

Our dependent variables include self-estimates of wakefulness and mood, measures of mental performance and of physiological functions. One major problem concerns the quantitative relationship between these three sets of variables as well as their interaction with the constitutional characteristics of the individual and environmental factors.

Among physiological parameters, the activity of the sympathetic-adrenal medullary system—as reflected in the urinary excretion of adrenaline and noradrenaline—is well suited for our purposes (see reviews by Frankenhaeuser 1971,

1975*a,b*; Levi, 1972). Urine samples can easily be obtained under field conditions without interfering with the subject's daily routine. A small fraction of the liberated amines is excreted in urine as free adrenaline and noradrenaline, and this fraction can be estimated quantitatively by spectrophotofluorometric methods (Euler and Lishajko, 1961).

Adrenaline excretion is a very sensitive indicator of behavioral arousal, reflecting both the effort that a person invests in what he is doing and the intensity of the feelings—pleasant as well as unpleasant—evoked by what is happening to him. Noradrenaline also reflects behavioral arousal, but the threshold for the release of this amine in response to psychological stimuli is much higher.

I shall use data from our experiments to illustrate, *first,* how measures of catecholamine excretion reflect the impact of the psychosocial environment, and *second,* how peripheral catecholamines affect mental functions.

THE IMPACT OF PSYCHOSOCIAL FACTORS

The "stress staircase" (Fig. 1) illustrates the levels of adrenaline secretion accompanying exposure to pressures typical of everyday life. The adrenaline level is low during night rest, doubles during daily routine activities, and rises three to five times above the resting level under conditions of "mild" to "moderate" stress. Severely stressful conditions—exceeding the everyday stresses pictured here—may induce a further pronounced rise in adrenaline secretion, as some of the data to be reported will show.

The terms understimulation and overstimulation refer to the inability of the central regulatory mechanisms to maintain an optimal arousal level at low and

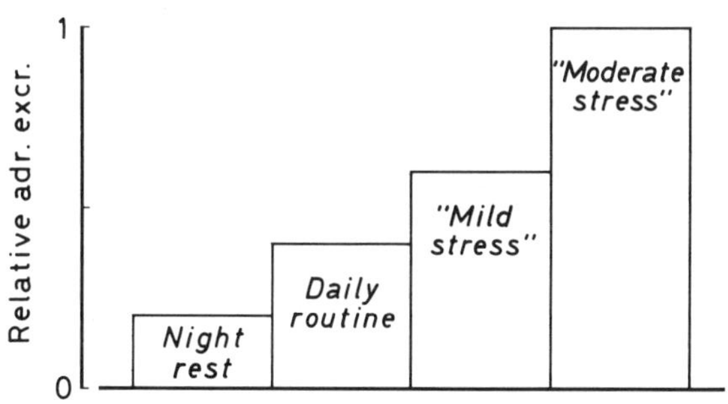

Fig. 1. Levels of adrenaline secretion during "the stress of everyday life." The level is low during night rest, doubles during daily routine activities, and rises 3–5 times above the resting level under conditions of "mild" to "moderate" stress.

high levels of stimulus input (Lindsley, 1961). The organism's adaptive resources are taxed at both ends of the stimulus continuum. However, tolerance is determined, not by the physical properties of the stimuli as such, but by the individual's cognitive appraisal of external conditions (Lazarus, 1966). Thus, cognitive and adreno-medullary functions interact in maintaining an optimal arousal level within a wide range of external stimulus conditions.

Data from a laboratory experiment (Fig. 2) show that understimulation—represented here by a repetitive, monotonous vigilance task—and overstimulation—represented by a complex audiovisual choice-reaction task requiring selective attention and rapid response—both induced an increase of catecholamine secretion as compared with a situation designed to match the "medium" input level of an "ordinary" environment (Frankenhaeuser *et al.,* 1971). Cognitively, understimulation and overstimulation were similar in that both situations were experienced as distressing and both required effort.

Any novel element in the environment elicits an increase of catecholamine secretion. As the individual's acquaintance with the situation grows, the output decreases. In situations where the subject plays the role of a passive observer, or "victim," there is a close relationship between the amount of adrenaline excreted and the intensity of the subjective reaction. This is seen in Fig. 3, where estimates of subjective stress, obtained by the method of ratio estimation, have been plotted against adrenaline excretion in each of six successive sessions involving exposure to gravitational stress in a human centrifuge (Frankenhaeuser

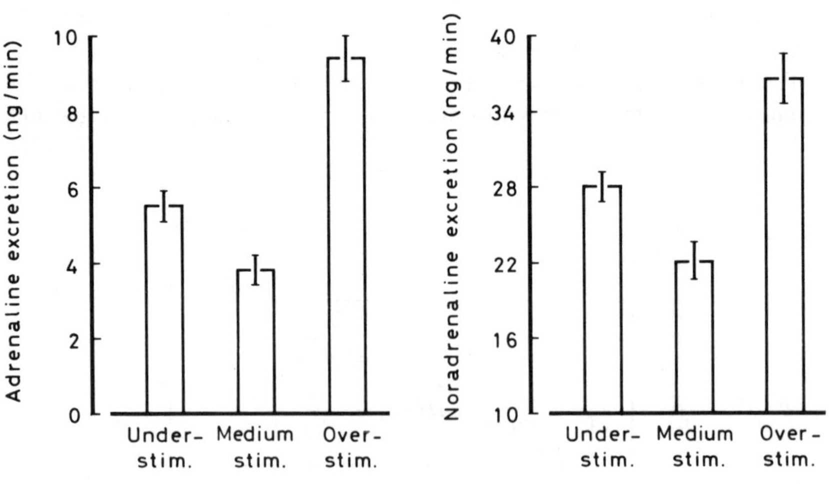

Fig. 2. Means and standard errors for adrenaline and noradrenaline excretion under conditions of understimulation, medium stimulation, and overstimulation. [Based on Frankenhaeuser *et al.* (1971). *Acta Psychol. 35,* 298–308.]

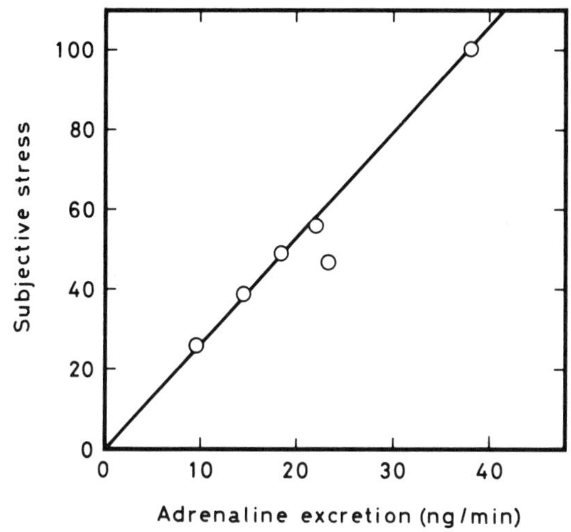

Fig. 3. **Mean estimates of subjective stress plotted against mean adrenaline excretion during each of six identical sessions involving exposure to gravitational stress in a human centrifuge.** Each point corresponds to one session. [Reprinted by permission of the publisher. Frankenhaeuser *et al.* (1962). *Percept. Mot. Skills, 15,* 63–72.]

et al., 1962). It should be noted, however, that repeated exposure to one and the same arousing stimulus is accompanied by a decrease in adrenaline secretion only if there is a concomitant decrease in subjective arousal. Thus, a stimulus which retains its stressful character will continue to induce a high rate of adrenaline secretion. The key characteristics of the cognitive-affective states under which a rise in catecholamine secretion occurs are effort, distress, and excitement. Other emotions, such as joy, anger, and fright—when intense—may also be accompanied by increased secretion of both adrenaline and noradrenaline. These conclusions are based on a series of studies in which various scaling and rating techniques have been used for quantifying subjective states.

When the subject is actively engaged in dealing with the stressor, his mastery and control of the situation will modify his catecholamine response. Figure 4 shows data from a laboratory experiment in which the level of control was varied systematically (Frankenhaeuser and Rissler, 1970). In Session I, where the subjects were exposed to uncontrollable electric shocks, adrenaline excretion was about three times as high as during relaxation (Session IV). Increasing the subject's control over the situation counteracted this rise. In Sessions II and III the subject performed a choice-reaction task and quick performance reduced the punishment. Session III was designed so that the subject exercised almost complete mastery. As the left-hand diagram shows, adrenaline output decreased

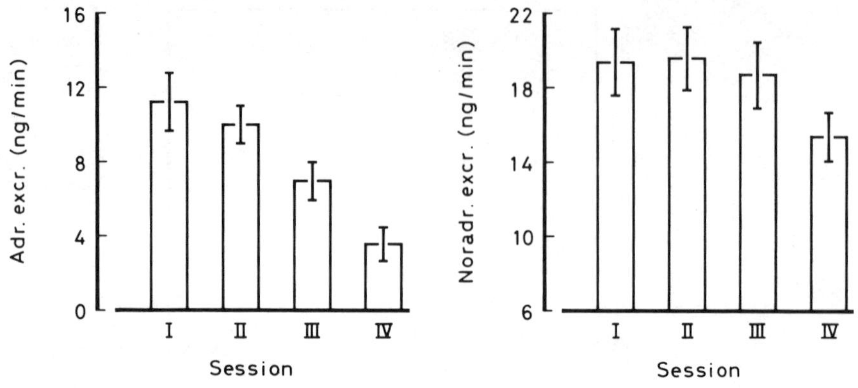

Fig. 4. Means and standard errors for adrenaline and noradrenaline excretion under four conditions. In Sessions I, II, and III the subject's control of the situation was successively increased. Session IV was a control condition. [Redrawn and reprinted by permission of the publisher. Frankenhaeuser and Rissler. (1970). *Psychopharmacologia 17,* 378–390.]

successively as the degree of control increased from a state of helplessness to an ability to master the disturbing influences. Noradrenaline excretion (right-hand diagram) was not much affected by the degree of control, but remained slightly elevated as long as the subject was engaged in the attention-demanding activity.

A further illustration of the modifying influence of control over the environment is provided by a study of urban commuting, initiated by Singer (State University of New York at Stony Brook) during his stay in Stockholm and followed up by Dr. Ulf Lundberg in my laboratory. Our aim was to identify the stressful aspects of daily commuting by train between a suburban home and a central-city job. A brief period of gas rationing during the oil shortage in the winter 1973–74 provided an opportunity to examine a group of passengers who made the same journey under different conditions of crowdedness, i.e., before and during the rationing period (Lundberg, 1976). Although the number of passengers per car increased by 10% only during rationing, adrenaline excretion (Fig. 5, left-hand diagram) was significantly higher on this occasion (values are expressed as a percentage of Sunday values, when the subjects stayed at home).

As to the actual traveling conditions, the most pronounced difference between the two trips was that the possibility to select seats and to choose one's own company was much more restricted on the second occasion. In an earlier study (Singer *et al.,* 1976) we found that subjects who boarded the train at the first stop secreted less adrenaline than those who boarded midway, when the train was more crowded. We interpret these results as indicating that the social and ecological circumstances, including controllability, are more important determinants of stress than the length and duration of the trip.

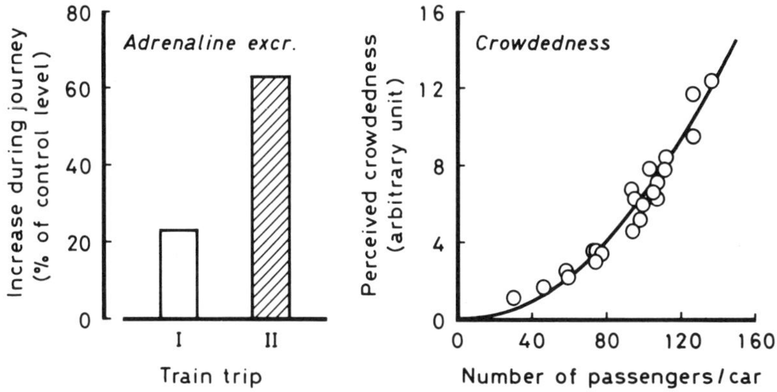

Fig. 5. Mean increase of adrenaline excretion in percent of control level during each of two train trips made under different conditions of crowdedness (left-hand diagram) and ratings of perceived crowdedness at different points on the journey (right-hand diagram). [Based on Lundberg. (1976.) *J. Human Stress* (in press).]

Perceived crowdedness, according to self-ratings made at successive points on the trip, increased as the square of the number of passengers. Figure 5 (right-hand diagram) shows mean data of a larger number of passengers, used here mainly to illustrate the ability of "the man in the street" to quantify his experiences.

THE ACTION OF PERIPHERAL CATECHOLAMINES

By varying the level of task demand, we can vary the individual's investment of effort and the rate of his catecholamine secretion (Frankenhaeuser and Johansson, 1976). Figure 6 shows adrenaline excretion in a group of subjects who performed the same color-word task under conditions of "single" and "double" conflict. In the latter case, when interfering auditory color words were added to those presented visually, adrenaline excretion was significantly higher and remained so during a subsequent arithmetic task.

A striking result was the ability of subjects to maintain the same performance level when the task demand was increased. Thus, the greater effort invested, "raising the body's thermostat for defense" (Selye, 1974), leads to a compensatory increase in adrenaline secretion, and performance remains intact.

There are considerable interindividual differences in catecholamine secretion, and by relating these to behavioral parameters we can analyze the action of peripheral catecholamines on psychological functions. Our analyses show that,

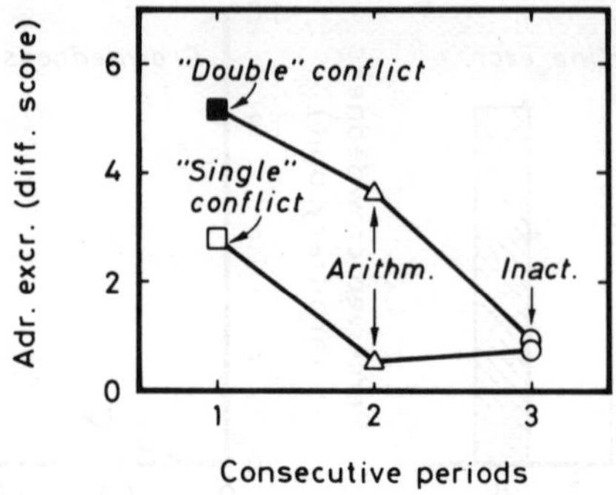

Fig. 6. Mean changes in adrenaline excretion during a color–word task performed under "single" and "double" conflict conditions, during a subsequent arithmetic task, and during inactivity. [Frankenhaeuser and Johansson. (1976). *J. Human Stress 2,* 15–23. Reprinted by permission of the publisher.]

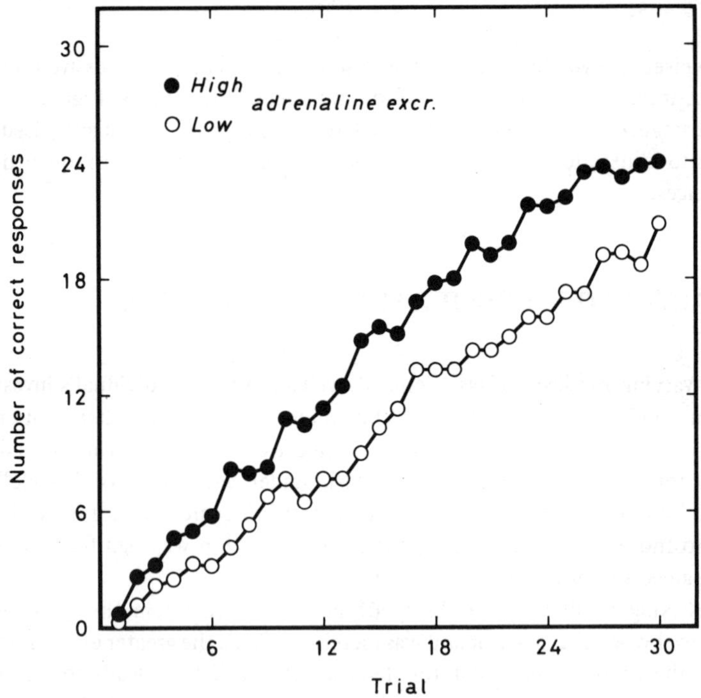

Fig. 7. Mean performance in a verbal rote learning task in subjects with high (above median value) and low (below median value) excretion rates of adrenaline. [Reprinted by permission of the publisher. Frankenhaeuser and Andersson. (1974). *Percept. Mot. Skills 38,* 557–558.]

among normal healthy individuals, those who secrete relatively more catecholamines tend to perform better in terms of speed, accuracy, and endurance than those who secrete less. This relationship is particularly marked for adrenaline secretion under low to moderate stimulation, but seems to hold for noradrenaline, too. The example given in Fig. 7 (Frankenhaeuser and Andersson, 1974) shows that performance in a learning task was consistently superior in high-adrenaline subjects to that of low-adrenaline subjects (i.e., subjects above and below the median adrenaline-excretion value).

When the conditions of stimulus load were varied (Frankenhaeuser *et al.*, 1971), subjects with relatively high adrenaline levels were found to perform better under monotonous conditions, whereas subjects with relatively low levels performed better under conditions of audiovisual overload requiring selective attention (Fig. 8). The impaired ability of high-adrenaline subjects to select relevant signals is consistent with an interpretation of Yerkes-Dodson's law in terms of a narrowing of the range of cues utilized at high arousal levels (Easterbrook, 1959).

The positive relationship between adrenaline secretion and psychological efficiency at low to moderate stimulus levels is not restricted to acute situations but involves cognitive functions in general. For example, studies of children show that school achievement and measures of intelligence correlate positively

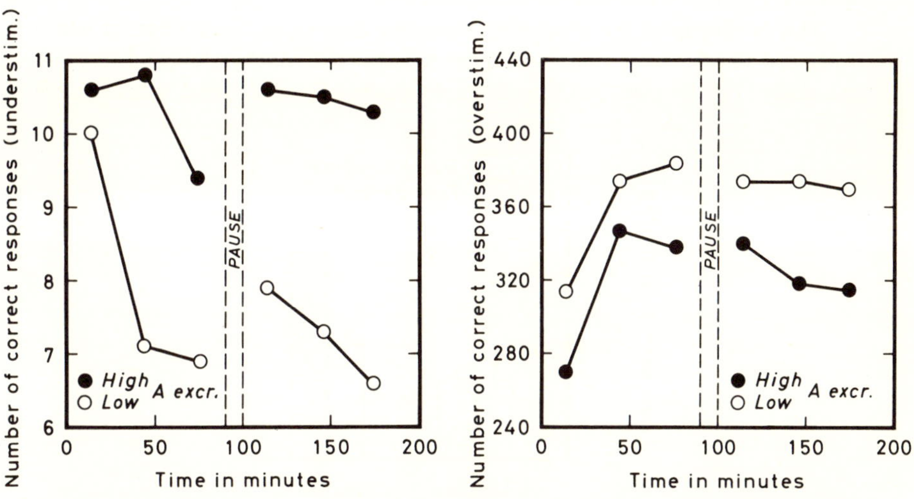

Fig. 8. Mean performance of subjects with high (above median value) and low (below median value) excretion rates of adrenaline (A) during understimulation (left-hand diagram) and overstimulation (right-hand diagram). [Reprinted by permission of the publisher. Frankenhaeuser *et al.* (1971) *Acta Psychol. 35,* 298–308.]

with catecholamine secretion (Johansson, *et al.,* 1973). Moreover, according to teachers' ratings and self-ratings, the high-adrenaline children are happier, livelier, and better adjusted to the school environment than their low-adrenaline peers.

Thus, interindividual differences in the capacity to regulate catecholamine release to suit environmental demands may account, in part, for differences in the ability to tolerate conditions characterized by low and high stimulus loads. We do not yet have a precise knowledge of the mechanisms by which circulating catecholamines modify mental capacity. However, the available evidence indicates that the catecholamines cross the blood-brain barrier in some regions only, but presumably penetrate sufficiently to exert a central effect. In addition, the perception of the peripheral changes accompanying catecholamine release may have an alerting effect.

SEX DIFFERENCES

The data reported so far emanate from male subjects and hold only in part for females. During rest and relaxation the two sexes do not differ in their catecholamine secretion (when body weight is taken into account). But psychosocial stress conditions produce a different picture, suggesting that the adrenal-medullary system is less reactive in females.

This is illustrated by adrenaline-excretion data from two studies, in which males and females were examined under stressful and nonstressful conditions. In one study (Fig. 9) male and female employees in a Swedish metal industry were

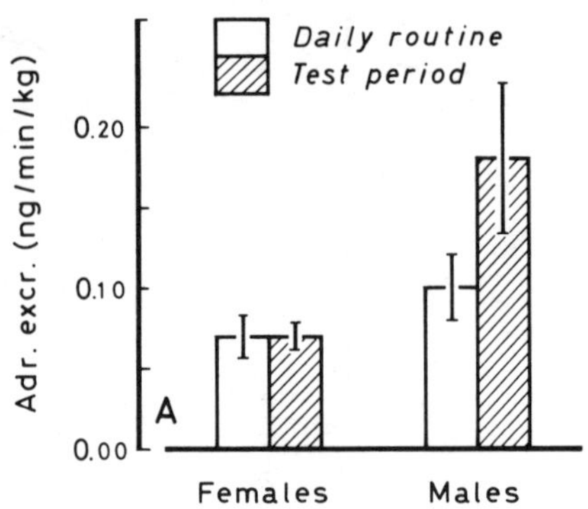

Fig. 9. Means and standard errors for adrenaline excretion (expressed in relation to body weight) in adult female and male subjects during daily routine activities and during intelligence testing. [Redrawn and reprinted by permission of the publisher. From Johansson and Post. (1974). *Acta Physiol. Scand. 92,* 557–565.]

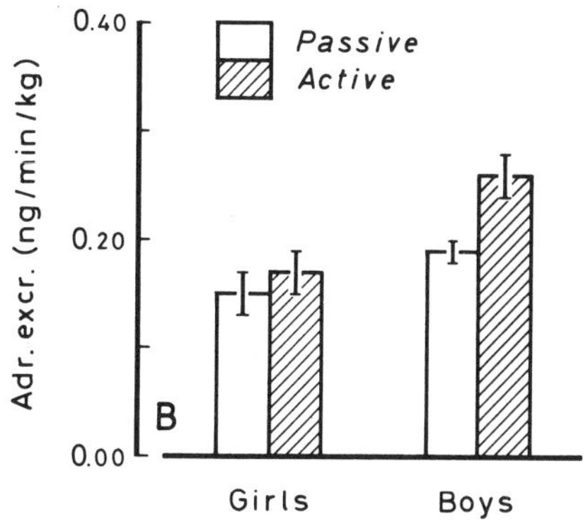

Fig. 10. Means and standard errors for adrenaline excretion (expressed in relation to body weight) in 12-year-old girls and boys during a passive and an active period. [Redrawn and reprinted by permission. From Johansson. (1972). *Acta Physiol. Scand. 85,* 569–572.]

compared while carrying out their daily routine activities and while performing an intelligence test under time pressure (Johansson and Post, 1974). In the female group, adrenaline output was the same in both sessions, whereas it increased significantly for the males during testing.

Similar results (Fig. 10) were obtained when 12-year-old boys and girls were compared in a passive situation, watching a movie, and an active, doing mental arithmetic (Johansson *et al.,* 1973). For the girls, adrenaline excretion did not rise significantly during the active period, whereas for the boys the increase was significant.

Since the females performed just as well or slightly better than the males in both studies, it seemed unlikely that the difference in adrenaline secretion could be associated with a difference in effort. However, the nature of the demand might be of significance. To examine the possibility that stressors other than mental work would elicit the typical male response in females, students of both sexes were exposed to two different stress situations, in one of which they were given a passive role, in the other an active (Frankenhaeuser *et al.,* 1976). In the passive situation, stress was induced by repeated venipuncture, a procedure that is generally considered somewhat stressful by members of both sexes. In the active situation, the subjects performed a cognitive task. Figure 11 shows that, in the female group, adrenaline excretion during both stress conditions was the same as during relaxation, whereas in the male group each of the stressors induced a significant increase.

Thus, females do not show the same readiness as males to respond to environmental demands by adrenaline release, regardless of whether the situation requires passive acceptance or active effort. Studies of individual differences among women may throw light on the mechanisms underlying this sex differ-

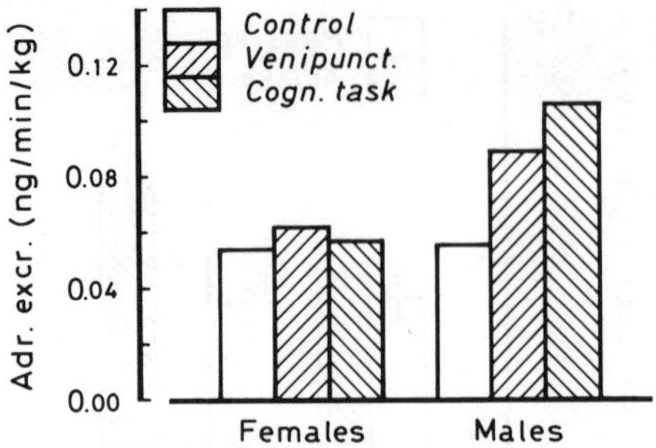

Fig. 11. Mean adrenaline excretion (expressed in relation to body weight) in female and male university students during a control condition, during repeated venipuncture, and during a cognitive task. [Based on Frankenhaeuser *et al.* (1976). *Psychopharmacology, 47,* 1–5.]

ence. Such studies are under way and we have already obtained some striking data from female members of our research team, showing a very high adrenaline release under conditions of intense examination stress.

A tentative hypothesis, on which our forthcoming investigations will be based, is that the tendency to respond by adrenaline release to requirements of the psychosocial environment is linked, not to sex per se, but to a behavior pattern which is more common in males in Western society. Type-A behavior, i.e., a behavior pattern correlated with coronary heart disease (cf. Friedman, 1969), seems to answer this description. Its characteristic features–hard-driving competitiveness, a sense of time urgency and impatience, constant struggling to meet deadlines, a strong need to be in control of life events–are all likely to be products of attitudes, expectations, and pressures of the early social environment. Hence, one might speculate about the possibility that the current change in sex-role patterns will lead to a growing proportion of type-A women. This, in turn, may lead to a decrease of the sex difference in catecholamine secretion and a concomitant decrease in the difference between the sexes in their susceptibility to diseases associated with the action of peripheral catecholamines.

ADAPTATION AND MALADAPTATION

These speculations pose the question whether catecholamine-mediated efforts to adapt to the stressors of everyday life may have aftereffects, leaving the individual less fit to cope with subsequent demands and more susceptible to disease.

A case in point is the duration of the catecholamine secretion elicited by a temporary stressor. Since the speed at which baseline levels are regained may determine the relative potency of harmful versus beneficial consequences, it is interesting that individuals differ with regard to the temporal pattern of catecholamine release. Figure 12 shows data from two groups of subjects, classified as rapid or slow "decreasers" depending upon the time taken for their adrenaline excretion to return to baseline after short-term exposure to mental overload (Johansson and Frankenhaeuser, 1973). It is seen that the rapid "decreasers" differ from the slow "decreasers" in that their adrenaline output was higher during inactivity, they performed better in a sensorimotor task, and had lower neuroticism scores on the Eysenck Personality Inventory.

Further support for the assumption that a rapid return to a baseline catecholamine level is indicative of well-being was obtained in a field study of employees in a Swedish metal industry, who were exposed to a stressful task before and after their summer vacation (Johansson, 1973). After the vacation, which had improved the subjects' physical and mental condition, the duration of catecholamine arousal following a stressful task was shorter than before the vacation.

A rapid drop in catecholamine secretion as demands decrease–quick "unwinding"–implies an "economic" mode of response, whereas a slow drop indicates poor adjustment in the sense that the organism "overreacts" by mobilizing resources which are no longer needed. When interpreted in terms of our cognitive model, a slow return to catecholamine baselines may either reflect a slowness in the reevaluation of environmental demands, or it may be part of a conscious effort to maintain a wide margin of safety.

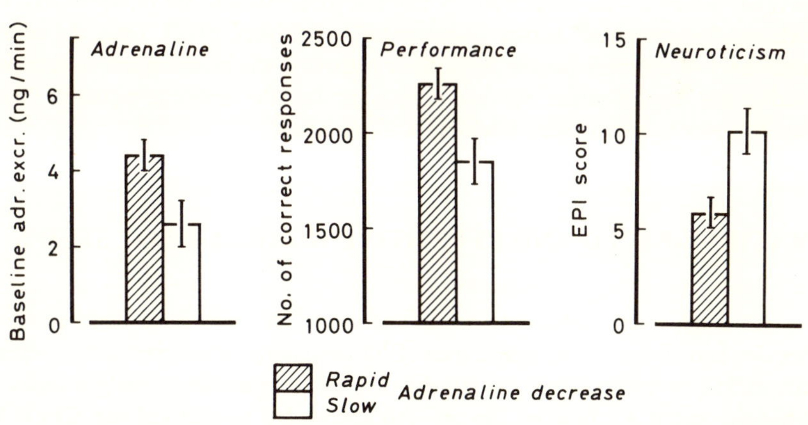

Fig. 12. Means and standard errors for adrenaline excretion during inactivity, performance scores in a sensorimotor task, and neuroticism scores (Eysenck's Personality Inventory). [Redrawn and reprinted by permission. From Johansson and Frankenhaeuser. (1973). *Biol. Psychol. 1,* 63–73.]

According to this line of reasoning, a mismatch between environmental demands and catecholamine secretion reflects a discrepancy between the actual and the expected capacity requirements. In other words: when the situation is misinterpreted, the message from the brain to the periphery will not contain adequate information about energy needs.

A striking example is the "paradoxical" reaction, i.e., a decrease of adrenaline secretion evoked by a stressor, a phenomenon observed from time to time in some apparently healthy, normal subjects (cf. Frankenhaeuser, 1975*b*). Such a reaction may reflect a denial of environmental demands, or, alternatively, it may be part of a conscious "energy-conserving" strategy. In either case, the failure has a cognitive rather than a peripheral origin. Regardless of which particular link in the chain is the weak one, adaptation will suffer.

As to possible practical consequences, the shift worker is a case in point. He is forced to change his work-sleep cycle in a way that does not coincide with the circadian rhythm of his bodily processes. Possible implications should be considered in the light of research on sleep deprivation which shows that peaks of adrenaline secretion and alertness tend to coincide during the 24-hr cycle (Fröberg *et al.,* 1975). In a worker who changes between day and night shifts at frequent intervals, the circadian rhythm of catecholamine secretion generally does not become reversed. Instead the function flattens, implying that he may be too aroused to sleep and too sleepy to work. This is likely to add to the cost of adapting to shift work and may account in part for the health hazards involved.

A point of interest is that the circadian rhythm of adrenaline output differs between persons who hold ordinary daytime jobs, but who are differentiated by extreme "morningness" versus "eveningness" (cf. Pátkai, 1971). Thus, the adrenaline level of morning people tends to be highest in the morning and to decrease during the day, whereas evening people display the opposite pattern. These results point to advantages of flexible working hours.

MACHINE-PACED AND REPETITIVE WORK: A PILOT STUDY

Although many questions remain unresolved, we now have fairly comprehensive knowledge of the significance of the peripheral catecholamines in human adaptation to psychosocial conditions. The most immediate question today is whether, and if so, how we can put this knowledge to practical use. Can it be applied to problems in real life? Can it be used as a basis for changing adverse work conditions?

The answer is that as isolated observations our data can carry little weight. It is not realistic to expect results from studies such as ours to have any real impact until we succeed in integrating them with other relevant knowledge, derived from studies of social psychology, ergonomics, and the medical sciences. Moreover, we have to learn how to communicate, at different stages of our research, with those involved in practical-political work.

With these considerations in mind, we have joined forces with a team of social psychologists, headed by Gardell, a prominent spokesman for ideas emphasizing autonomy, participation, and control over one's own work as prime determinants of job satisfaction, health, and well-being (Gardell, 1976). In a long-term project we are focusing on psychophysiological reactions to understimulation and overstimulation in working life (Frankenhaeuser and Gardell, 1976). As our first target we have selected the highly automated sawmill industry. Data from a recent large-scale ergonomics study of sawmills in Sweden—based on experts' ratings, health surveys, and interviews—show that work in the sawmill is characterized by severe physical strain and restriction of social interaction and movement (Ager, 1975). The workers report feelings of extreme monotony, repetitiveness, nonparticipation, and coercion. The work is machine-paced, and the monitoring process demands unfailing attention, since skilled judgments of timber quality have to be made at very short intervals. A striking feature is the short work cycle, in some cases less than 10 sec. Great effort is required to maintain a high level of performance and attention under conditions as completely lacking in variety as these. In the case of the sawmill worker, these factors are combined with a high pace of work—set by the machine—responsibiltiy for correct judgments, piecework rush, and a high noise level in the sawhouse. In other words: the worker is exposed to conditions characterized by a pronounced lack of personal control combined with elements of both understimulation and overstimulation, i.e., exactly those conditions which have been shown to elicit sympathetic-adrenal medullary arousal. Moreover, he is a shift worker, which requires adaptation to repeated changes in the work-sleep cycle, and he is paid on a piecework basis, which adds to the rush and time pressure.

For our pilot study we selected a group consisting of edgermen, graders, and sawyers, all judged as high-risk workers on the basis of available data. Somatic and psychosomatic symptoms (e.g., high blood pressure, gastrointestinal disorders, back pain, headache, unspecific nervous disorders) are exceptionally high in this group, not only in relation to other workers in the sawmill but to the total population of Swedish workers as well. The work performed by all members of the high-risk group is distinguished from other jobs in the sawmill by a work cycle of less than 1 min. For our control group we selected grinders and stickers, whose work is not as constricted physically and mentally, and who have a work cycle of several minutes.

Figure 13 shows four measurements of catecholamine excretion, taken during an 8-hr work shift and expressed in percent of baseline values obtained

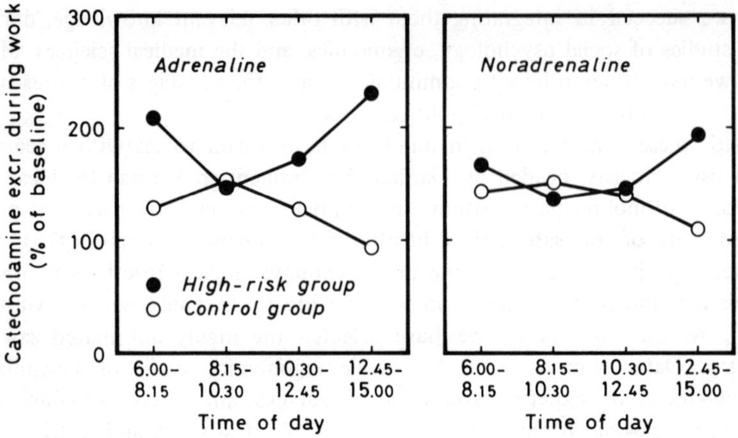

Fig. 13. Mean changes in the excretion of adrenaline (left-hand diagram) and noradrenaline (right-hand diagram) during an 8-hr work shift (in relation to baseline levels) in two groups of sawmill workers: a high-risk group and a control group. [Reprinted by permission of the publisher Frankenhaeuser and Gardell. (1975). *J. Human Stress, 2* (in press).]

under nonwork conditions. The average adrenaline excretion was consistently higher in the high-risk group than in the controls. Furthermore, the time course was strikingly different for both amines, catecholamine release decreasing toward the end of the work day in the control group, but increasing in the high-risk group. The differences between the last measurements of the day were significant for both amines. Such a buildup of catecholamine arousal during an 8-hr work day should be regarded as a warning signal indicating that the organism is forced to mobilize "reserve capacity" which in the long run is likely to add to its wear and tear. In other words: the cost of adaptation may be exceedingly high.

As yet, we have not been able to obtain the 24-hr catecholamine measures needed to provide information about the aftereffects of a day's work. But interview data indicate that an inability to relax after work is a common complaint among our high-risk workers, since—to quote them—"it takes hours for the noise and the machine-paced tempo of the mill to disappear from mind and body." Moreover, they complain of being too tired even to speak to their wife and children for several hours after returning home.

According to our tentative interpretation, the common origin of the high catecholamine level and the high frequency of psychosomatic symptoms in our high-risk group, is the monotonous, coercive, machine-paced nature of their work. If this interpretation is correct, conditions could be improved by increasing personal control, letting the individual himself decide how, in the course of work, to allocate his resources.

Research along these lines is being facilitated by a temporary cutback at Swedish sawmills, which will leave more time for testing new ideas in advanced field experiments involving, e.g., systematic variations in the speed of the assembly line, in the degree of personal control at various stages of the work process, and changes from piecework wages to fixed wages, etc. In parallel laboratory experiments we are focusing on various aspects of man-paced versus machine-paced work. By degrees, the number of parameters is being stepped up, so as to enable us to study patterns of profiles of multiple hormonal responses to psychosocial stimuli (cf. Mason, 1975).

In due course, we hope to contribute to the knowledge on which action can be based for eliminating adverse factors associated with automation, thereby creating work conditions that are more in harmony with the goals and aspirations of the welfare state.

ACKNOWLEDGMENT

The research reported in this paper had been financed by grants from the Swedish Medical Research Council (Project No. 997), the Swedish Social Science Research Council, and the Swedish Work Environment Fund (Project No. 73/55:2).

REFERENCES

Ager, B. (1975). Arbetsmiljön i sågverk (Work environment of sawmills). In *Arbetarskyddsstyrelsens Undersökningsrapport AM 101/75.* Stockholm. Chapter II, pp. 2–22.

Easterbrook, J. A. (1959). The effect of emotion on cue utilization and the organization of behavior. *Psychol. Rev. 66,* 183–201.

Euler, U. S. v., and Lishajko, F. (1961). Improved technique for the fluorimetric estimation of catecholamines. *Acta Physiol. Scand. 51,* 348–355.

Frankenhaeuser, M. (1975*a*). Sympathetic adrenomedullary activity, behaviour and the psychosocial environment. In *Research in Psychophysiology* (P. Venables and M. Christie, Eds.). New York, London, and Sydney: Wiley, Chapter 4, pp. 71–94.

Frankenhaeuser, M. (1975*b*). Experimental approaches to the study of catecholamines and emotion. In *Emotions–Their Parameters and Measurement* (L. Levi, Ed.). New York: Raven Press, 209–234.

Frankenhaeuser, M. (1971). Behavior and circulating catecholamines. *Brain Res. 31,* 241–262.

Frankenhaeuser, M., and Andersson, K. (1974). Note on interaction between cognitive and endocrine functions. *Percept. Mot. Skills 38,* 557–558.

Frankenhaeuser, M., Dunne, E., and Lundberg, U. (1976). Sex differences in sympathetic-adrenal medullary reactions induced by different stressors. *Psychopharmacology, 47,* 1–5.

Frankenhaeuser, M., and Gardell, B. (1976). Underload and overload in working life: Outline of a multidisciplinary approach. *J. Human Stress* (in press).

Frankenhaeuser, M., and Johansson, G. (1976). Task demand as reflected in catecholamine excretion and heart rate. *J. Human Stress 2,* 15–23.

Frankenhaeuser, M., Nordheden, B., Myrsten, A. L., and Post, B. (1971). Psychophysiological reactions to understimulation and overstimulation. *Acta Psychol. 35,* 298–308.

Frankenhaeuser, M., and Rissler, A. (1970). Effects of punishment on catecholamine release and efficiency of performance. *Psychopharmacologia 17,* 378–390.

Frankenhaeuser, M., Sterky, K., and Järpe, G. (1962). Psychophysiological relations in habituation to gravitational stress. *Percept. Mot. Skills 15,* 63–72.

Friedman, M. (1969). *Pathogenesis of Coronary Artery Disease.* New York: McGraw-Hill.

Fröberg, J. E., Karlsson, C. -G., Levi, L., and Lidberg, L. (1975). Circadian rhythms of catecholamine excretion, shooting range performance and self-ratings of fatigue during sleep deprivation. *Biol. Psychol. 2,* 175–188.

Gardell, B. (1971). Alienation and mental health in the modern industrial environment. In *Society, Stress and Disease, Vol. I :The Psychosocial Environment and Psychosomatic Diseases* (L. Levi, Ed.). London: Oxford University Press, pp. 148–180.

Gardell, B. (1976). Autonomy and participation at work. In *Society, Stress and Disease, Vol. IV: Working Life* (L. Levi, Ed.). London: Oxford University Press (in press).

Johnasson, G. (1973). Intraindividual variation in the temporal pattern of sympathetic-adrenal medullary activity. Reports from the Psychological Laboratories, University of Stockholm, No. 389.

Johansson, G., and Frankenhaeuser, M. (1973). Temporal factors in sympatho-adrenomedullary activity following acute behavioral activation. *Biol. Psychol. 1,* 63–73.

Johansson, G., Frankenhaeuser, M., and Magnusson, D. (1973). Catecholamine output in school children as related to performance and adjustment. *Scand. J. Psychol. 14,* 20–28.

Johansson, G., and Post, B. (1974). Catecholamine output of males and females over a one-year period. *Acta Physiol. Scand. 92,* 557–565.

Lazarus, R. S. (1966). *Psychological Stress and the Coping Process.* New York: McGraw-Hill.

Levi, L. (1972). Stress and distress in response to psychosocial stimuli. *Acta Med. Scand. 191,* Suppl. 528.

Lindsley, D. B. (1961). Common factors in sensory deprivation, sensory distortion, and sensory overload. In *Sensory Deprivation* (P. Solomon, P. H. Kubzansky, J. H. Leiderman, J. H. Mendelson, and D. Wexler, Eds.). Cambridge, Mass.: Harvard University Press, pp. 174–194.

Lundberg, U. (1976). Urban commuting: Crowdedness and catecholamine excretion. *J. Human Stress* (in press).

Mason, J. W. (1975). Emotion as reflected in patterns of endocrine integration. In *Emotions–Their Parameters and Measurement* (L. Levi, Ed.). New York: Raven Press, pp. 143–181.

Pátkai, P. (1971). Interindividual differences in diurnal variations in alertness, performance and adrenaline excretion. *Acta Physiol. Scand. 81,* 35–46.

Selye, H. (1974). *Stress without Distress.* Philadelphia and New York: Lippincott.

Singer, J. E., Lundberg, U., and Frankenhaeuser, M. (1976). Stress on the train: A study of urban commuting. *Advances in Environmental Research* (in press).

Resistance and Overmotivation in Achievement-Oriented Activity

JOHN W. ATKINSON

What is stressful in the everyday effort to achieve? We get two answers from experimental work that has been designed to sharpen our conception of how individual differences in personality influence motivation and its expression in action: viz., *resistance* and *overmotivation.*

The observed pattern of achievement-oriented action in an individual who is positively motivated to achieve can be taken as a norm in relation to which the behavioral manifestations of fear of failure, and even fear of success, may often be described as maladaptive trends. These fears produce resistance to achievement-oriented action (Atkinson and Birch, 1970, 1974). The behavioral effects of resistance are, in brief: unrealistic levels of aspiration; unwarranted persistence in the face of continued failure in pursuit of impossible goals; paradoxical changes in the level of aspiration following success and failure; a substantial dampening of interest following a failure; and a paradoxical dread and deadening of interest when some immediate task is clearly perceived as having important future implications for the person.

These maladaptive trends, observed in studies of achievement motivation among college students, represent the psychopathology of everyday achievement-related action. Undoubtedly, they occur in more extreme form in other populations.

JOHN W. ATKINSON • The University of Michigan, Ann Arbor, Michigan.

Now we are beginning to appreciate another paradoxical effect of resistance. The dampening of interest produced by the strength of one's tendency to avoid failure can sometimes have an enhancing effect on performance. It can prevent a person from becoming so enthusiastically motivated by the prospects of the various positive consequences of an activity that performance suffers the detrimental effect of inefficiency caused by overmotivation.

First—a little history. Early experiments using thematic apperception to assess individual differences in strength of motive to achieve (Atkinson, 1953; Moulton *et al.,* 1958) showed two things: that the motivational state must involve an interaction of personality disposition and immediate situational factors, and that motivational tendencies having exactly opposite behavioral implications are aroused when success and failure are at stake. We soon came to appreciate that the behavioral effects of differential anxiety, then being studied at Yale (Mandler and Sarason, 1952) and at Iowa (Taylor and Spence, 1952), were exactly opposite to those of differential achievement motivation across treatments. Those scoring high in anxiety, measured by self-descriptive tests, behaved like those scoring low in *n* Achievement measured by content analysis of imaginative behavior. Yet our studies, which began with Raphelson's (1957) use of both diagnostic tests to show diametrically opposite trends in the relation of the two measures to skin conductance, failed to show any correlation between strength of *n* Achievement and Test Anxiety within individuals when both were diagnosed under neutral conditions (Atkinson and Feather, 1966, p. 341). This implied that we were dealing with two variables which apparently influence achievement-related behavior in diametrically opposite ways.

Studies showing greater preference and higher level of performance at moderately difficult tasks among those strongly motivated for achievement suggested our initial formulation of a theory of achievement motivation (Atkinson, 1957, 1958). This scheme makes coherent sense out of the pattern of reactions we attribute to resistance among persons more fearful of failure than positively motivated to achieve. It provides the foundation of more recent work. My presentation is a survey of the old (Atkinson and Feather, 1966) and the new (Atkinson and Raynor, 1974).

In brief, it says that whenever an individual undertakes an activity knowing that he/she is responsible for the outcome, that skill will be evaluated, and there is uncertainty about the outcome, both a tendency to achieve success (in reaction to the challenge) and a tendency to avoid failure (in reaction to the threat) influence behavior. One instigates the activity. The other produces resistance to the activity. One constitutes what is called achievement motivation. The other constitutes the behavioral effect of what is called fear of failure, or anxiety about failure.

Figure 1 provides a succinct description of how personality and subjective probability of success, as defined by the difficulty of a task, interact to

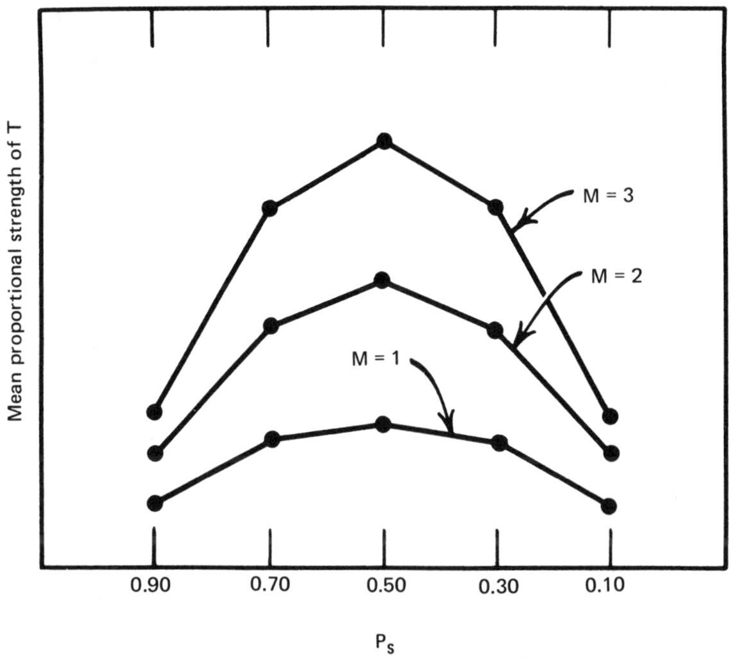

Fig. 1. Graphic representation of assumptions stated in text that strength of both the tendency to achieve success and strength of tendency to avoid failure (resistance) depend upon strength of relevant motive (personality) and subjective probability of success (situation). (Adapted from Atkinson and Feather, 1966.)

determine the strength of tendency to succeed. If the strength of a person's motive to achieve, the subjective probability of success in the activity, and incentive value of success combine multiplicatively to determine the strength of positive motivation, and if incentive value of success is greater the more difficult the task ($I_s = 1 - P_s$), then motivation to achieve will be strongest where P_s is 0.50 and when the motive is strong.

Figure 1 presents an analogous description concerning the determinants of the strength of the tendency to avoid failure, the other outcome inherent in any test of competence. Here the fundamental assumption is that it hurts more to fail the easier the task (that is $I_f = -P_s$). So the tendency to avoid failure based on subjective probability of failure ($P_f = 1 - P_s$) is an avoidance tendency, a tendency not to undertake an activity, a negation tendency, the source of resistance.

If the two tendencies combine algebraically, as supposed, then the tendency to avoid failure always will oppose, dampen, subtract from the positive tendency to reduce the strength of the resultant tendency to achieve that is expressed in behavior. If the motive to achieve (M_S) is the stronger within a person, the resultant tendency is always positive and the implications of positive motivation are expected in behavior. If the motive to avoid failure (M_{AF}) is the stronger within a person, the resultant tendency is always negative and the implications of greater resistance are expected in behavior.

To illustrate the way motives to achieve and to avoid failure influence level of aspiration, and risk-taking, a well-documented behavioral phenomenon (e.g., Hamilton, 1974), we recall Mahone's (1960) results concerning vocational aspiration. Both the unrealistically high aspirations said to characterize neurasthenics and the unrealistically low aspirations said to characterize hysterics (Eysenck and Himmelweit, 1946; Miller, 1951) were evident among those scoring low in *n* Achievement and high in Text Anxiety. Realism of aspiration was judged by clinicians and from an index of goal discrepancy, the difference between ability required for the chosen vocation and the subject's own level of ability.

Hardest to come by among college students, for whom motive to achieve success may be stronger than motive to avoid failure in most or all persons in the sample studied, is evidence that the level of performance is never actually depressed by resistance among those low (below the median) in *n* Achievement and high (above the median) in anxiety when the task is moderately difficult. For this reason, results obtained by Karabenick and Youssef (1968) at a less academically demanding college than the University of Michigan are of special interest. There it might be assumed, the balance of strength of motives to achieve success and to avoid failure might be different. Those in whom the motive to avoid failure is dominant would be more frequent in the population of subjects. Equally difficult verbal paired-associates were presented to subjects as very easy, intermediate in difficulty, or very difficult in terms of the proportion of students who could be expected to learn them. The result shown concerns

Table 1. Realistic (Intermediate) and Unrealistic Vocational Aspiration in College Men (N = 135) According to Strength of Achievement Motive (M_s) and Motive to Avoid Failure (M_{AF})[a]

			Clinical judgments		Goal discrepancy	
n Achievement	Test anxiety score	*N*	Realistic	Unrealistic	Intermediate third	Extreme thirds
High	Low	36	75%	25%	50%	50%
High	High	31	48	52	30	70
Low	Low	40	68	32	38	62
Low	High	28	39	61	18	82

[a]After Mahone (1960), with permission of the author and publisher, American Psychological Association.

Table 2. Means for all Groups on Total Number Correct of 10 PA Trials[a]

TAT-TAQ	Supposed difficulty of word pairs		
	Easy	Intermediate	Difficult
High–Low	24.71	27.71	26.50
High–High	22.71	22.84	22.79
Low–Low	24.09	23.92	24.04
Low–High	23.21	20.00	25.43

[a]From Karabenick and Youssef (1968).

level of performance after ten trials. It illustrates both the minor differences expected between extreme motivational groups divided at the median on TAT *n* Achievement and Test Anxiety when the task is very easy or very difficult and the depressed level of performance where resistance should be greatest.

Change in level of aspiration is pretty straightforward among those who are positively motivated. Success raises and failure lowers the subsequent probability of success at the same and similar tasks. This change in P_s produces a comparable change in I_s and the kind of changes in motivation that are shown in Fig. 2. So it is generally expected that the "typical" or "normal" shift in level of aspiration is to raise it following success and to lower it following failure. But now consider the figure as referring to the resistance which characterizes someone in whom the motive to avoid failure is dominant. It suggests the paradoxical expected consequences for more anxious persons whose initial choice of task (either very easy or very difficult) is determined by default since there is strongest resistance

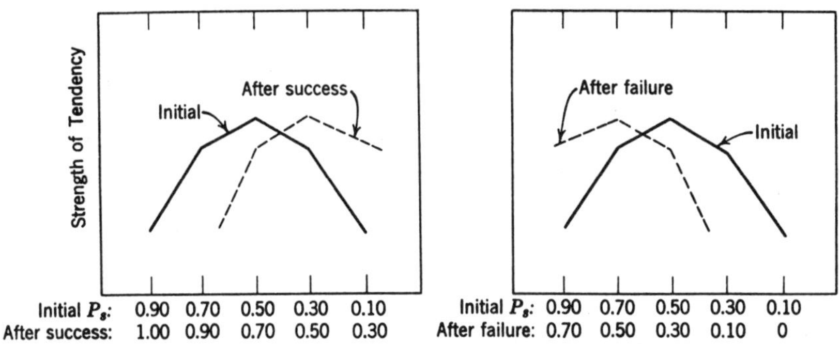

Fig. 2. Changes in level of aspiration following success and failure express the change in motivation produced by a change in subjective probability of success. The curves describe resultant achievement motivation when $M_S > M_{AF}$ and resistance when $M_{AF} > M_S$. (Adapted from Atkinson and Feather, 1966.)

to the moderately difficult alternatives. If the person dominated by a tendency to avoid failure succeeds when the task is easy, it then seems easier. If he fails when the task is very difficult, it then seems even more difficult. Both of these highly probable outcomes reduce resistance. Yet neither provides any dynamic for a change in aspiration. But should the very unlikely outcomes ever occur, namely, success at the very difficult task or failure at the very easy task, then both tasks will later seem more like moderately difficult challenges, for the P_s has changed. The new pattern of resistance favors avoidance of the old task and such "atypical" shifts in the level of aspiration as raising it substantially after failure at an easy task and lowering it substantially after success at a very difficult task.

By an ingenious experimental procedure, Moulton (1965) provided some illustrative evidence of these defensive shifts in aspiration. After obtaining the predicted initial preference of high school boys for an easy, moderately difficult, or very difficult task, he requested that each perform the moderately difficult task in the set first. He controlled the outcome to see whether a new choice of task after success or failure would manifest the typical or atypical shift in level of aspiration. Eleven out of 31 of his most anxious high school aged subjects exhibited either an apparently irrational increase in level aspiration after failure or a decrease in level of aspiration after success. All but one of 31 positively motivated boys displayed the typical shifts in aspiration.

A striking maladaptive trend in persistence following failure is shown when the subject confronts failure in a series of repeated trials trying to solve a puzzle that is actually insoluble. The puzzle has initially been presented in one condition as fairly easy (with a stated 70% chance of success) and in another condition as very difficult (with a stated 5% chance of success). How many trials will a person spend trying to solve the puzzle before quitting to move on to an alternative activity?

If failure lowers P_s, then those who began thinking the task is easy are soon confronted with a moderately difficult task. This should heighten motivation and encourage persistence in those who are primarily motivated to achieve but should greatly increase resistance and cause early termination of activity among those who are more anxious about failure. In contrast, continual failure at a problem already perceived as very difficult should immediately reduce motivation as P_s drops still further among those who are positively motivated but reduce resistance and encourage continuance among those who are most anxious. And so it came out in Feather's (1961) incisive demonstration of what it means to say that motivation depends upon interaction of personality and situational influences. The more anxious subjects were doggedly persistent in the face of failure, or at least continued to go through the motions, when told a task is very difficult but not when told it is comparatively easy at the outset.

Still other evidence of differential motivational effects attributable to a

Table 3. Persistence Among College Men (N = 35) in the Face of Continued Failure as a Function of Personality and Initial Difficulty of Task[a]

		Percent above median in persistence	
n Achievement	Test Anxiety	Task seen initially as easy (P_s = 0.70)	Task seen initially as difficult (P_s = 0.05)
High	Low	75	22
Low	High	33	75

[a]After Feather (1961), with permission of the author and of the publisher, American Psychological Association.

change in probability of success comes from a study in which students from a traditionally heterogeneous fifth-grade class were grouped according to ability in sixth grade. Very bright students, for whom P_s is usually very high in classroom, must now face more of a competitive challenge when grouped with equally able peers. Very dull students move from a situation having consistently low probability of success for them to a more realistic challenge. The change should be responded to positively by some students and negatively by others. There was greater academic growth in ability-grouped classes than traditionally heterogeneous control classes among those who were high in resultant achievement motivation no matter what the level of ability of the students. We expected, but did not find, that ability grouping might have a detrimental effect among those relatively more anxious about failure. They held their own when compared with the control group. Opposite effects were evident, however, in the subjective reactions to schoolwork of the differently motivated students in the ability-grouped as compared to control classes at the end of sixth grade (Table 4).

Among the more interesting recent developments is Horner's identification of "fear of success" among college women (Horner, 1974*a*). In thematic apper-

Table 4. Percentage of Students High, Moderate, and Low in Resultant Achievement Motivation Reporting Above Median Interest in Sixth as Compared with Fifth Grade

	Homogeneous classes		Heterogeneous classes	
n Achievement–Test Anxiety	N	Percent above median	N	Percent above median
High	(18)	78	(78)	56
Moderate	(22)	41	(82)	43
Low	(22)	36	(73)	52

[a]From O'Connor, Atkinson, and Horner (1966).

ception, this fear was blatantly expressed when women were asked to tell a story in response to structured verbal cues such as this: "Nancy and the boy she has been dating for over a year have both applied to the same highly selective university;" or "After the first term finals, Anne finds herself at the top of her med school class." Negative consequences associated with success were introduced by 62% of the women but only 9% of the men in Horner's first study of this problem in the winter of 1965. For example: "Anne . . . is pretty darn proud of herself. But everyone hates and envies her."

In contrast, the imaginative stories of men were most frequently variations on this theme: "John is very pleased with himself and realizes his efforts have been rewarded, he has finally made the top of his class. . . . He will go on in med school making good grades and be successful in the long run."

Is this an additional source of resistance among women? In Horner's study, 67% of the men for whom the comparison could be made performed an intellective task better under conditions of interpersonal competition than when working alone. Among females, it clearly depended upon the presence (24%), or not (92%), of evidence of fear of success.

This problem is now being widely investigated (Hoffman, 1974). It may provide one key to understanding the relative lack of achievement even among educated women in the past.

Another fundamentally important new development is Raynor's (1969, 1974) analysis of what happens when an individual perceives the immediate task as instrumental to future opportunities, as a step in a longer path leading on to one or more future consequences and not merely as an isolated task with no future implications. Perceiving that some future opportunity, and the consequences of it, are contingent upon success in the present activity should intensify the characteristic achievement-related motivation of a person. The strength of tendency to achieve success in a present activity will depend upon two components, the strength of the tendency to succeed in reaching the more distant (*d*) expected goals, when there is expectation of one (or more), in addition to the strength of tendency to succeed in the immediate task (*i*).

The same should apply to the strength of tendency to avoid failure that produces resistance to the present activity. It, too, should be a summation of separate components, one referring to immediate failure (*i*) and the other to failing to achieve more distant goals (*d*).[1]

The effect has been shown experimentally by Raynor and Rubin (1971) by presenting male college students with several tests. In one condition, subjects were told that the opportunity to take the next test in the series depended on how well they did in the previous test. In another condition, there was no

[1] Extending the logic of Fig. 1, (a) $T_S = T_{S_i} + T_{S_d} = M_S\,(P_{s_i}I_{s_i} + P_{s_d}I_{s_d})$; (b) $T_F = T_{F_i} + T_{F_D} = M_{AF}\,(P_{f_i}I_{f_i} + P_{f_d}I_{f_d})$; so (c) $T_{Res} = (M_S - M_{AF})\,(P_{s_i}I_{s_i} + P_{s_d}I_{s_d})$.

Table 5. Mean Number of Problems Attempted as a Function of Motive Groups and Experimental Conditions[a]

	Condition			
	Noncontingent		Contingent	
Motive group	*N*	*M*	*N*	*M*
High–Low	8	15.63	7	18.43
High–High	6	11.67	6	14.17
Low–Low	10	14.40	6	12.67
Low–High	7	14.14	8	8.38

[a]From Raynor and Rubin (1971).

contingency. They were told that they would have an opportunity to take each of four tests regardless of their performance on any one of them. Table 5 shows evidence of intensification of both positive motivation and resistance when the future opportunity is contingent upon success in the immediate task.

A similar result, paradoxical as it may seem, was obtained by Raynor (1970) in one study of academic achievement in an introductory psychology course (Table 6). Students had been asked how important getting a good grade was for having their career plans work out. Those who replied affirmatively, and were positively motivated, performed better. In contrast, those more strongly motivated to avoid failure performed worse when they viewed the course as instrumental to future goals than when they did not.

In another study (Raynor *et al.,* 1974), students came early before a final exam in a college course and were asked to introspect, to describe their thoughts and feelings in common everyday language. They were also asked to indicate on a rating scale to what extent they felt their exam performance was related to their own future goals. For simplicity, only the subjective reactions of those in the highest and lowest thirds concerning perceived instrumentality of the exam grade are presented in Table 7.

Table 6. Mean Grades in Introductory Psychology as a Function of Achievement-Related Motives and Perceived Instrumentality[a]

	Perceived instrumentality (PI-Psych)			
n Achievement–Test Anxiety	*N*	Low	*N*	High
High–Low	15	2.93	19	3.37
Low–High	11	3.00	17	2.59

[a]From Raynor (1970).

Table 7. Subjective States Described Immediately Before a Final Exam in a College Course[a]

Concern about Doing Well and/or Getting a Good Grade

	Perceived instrumentality of exam			
	N	Low (%)	N	High (%)
n Achievement				
High	(24)	54	(31)	*83*
Low	(27)	56	(24)	63
Text Anxiety				
High	(20)	60	(34)	65
Low	(31)	52	(21)	*86*
n Achievement–Test Anxiety				
High–Low	(15)	47	(16)	*94*
Low–High	(11)	55	(19)	63
	Expression of anxiety			
n Achievement				
High	(24)	33	(31)	56
Low	(27)	33	(24)	*71*
Test Anxiety				
High	(20)	40	(34)	*74*
Low	(31)	40	(21)	38
n Achievement–Test Anxiety				
High–low	(15)	27	(16)	38
Low–high	(11)	36	(19)	*79*

[a]From Raynor, Atkinson, and Brown (1974).

There are no surprises in the descriptive characteristics of the subjective state of persons who differ in achievement-related motivation. The intensified subjective reaction is italicized in the right-hand column. Here we have the inside view of the effect of future orientation to complement the overt behavioral evidence. An intensified positive resultant tendency to achieve is accompanied by subjective concern about doing well, an intensified resistance by the experience of anxiety.

Another direction of recent research has focused on the differential *immediate* motivational consequences of success and failure that are attributable to the consummatory character of a success as distinguished from the nonconsummatory and possibly punitive character of a failure. Weiner (1965) ingeneously arranged procedures for disentangling effects mediated by a change in subjective probability of success (P_s), given the logic of theory of achievement motivation, from the *immediate* motivational effects of the success and the failure. He allowed some subjects to succeed when they believed the probability of success was 0.70 and arranged for others to fail when they believed the probability of success was 0.30. Both the initial levels of motivation (where $P_s = 0.70$ and 0.30) and those consequent to the equivalent changes in P_s produced by success at the easier task and failure at the more difficult one should be the same *unless,* as Weiner proposed, success immediately reduces the characteristic motivation of a person but failure does not. In the latter case, the characteristic motivation would persist and influence any immediately subsequent performance. This means we should expect persistent positive motivation and enhancement of performance when motive to achieve is dominant but persisting resistance and dampening of subsequent performance when motive to avoid failure is dominant. This pattern, shown in a number of Weiner's studies (Weiner, 1966, 1972), is illustrated in Table 8. It shows speedier performance of 60 digit-symbol substitutions following failure than success among the more positively motivated and just the opposite among the more fearful persons.

So much of what I have said concerning the resistance produced by the tendency to avoid failure is negative in its implication concerning the effectiveness of an individual, I am moved to end on a somewhat more optimistic note. Let us consider the effect of intensity of motivation on efficiency of performance. Recent evidence in our work confirms the view advanced by others that the relationship between strength of motivation and efficiency of intellective performance is nonmonotonic and described by an inverted U-shaped curve as in Fig. 3 (e.g., Broadhurst, 1959; Eysenck, 1966).

This means that a surfeit of positive motivation (overmotivation), in addition to resistance attributable to fears of failure or success which dampen resultant motivation (undermotivation) can produce less than optimally efficient perfor-

Table 8. Time (in Seconds) to Complete 60 Digit-Symbol Substitutions on Trial 1 and Trial 2

	Motive Classification			
	$M_S > M_{AF}$		$M_{AF} > M_S$	
Condition	Trial 1	Trial 2	Trial 1	Trial 2
N	19	19	12	12
Success *M*	60.10	66.63	60.58	62.75
N	14[a]	14[a]	14	14
Failure *M*	60.93	64.33	60.93	70.43

[a]Level of performance for one *S* could not be accurately evaluated. From Wiener (1965).

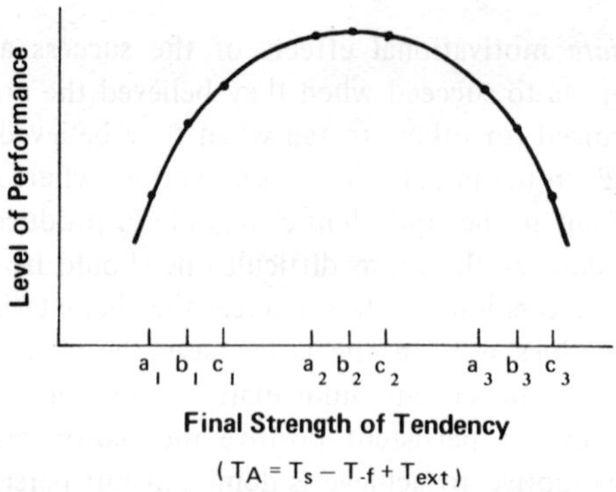

Fig. 3. The assumption that efficiency of performance (and therefore its level holding ability constant) increases up to some optimal level after which efficiency decreases as motivation increases still further. (Adapted from Atkinson, 1974*a*.)

mance. If the hypothesis is correct, any single determinant of the strength of tendency to perform a task, such as *n* Achievement, should be positively related to level of performance when there are few or no other situational incentives to perform and the overall strength of motivation falls in the relatively low range 1 of the figure. When other situational incentives bring the overall strength of tendency to the middle range 2, the correlation between *n* Achievement and level of performance should be zero. And given sufficiently strong extrinsic incentives for performance, or the kind of intensification of motivation that is attributable to future orientation, the correlation of need for achievement and performance should be negative when the total strength of tendency is brought into the upper range 3 beyond what is optimal for the task. Corresponding to these three hypotheses is an exactly opposite one concerning the correlation between strength of motive to avoid failure (as measured by Test Anxiety) and performance. The nonobvious hypothesis is that when positive motivation is likely to be more than optimal for the task, the dampening effect of resistance produced by expectation of negative outcomes should enhance performance.

In a series of studies, all of which confirm the hypothesis of a curvilinear relationship between the overall or final strength of tendency and efficiency of performance (see Atkinson, 1974*a*), some results obtained by Smith (1966) illustrate the enhancing effect of resistance when positive motivation is too strong (Table 9).

Subjects worked at complex, three-step arithmetic problems for 14 min. In the *relaxed* condition, the importance of the task was minimized. In the *extrinsic* condition, the only incentive was to finish quickly in time for the evening meal. In the *achievement* condition, the task was presented as a test of ability and each subject worked alone in a small room. In the *multi-incentive* condition, the subjects worked in a competitive coacting group situation, with

Table 9. Mean Correct Arithmetic Performance (14 Minutes) According to Strength of Achievement-Related Motives Under Various Conditions[a]

Mean post-test rating by *Ss*: how hard they had worked	Relaxed (3.41)	Extrinsic (3.97)	Achievement (4.00)	Multi-incentive (4.39)
n Achievement				
High[b]	52.07	56.47	70.69	58.39
Low	51.43	55.93	64.93	76.94
Difference	.63	.54	6.36	−18.55
Test Anxiety				
High[c]	48.33	56.00	58.92	70.31
Low	55.43	56.38	75.71	64.74
Difference	− 7.10	− .38	−16.79	5.57
n Achievement–Test Anxiety				
High–low[d]	53.0	51.7	78.8	55.3
High–high	55.1	60.6	66.0	58.7
Low–low	56.4	60.0	71.3	71.5
Low–high	34.8	49.8	55.8	85.3

[a]After Smith (1961, 1966), with permission of the author and of the publisher, John Wiley and Sons. Differences attributable to *n* Achievement here are smaller than when one picture of questionable validity is eliminated from the set of six (Smith, 1961, Table 15). This picture is removed for the joint classification based on Smith (1961, Table 58). Generally the differences are somewhat smaller but the pattern is unchanged with quantitative ability controlled statistically (Smith, 1961). Data from Atkinson (1974*a*).
[b]N_s = 14 to 18 in each subgroup.
[c]N_s = 13 to 19 in each subgroup.
[d]N_s = 4 to 10 in each subgroup.

achievement orientation, the promise of a $5.00 prize for the best performance, and in the presence of two proctors who walked around as potential agents of social approval/disapproval, looking directly into the eyes of anyone who slackened in the work.

The conditions are arranged in order according to an a priori appraisal of the incentives offered and also according to ratings made by subjects immediately after the experiment in response to the question of how hard they had worked at the task. Look particularly at the levels of performance of the subgroups considered most positively motivated, and those most likely to be more strongly motivated to avoid failure. One sees the hint of the enhancement of efficiency of performance by resistance when, without it, one might suffer a decrement in performance attributable to overmotivation.

Similar evidence of inefficiency in performance attributable to excessive positive motivation is shown in a similar study by Horner (1974*b*) where each male subject worked alone under achievement-orientation, or in direct competition with a female, or with a male. In the last two competitive conditions, subjects were told that it would be announced which of them had done the best (Table 10).

Table 10. Mean Anagrams Performance of Men Working Alone and in Competition According to Resultant Achievement Motivation and Affiliative Motivation[a]

	Alone		Competition with female		Competition with male	
	N	score	*N*	score	*N*	score
High resultant *n* Achievement (*n* Achievement–Test Anxiety)						
High *n* Affiliation	(10)	46.5	(7)	53.9	(7)	48.4
Low *n* Affiliation	(8)	48.4	(6)	53.4	(5)	53.7
Low resultant *n* Achievement (*n* Achievement–Test Anxiety)						
High *n* Affiliation	(6)	41.8	(7)	53.6	(6)	56.1
Low *n* Affiliation	(6)	40.8	(10)	47.7	(10)	46.7

[a]After Horner (1968), with permission of the author. Raw scores on 10 min anagrams task were normalized to yield a mean of 50 and an *S.D.* of 10.

Here, the three treatments are ordered in terms of the absence versus presence of an additional incentive for performance (social approval) and according to evidence from earlier work showing that the subgroup whose performance level is monotonically sensitive to the introduction of an incentive for approval is the one classified low in resultant achievement motivation (i.e., low *n* Achievement–high Test Anxiety) and high in *n* Affiliation (i.e., concern about positive affective reactions from others).

Given this justifiable ordering of the three treatments, the pattern of results is entirely consistent with the supposition of a nonmonotonic relationship between strength of motivation and efficiency of performance. Among those low in resultant achievement motivation, those also low in tendency for social approval improve somewhat when an incentive for approval is offered but not as much as those who are highly motivated for approval. Here we are talking about the low-to-moderate range of motivation.

Among those high in resultant motivation to achieve, who are already performing at a reasonably high level (suggesting at least moderate motivation) when working alone, both those low and high in tendency for approval show equivalent increases in performance in competition with a female. But one of these groups should be more strongly motivated. Why is the level of performance the same? An answer, consistent with the curvilinear hypothesis is that those high in *n* Affiliation are already beyond the optimal level of motivation when they score 53.9 and those low in *n* Affiliation have not yet reached the optimal level. Given this presumption, the level of performance of the most positively motivated group must be still lower in competition with a male, and it is. And the similar level of performance of the low *n* Affiliation group in competition with a male is expected if the strength of motivation has moved from just below to just above the optimal level. The pieces of the puzzle fall into place and the curve describing efficiency in performance in relation to strength of motivation can be seen in two places. First, it is evident across treatments for those who are predisposed to overmotivation by being unambiguously strongly positive in both motivation for achievement and social approval. Second, it is evident across different personalities within the treatment that offers the strongest combination of incentives.

It is clear that the same person (or kind of person) is not the most efficient performer in all situations. And here we have suggestive evidence that the same resistance which motivates so many defensive and maladaptive trends in achievement-oriented activity may at least sometimes, when it dampens enthusiasm, have a more positive effect on behavior.

Consistent evidence of a curvilinear relationship between strength of motivation at the time of a test and intellective performance allows a motivational interpretation of some of the differences in level of intellective test performance that have traditionally been called "IQ" and used to define the level of ability (Atkinson, 1974*b*).

What are the sources of stress in achievement-oriented action? Our work suggests two: the resistance produced by dread of negative consequences that must be overcome to perform at all; and the excessive motivation that causes inefficiency in execution among those who are sometimes too excited by multiple positive prospects.

REFERENCES

Atkinson, J. W. (1953). The achievement motive and recall of interrupted and completed tasks. *J. Exp. Psychol. 46,* 381–390. Also in *Studies in Motivation* (D. C. McClelland, Ed.). New York: Appleton-Century-Crofts (1955).

Atkinson, J. W. (1957). Motivational determinants of risk-taking behavior. *Psychol. Rev. 64,* 359–372.

Atkinson, J. W. (1958). Toward experimental analysis of human motivation in terms of motives, expectancies, and incentives. In *Motives in Fantasy, Action and Society* (J. W. Atkinson, Ed.). Princeton: Van Nostrand (1958).

Atkinson, J. W. (1974*a*). Strength of motivation and efficiency of performance. In *Motivation and Achievement* (J. W. Atkinson and J. O. Raynor). Washington, D. C.: V. W. Winston and Sons, pp. 193–218.

Atkinson, J. W. (1974*b*). Motivational determinants of intellective performance and cumulative achievement. In *Motivation and Achievement* (J. W. Atkinson and J. O. Raynor). Washington, D. C.: V. W. Winston and Sons, pp. 389–410.

Atkinson, J. W., and Birch, D. (1970). *The Dynamics of Action.* New York: Wiley.

Atkinson, J. W., and Birch, D. (1974). The dynamics of achievement-oriented activity. In *Motivation and Achievement* (J. W. Atkinson and J. O. Raynor). Washington, D. C.: V. W. Winston and Sons, pp. 271–326.

Atkinson, J. W., and Feather, N. T. (Eds.). (1966). *A Theory of Achievement Motivation.* New York: Wiley.

Atkinson, J. W., and Raynor, J. O. (1974). *Motivation and Achievement.* Washington, D. C.: V. W. Winston and Sons. (Distributed by Halsted Press/Wiley.)

Broadhurst, P. L. (1959). The interaction of task difficulty and motivation: The Yerkes-Dodson Law revived. *Acta Psychol. 16,* 321–338.

Eysenck, H. J. (1966). Personality and experimental psychology. *Brit. Psychol. Soc. Bull. 19,* 62, 1–28.

Eysenck, H. J., and Himmelweit, H. T. (1946). An experimental study of the reactions of neurotics to experiences of success and failure. *J. Gen. Psycho. 35,* 59–75.

Feather, N. T. (1961). The relationship of persistence at a task to expectation of success and achievement related motives. *J. Abnorm. Soc. Psychol. 63,* 552–561.

Hamilton, J. O. (1974). Motivation and risk taking behavior. A test of Atkinson's theory. *J. Person. Soc. Psychol. 29,* 856–864.

Hoffman, L. (1974). Fear of success in males and females: 1965 and 1972. *J. Consult. Clin. Psychol. 42,* 353–358.

Horner, M. S. (1974*a*). The measurement and behavioral implications of fear of success in women. In *Motivation and Achievement* (J. W. Atkinson and J. O. Raynor). Washington, D. C.: V. H. Winston and Sons, 91–120.

Horner, M. S. (1974*b*). Performance of men in noncompetitive and interpersonal competitive achievement oriented situations. In *Motivation and Achievement* (J. W. Atkinson and J. O. Raynor). Washington, D. C.: V. W. Winston and Sons, pp. 237–254.

Karabenick, S. A., and Youssef, Z. I. (1968). Performance as a function of achievement motive level and perceived difficulty. *J. Personal. Soc. Psychol. 10,* 414–419.

Mahone, C. H. (1960). Fear of failure and unrealistic vocational aspiration. *J. Abnorm. Soc. Psychol. 60,* 253–261. Also in *A Theory of Achievement Motivation* (J. W. Atkinson and N. T. Feather, Eds.). New York: Wiley (1966).

Mandler, G., and Sarason, S. B. (1952). A study of anxiety and learning. *J. Abnorm. Soc. Psychol. 47,* 166–173.

Miller, D. R. (1951). Responses of psychiatric patients to threat of failure. *J. Abnorm. Soc. Psychol. 46,* 378–387.

Moulton, R. W. (1965). Effects of success and failure on level of aspiration as related to achievement motives. *J. Personal. Soc. Psychol. 1,* 399–406. Also in *A Theory of Achievement Motivation* (J. W. Atkinson and N. T. Feather, Eds.). New York: Wiley (1966).

Moulton, R. W., Raphelson, A. C., Kristofferson, A. B., and Atkinson, J. W. (1958). The achievement motive and perceptual sensitivity under two conditions of motive-arousal. In *Motives in Fantasy, Action, and Society* (J. W. Atkinson, Ed.). Princeton: D. Van Nostrand, pp. 350–359.

O'Connor, P., Atkinson, J. W., and Horner, M. (1966). Motivational implications of ability grouping in schools. In *A Theory of Achievement Motivation* (J. W. Atkinson and N. T. Feather, Eds.). New York: Wiley, pp. 231–250.

Raphelson, A. C. (1957). The relationships among imaginative, direct verbal and physiological measures of anxiety in an achievement situation. *J. Abnorm. Soc. Psychol. 54,* 13–18.

Raynor, J. O. (1969). Future orientation and motivation of immediate activity: An elaboration of the theory of achievement motivation. *Psychol. Rev. 76,* 606–610.

Raynor, J. O. (1970). Relationships between achievement-related motives, future orientation, and academic performance. *J. Personal. Soc. Psychol. 15,* 28–33.

Raynor, J. O. (1974). Future orientation in the study of achievement motivation. In *Motivation and Achievement* (J. W. Atkinson and J. O. Raynor). Washington, D. C.: V. W. Winston and Sons, pp. 121–154.

Raynor, J. O., Atkinson, J. W., and Brown, M. (1974). Subjective aspects of achievement motivation immediately before an examination. In *Motivation and Achievement* (J. W. Atkinson and J. O. Raynor). Washington, D. C.: V. H. Winston and Sons, pp. 155–172.

Raynor, J. O., and Rubin, I. S. (1971). Effects of achievement motivation and future orientation on level of performance. *J. Personal. Soc. Psychol. 17,* 36–41.

Smith, C. P. (1966). The influence of testing conditions and need for achievement scores and their relationship to performance scores. In *A Theory of Achievement Motivation* (J. W. Atkinson and N. T. Feather, Eds.). New York, Wiley, pp. 277–297.

Taylor, J. A., and Spence, K. W. (1952). The relationship of anxiety level to performance in serial learning. *J. Exp. Psychol. 44,* 61–64.

Weiner, B. (1965). The effects of unsatisfied achievement motivation on persistence and subsequent performance. *J. Personal. 33,* 428–442.

Weiner, B. (1966). Role of success and failure in the learning of easy and complex tasks. *J. Personal. Soc. Psychol. 3,* 339–344.

Weiner, B. (1972). *Theories of Motivation.* Chicago: Rand-McNally.

From the Dynamics of Conscience to Contract Psychology

Clinical Theory and Practice in Transition

O. HOBART MOWRER

At the outset I should explain how the rather strange sounding expression, "the dynamics of conscience," will be used in this paper. I shall use it, first of all, to convey the essential aspects of Sigmund Freud's views concerning the relationship between what he called the superego and the id, ego, and external reality. Then, retaining the same psychic topography as Freud, I shall suggest another, and more fruitful, conception of personality functioning and malfunctioning. Finally, I shall indicate how a further advance can be made, not by entirely rejecting this second conceptual scheme, but by extensively supplementing it with–indeed, one might say, transposing it into–the theory and pragmatics of social contracts, agreements, promises, commitments, explicit or implicit. We then discover that we have here an ethic that is not only without serious scientific or, for that matter, theological, objection in our own culture; once fully understood, it is a social or cultural universal.

PRELIMINARY CONSIDERATIONS

Recently a psychiatrist friend told me an "in" joke: "Psychiatry isn't a profession–it's a diagnosis!" It is a common supposition that psychologists also often

---

O. HOBART MOWRER • University of Illinois, Evanston, Illinois.

enter their profession in a search for solutions to "personal problems." Without attempting to speak for others, I readily concede that my entering psychology as a professor represented an attempt to find "psychological salvation."

Beginning early in my freshman year in high school, I woke up one morning with a pervasive and unrelenting sense of unreality and depersonalization, harbingers of a relatively severe depression, which lasted about 18 months and was only the first of some six or seven more that I have since experienced. Elsewhere (Mowrer, 1966, 1974) I have, in retrospect, been able to identify a multiplicity of "causes" for my predisposition to depression, and my family and I, in response to the first one, resorted to innumerable forms of treatment, some of which now seem quite ludicrous.

Suffice it to say that when, in the Fall of 1925, I entered the University of Missouri, it was with the hope and expectation that I would find out what had happened to me four years earlier–and what kind of really effective treatment was available for such an affliction. We were then in the heyday of Watsonian Behaviorism, and the textbook in our introductory psychology course was entitled, *The Psychology of the Other-One* (Meyer, 1922). What *I* needed and wanted was to find out about *me,* so my collegiate years did not bring me significantly nearer to my objective. But then I learned that you shouldn't really expect too much from undergraduate psychology; if you wanted to get to the heart of this discipline, you went to a graduate school, which I duly did, 1929–1932, at Johns Hopkins University, in Baltimore, Maryland.

At Hopkins there was some talk about the then new and quite radical conception of "neurosis" known as psychoanalysis which had been promulgated by Sigmund Freud; but all my instructors were very critical of this approach and had neither the inclination nor training to "practice" it. However, after I had done a little independent reading, I concluded that here was a lead that might be well worth exploring; and I was able to locate a woman analyst with an office only a few blocks from where I was living. Accordingly, in the Fall of 1929 I entered analysis and was in and out of this type of psychotherapy over a period of some 15 years, for a total of about 700 contact hours. I became deeply immersed in this movement, read extensively, wrote some papers articulating analysis and learning theory, and reported several experimental analogues of certain Freudian "dynamisms" with animal subjects (Mowrer, 1950).

I very much wanted psychoanalysis to work for me and to *be true,* but I found that after a period of analysis I would experience less anxiety and depression for a while, only to have, in a few years, a recurrence of my difficulties. Gradually I became discouraged and began to look elsewhere for a different approach to psychotherapy and source of personal "help."[1]

[1] The original draft of this paper was too long, and the foregoing paragraphs represent a drastic condensation of the first two sections entitled, "Personal Factors in the Choice of a Profession" and "The Dynamics of Conscience, Freudian Style."

THE DYNAMICS OF CONSCIENCE REINTERPRETED

In the Spring of 1944, I was invited to serve as an assessment psychologist in the Office of Strategic Services, and soon, with my wife and two small children, moved to Washington, D. C. By a great stroke of luck, Harry Stack Sullivan, today regarded by many as the founder of social psychiatry in this country, was then living in Washington, and practicing, consulting, writing, and lecturing there, in New York City, and elsewhere. From some of my colleagues in OSS who had been or were then students or associates of Sullivan, I learned that he had founded and was still the director of the Washington School of Psychiatry. Only medical students or licensed physicians were then, as now, ordinarily able to get instruction and training in the Departments of Psychiatry associated with medical schools. And independent psychoanalytic institutes also tended to limit their training to physicians. Sullivan believed there ought to be a different approach in this area. He felt that the social sciences are just as important in psychiatry as are the biological and physical ones; and he wanted his type of theory and training to be available to a broad professional spectrum: physicians, of course, but also psychologists, social workers, clergymen, school counselors, and still others.

In any case, it was clear that it would be possible for me to take courses at the Washington School, most of which were held in the evenings; so in the 1945 Spring Semester, I enrolled for a seminar with Sullivan and one with Dexter Bullard, then Director of Chestnut Lodge, a private psychiatric facility in nearby Rockville, Maryland, which operated according to Sullivanian precepts and procedures. I don't remember much about the rest of Sullivan's course; but I had heard only one or two of his lectures, when I became galvanized with interest in what was to me a fascinating new idea.

In order to explain why this happened as it did, let me say that during the preceding year (1944), I had finally lost my faith in both the theoretical soundness and practical efficacy of Freudian analysis. As early as 1934, while a National Research Fellow at Princeton University, I happened upon an issue of the *Journal of Mental Hygiene* (founded and long edited by Clifford Beers), where there was an article in which the therapeutic results obtained in various psychoanalytic institutes throughout the world were reported in tabular form. To my dismay there were very few entries in the "Cured" column, a somewhat larger number in the "Improved" column, but there was also about the same number in the "Unimproved" column, and, as the last column of the table indicated, some patients had evidently gotten worse during analysis. However, the impact of these dismal findings were largely "washed out" during my ebullient days at the Yale Institute of Human Relations; and it was only after I had moved to Cambridge, had further analysis with Hanns Sachs, and gone to Washington, D. C., and there found that I was still having trouble with anxiety and depression, that I again began to have doubts, doubts which were intensified by a growing awareness of other persons who had likewise not benefited anymore than I had from analysis. Eyesenck's controversial, but probably sound article suggesting that individual psychotherapy does not produce a rate of recovery any higher than "spontaneous remission" and that psychoanalyzed persons may, on the average, be worse off afterwards than before analysis, was not to appear until 1952 (cf. Bergin, 1971); but as early as 1944–1945, Freudian psychoanalysis was beginning to have considerable "face invalidity," and I was certainly looking for something different and, hopefully, better to take its place (cf. Jurjevich, 1974; Mowrer, 1975).

Sullivan's thrust aroused new hope. A disenchanted Freudian himself, he was now saying something like this: psychopathology is not to be understood and treated primarily in terms of intrapsychic complexes and dynamics but rather as parataxic distortions (misperceptions) and malfunctioning in interpersonal relations. The departure from Freud was thus relative rather than absolute, but it was sufficient to start me thinking in an entirely new frame of reference. As an undergraduate at the University of Missouri, I had taken a major in sociology and was familiar with such expressions as "social action and reaction," "social interaction," "roles," and "role-conflict"; but there was a link missing here, which Sullivan seemed to be supplying, namely, *the connection* between one's adequacy or inadequacy in the interpersonal sphere and psychopathology!

Perhaps the reason I don't remember much about Sullivan's ensuing lectures and have never really studied his voluminous writings and become an avowed follower of his is that I became so preoccupied with a somewhat independent line of thought and action which Sullivan's introductory lectures had suggested to me. Although then a man 38 years old, this was, I think, the first time I had ever thought at all carefully about the communal aspects of human nature and nurture. I have already indicated the lack of any very serious interpersonal emphasis in my upbringing (although there was a lot of talk about both God and the Devil—in fact, we had a dramatically illustrated book about the latter in our meager family library); and even before I entered high school I had picked up such shibboleths as: "What you do is your own business" and "What others don't know won't hurt them." And in college I purchased and hung immediately above my desk a burnt-wood motto by Elbert Hubbard: "Think so well of yourself it doesn't matter what others think of you."

Sullivan was the catalyst that started an intellectual and emotional ferment of a far more profound nature than anything that had ever happened as a result of my association with Freudian psychoanalysis. Here, as I now saw in retrospect, I had spent a lot of time and money only to have my self-isolating and self-defeating tendencies reinforced. All three of my analysts were immediately interested whenever I said anything about how badly I had been treated by others, particularly by my parents during childhood; but I can recall only two occasions when they expressed surprise or concern over something *I* had done. And to this day, many schools of psychotherapy, although perhaps not using specifically Freudian treatment procedures or terminology, are nevertheless Freudian in general orientation. Transactional Analysis, which at the present time is enormously popular, is a case in point. Although its official lineage goes back through Eric Berne to Sullivan, Thomas Harris, in *I'm OK, You're OK* (1967), minces no words in saying: "The question has always been how to get Freud off the couch and to the masses" (p. 15).

Under the impulsion of an idea that had been suggested to me by Sullivan in his Washington lectures (but which, so far as I can detect, he himself never explicitly embraced or promoted), I had begun to reexamine my own life history and, without difficulty, found many instances when I had deceived, cheated, or otherwise injured others and had made not the slightest attempt at restitution or reconciliation. Small wonder, then, that I found intimacy with others difficult (dangerous–lest they find me out) and that my sense of identity was often weak (because I had so often denied who I really was). Integrity (a word not often heard in professional circles during the Freudian era) emerged as the key to both interpersonal intimacy (or the "psychological sense of community," as Sarason, 1974, has recently called it) and a solid basis for identity, for only through integrity can there be congruence between the person we appear to be to others and the person we, ourselves, know we are.

But now back to the evolution of the dynamics of conscience suggested in the title of this section. Freud had said, in essence, that we become neurotic because we have an excessively strong and implacable superego which takes the ego captive and forces it to repress biologically given, "natural" sources of instinctual pleasure (particularly in the realms of sex and aggression). And, under the pressure of instinctual forces thus denied outlet, neurotic anxiety develops, with ensuing symptom formation. It now seemed increasingly likely to me that what happens in neurosis (a very misleading and unfortunate term which, happily, is being gradually abandoned) is that the ego becomes id-dominated and permits actions to occur which precipitate severe conflict with the super-ego, or conscience, thus causing the ego to institute repression which is *directed toward the superego.* Instead of assuming, with Freud, that the neurotic is an unduly inhibited and "good" person, the alternative hypothesis holds that sociopaths and neurotics are both "offenders" but that the neurotic, being somewhat more socialized than the sociopath, has a stronger conscience which, even though repressed, is nevertheless able to assert itself, in the form of sometimes quite awesome symptoms. Some small-scale experimentation in my own life, involving self-disclosure and restitution, produced surprisingly good results and prompted me to pursue the matter further.

Almost immediately upon returning to my academic post, in the summer of 1945, I wrote a paper entitled "The Problem of Anxiety," which was delivered as an invited address at the 1947 (Chicago) meeting of the American Association for the Advancement of Science (Mowrer, 1950, Chapter 19). The crux of this paper's thesis is contained in the following quotation:

> In essence, Freud's theory holds that anxiety comes from evil wishes, from acts which the individual would commit if he dared. *The alternative view here proposed is that anxiety comes, not from acts which the individual would commit but dares not, but from acts which he has committed and wishes that he had not. It is, in other words, a "guilt theory" of anxiety rather than an "impulse theory"* (p. 537).

This basic thesis is elaborated and its clinical implications delineated in three subsequent books: *Psychotherapy–Theory and Research* (1953), *The Crisis in Psychiatry and Religion* (1961), and *The New Group Therapy* (1964). Those of us who have taken this modification of Freudian theory seriously have gotten a good deal of "mileage" out of it. For example, Freud and his followers held that most of the concerns which patients bring to analysis regarding guilt are just "guilt feelings" or "false fears"; and the troublesome problem of *scrupulosity,* so often encountered in the Catholic confessional, would seem, at least on first blush, to bear out the Freudian view. So I decided to take this matter to a psychologist-priest Fr. John Stafford, whom I knew and trusted. Dr. Stafford responded by saying, "I have written a paper on this very matter. It is entitled, 'Is Scrupulosity a Mark of Insincerity?' and I will send you a copy of it" (Stafford, 1950).

When the paper in question arrived, I found it involved the following "case history." A man one day appeared in Fr. Stafford's confessional booth and said: "Father, I have an awful fault, a terribly sinful habit. I constantly try not to swear, Father, but I just can't keep from doing it, all the time. Please absolve me, Father, and ask the Lord's help in overcoming this vice." The priest heard and responded to this confession in the usual way; but a week later, the man was back with the same complaint–now, if anything, more urgent than at first. Absolution and nominal penance were administered; but the man kept returning, week after week, with no mitigation in the swearing or the guilt felt in respect to it.

At this point, Dr. Stafford did a very unorthodox but canny thing: He found out the name of this penitent and discovered that he was a day laborer whose real, but previously unmentioned, "sin" was that, instead of buying food and clothing for his family and paying the rent, he spent most of his meager earnings in taverns. When confronted by this previously undisclosed aspect of his life, this man gave up the game of scrupulosity with respect to his swearing. So the Freudian hypothesis was not confirmed, after all. There was *real* guilt in this man's life, not just "guilt feelings."

A decade or so ago I became acquainted with the Daytop Village rehabilitation program for drug addicts in New York City and the other Eastern cities, and was eventually "welcomed into the family." I never once heard the word "scrupulosity" used at any of the Daytop facilities; but everyone knew what it meant, in group sessions, to "throw a bone." The same diversionary tactic is implied in both cases. I have discussed Dr. Stafford's views on scrupulosity more fully elsewhere (Mowrer, 1953, pp. 93, 318, 330, 358, and 362).

Recently a particularly dedicated and wise member of Alcoholics Anonymous called me from Chicago and said: "Look, I've been working with this alcoholic, and he's gotten sober but he's had to be admitted to the Elgin State Hospital. His main symptom is that he thinks he is being watched all the time and that other people can hear his thoughts. Should I continue to try to work the [12 A.A. Recovery] steps with him or not?" My reply went something like this:

> There is undoubtedly a genetic predisposition in some people to show "schizophrenic" thought patterns and bizarre behavior when they get under too much stress. And sometimes, medication is the only thing that seems to help them. But maybe it goes like this in the case of your friend. Perhaps he has a history of dishonesty and irresponsibility, for which he has every right to feel guilty. Suppose, further, that he became a compulsive drinker because he found that alcohol tended to deaden his guilt and anxiety, and that, now he's been sober for a while and without this crutch, the guilt is impinging upon him with unimpeded force and the resulting stress has caused him to react schizophrenically.
>
> Now, I continued, if our suppositions are true, this man's so-called delusions make a lot of sense. If he has, in fact, behaved in an untrustworthy manner, what is more appropriate than for his conscience to make him feel that he is being constantly watched; and if he has been secretive and dishonest, the belief that others can "hear" his thoughts would be an excellent way of taking his privacy, which he has abused, away from him.
>
> Paranoia has commonly been regarded as quite intractable to conventional therapeutic approaches; but, using the approach I've just suggested, we've had very good results with it in our Integrity Groups. I think the fact that other, more secure members share deeply out of their life experiences with such a person at his Intake, and in the regular group to which he is then assigned, engenders trust in a remarkably effective way and helps dissolve the paranoid's suspiciousness, and gives him enough confidence to begin telling the truth about himself, so that, as in the case of your man, he doesn't have to be *watched* anymore (because he has become more responsible) and other people don't need to *hear* his thoughts (because he is now honestly and accurately communicating with them).

My A.A. friend replied: "I guess it probably won't do any harm for me to continue working the Steps with this fellow." Anyone who is familiar with the 12-Step Recovery Program of Alcoholics Anonymous will have no difficulty seeing its congruence with the approach to paranoia just suggested.

Let me now give an example from my own recent experience. About a month ago I was working on an exciting idea I had for an article for another paper; but one evening, when I settled down to enjoy watching the CBS 5:30 News, I remarked to my wife that instead of having the feeling of relaxation I usually experience this time of day, I was a little anxious. My wife replied, "You know what's the matter with you, don't you?" I made some ambiguous gesture, and she said: "You ought to be finishing that paper for the New York Symposium instead of writing on something for which there is no deadline."

In this situation I had been rather successfully lying to myself. I had been saying, in effect: "Oh, the New York paper is already about one-fourth completed and I have the rest of it outlined in my head so completing it is just a matter of three or four more days of writing. Therefore, there is plenty of time for me to work on the other paper." But my wife, and conscience, were quite right in insisting that I put "first things first." This incident illustrates, I may add, the advantage of disclosing one's future plans, as well as one's past follies and weaknesses, to other persons. Without having previously informed my wife about the New York trip and the responsibilities associated with it, she could not have "pulled my covers," to use another Daytop saying, and made me admit to the truth, not to someone else, but to myself—and change my behavior accordingly.

"This kind of anecdote and clinical evidence is all well and good," someone may say, "but where is your hard supporting data?" Freud never had and was rarely asked for "hard data" to support his constructs, so why should this stricture be laid on us? Perhaps the very fact that we so readily took Freud's verbal persuasiveness in lieu of hard evidence is all the more reason others should not be falsely persuaded and disappointed by us. So we do not resent the question, and are happy to say there is now quite a lot of empirical support for the position just described. This has been described in a variety of other places (Mowrer, 1971, 1973*a*, 1973*b*, 1973*c*). In fact, as early as 1968 I published a paper reviewing ten empirical studies by others, all of which supported, not the Freudian "characterology," but, unequivocally, the one described above, yet I have not seen that paper cited a single time!

At about the same time, Johnson and associates (see Johnson *et al.*, 1972, Chapter 34) have approached the problem in a rather different way. They begin by making the assumption that the best index of ego strength is an individual's capacity to "resist temptation" (*i.e.*, the ability to forego immediately satisfying but ultimately self-defeating acts); and that the best measure of superego strength is the amount of remorse or guilt a person suffers after having yielded to temptation (*i.e.*, performed immediately satisfying but ultimately self-defeating acts). Johnson and collaborators, using scaled responses to events described in brief stories, found that neurotics and normal persons do not differ in superego strength (as Freud postulated) but do differ, dramatically, in ego strength, neurotics being markedly inferior to normals in this regard, thus justifying the lively contemporary interest in "ego strength" or "ego psychology" (Mowrer and Vattano, 1975).

Although the logic of the Johnson work is somewhat different from that of Rolls *et al.* (Mowrer, 1968), the end results are much the same. On a scale of socialization, Johnson's findings put normal persons above neurotics, for the reason that the latter more often "yield to temptation," and since guilt is experienced only in those instances in which an individual does not succeed in resisting temptation, neurotics experience more guilt and have more anxiety than normal persons. Johnson did not collect data bearing on the position of sociopaths, because virtually everyone agrees that sociopaths have weak egos and thus readily yield to temptation; but since they also have weak superegos, yielding to temptation produces little anxiety or guilt.

The reformulation of Freudian theory described in this section, despite all the clinical and empirical evidence in support of it, is only now beginning to attract the interest and attention it would seem to warrant. Part of the difficulty almost certainly is that, for many persons, this reformulation has "moralistic" overtones and thus, however sound it may prove to be, they find difficulty regarding it as "scientific." But this is a squeamish, rather than realistic, objection. In the facilities operated by Daytop Village, a great deal of attention is put upon honesty, responsibility, and concern for others (and oneself). However, I never once heard a resident called bad, immoral, evil, wicked, or sinful; if he is deficient in certain character traits—he is simply yelled at and called STUPID. Drug addicts have been called "bad" most of their lives, and such "scolding" has had little or no effect on them; but when it is pointed out to them by their peers that their asocial or antisocial behavior is unintelligent and self-defeating, this "gets to" them. Bad? Of course, but stupid? Surely not that! Look how clever they have been in so many ways; but they have to admit that their so-called cleverness has, in the end, not worked out very well for them. Why, otherwise, would they be where they are now?

In Integrity Groups our logic and approach to the behavior of others is essentially the same: We do not condemn them for having manifested poor integrity prior to coming into our groups; instead, we ask them how their lack of integrity has worked out for them? If they reply, "Just fine," then we ask them why they are trying to get into an Integrity Group. Usually, of course, they admit that their lack of integrity has caused them a great deal of trouble and they now wish to learn how to live with greater congruence between promises and performances, appearances and actions.

So we see no reason for rejecting the revised Freudian formulation on the grounds that it is "moralistic." Moreover, Henri Ellenberger—the same Canadian psychiatrist who wrote the widely acclaimed volume, *The Discovery of the Unconscious* (1970), somewhat earlier published a trenchant paper entitled "The

Pathogenic Secret and Its Therapeutics" (1966). Here extensive clinical and literary materials are reviewed which indicate the importance and relevance of integrity (truthfulness, responsibility, and concern). In fact, it seems that integrity is a cultural universal, *i.e.*, essential to the proper functioning of individuals and societies everywhere. The cry of "moralistic" is a red herring; and perhaps the time has come, as previously noted, to disregard it and begin approaching the phenomenon of morality as a social reality which can be dealt with in a completely naturalistic and humanistic way (Mowrer, 1971).

CONSCIENCE OR CONTRACT? TWO SIDES OF THE SAME COIN

We come now to the final, in some ways most difficult, but also most rewarding part of our discussion.

In the preceding section, we have seen that the hypothesis that repression of conscience as the "primal pathogenic act" has numerous advantages over Freud's supposition that repression of id impulses is the taproot of neurotic difficulties. In taking this position, we do not categorically maintain that Freudian-type repression cannot or does not sometimes occur—our position is simply that repression of conscience is by far the more common phenomenon and that this is the most useful assumption with which to start. Exceptions to this guiding principle can be made if and when the clinical facts require it; but attempts to help a neurotic person are, in most instances, more rewarding and carry us much further if we operate on the supposition that the individual is in trouble because he has repressed, not id, but conscience.

As indicated in one of the illustrations mentioned in the preceding section, the typical delusions of paranoid persons become immediately—and intuitively—intelligible if one posits a repressed, outraged conscience. For Freudians, on the other hand, paranoia has always been an enigma; and the basis for their conjecture that it represents repressed homosexuality has been as hard to fathom as their efforts to treat paranoia have been unavailing. We conjecture that the main reason Freudian analysts have always felt that they must proceed very slowly with paranoid persons—and that other, presumably less well-trained therapists should not even attempt to deal with such persons lest they precipitate an even more malignant type of psychosis—comes from their fundamental misunderstanding of paranoia and hence the very real likelihood that if they are "too active" they will indeed exacerbate client confusion and anxiety. Persons displaying paranoid characteristics are usually comforted by our perception of their difficulties—which suggests that, when all the facts are known, they are not nearly so "crazy" as they otherwise seem to be, both to themselves and to others. Our view of paranoia leads quickly to certain prescriptions which, if pursued in the context of a trusting and trusted group, can often dissipate delusional symptomology in a relatively short time.

It should also be added here that although we perceive and deal with paranoia and other forms of functional symptomology (with qualifications to be discussed presently) in the manner indicated, we do not by any means overlook or deny the role of constitutional (genetic) factors in determining a person's predisposition, when under stress, to schizophrenic, cyclothymic, and other forms of emotional and mental disorder. We follow with great interest the exciting advances presently being made in the field of brain chemistry (especially under the guidance of the catecholamine hypothesis); and we are grateful for those psychotropic drugs which often dramatically relieve symptoms by temporarily increasing stress tolerance (although we are aware that such drugs can be misused and abused). The approach followed in our Integrity Groups, which will be alluded to again shortly, aims at helping nervous, insecure individuals develop a new style of life which will be inherently less stress-provoking than the one they have previously been following (Mowrer, 1973*a,b*), although, here again, we are not unmindful that our whole civilization is sick (perhaps fatally so) and subjects us all to quite severe, and often intolerable, sources of stress.

As a means of now beginning to move more directly toward the central concerns of this last section, let us look at some of the difficulties associated with the concept of conscience, which has beset the Freudian theory of psychopathology as well as our own. Freudians, in assaulting a patient's absolutistic superego, commonly insisted that they knew precisely what kind of new superego, or value system, ought to be substituted during the transference and its resolution. We, on the other hand, have operated on the assumption that a person can most swiftly eliminate his neurotic suffering by coming to terms with his existing conscience, rather than trying to remodel it. We have known, of course, that individual consciences differ in both small and major ways; and our only recourse was to assume that, regardless of whatever "make" of conscience a person happens to have, he is stuck with it and has to adjust to it. Freudian psychoanalysis and our own approach have thus constituted the horns of a most troublesome dilemma: to try to change one's conscience, by means of psychoanalysis, with uncertain and often unsatisfactory results, *or* to keep the conscience one already has and try to do its bidding, even if it is in some respects archaic.

Perhaps a way out of this dilemma can be found if we reexamine the concept of conscience and possible alternatives. One of the most obvious difficulties with this concept is that it is so global, vague, subjective, and difficult to operationalize. Freud, in *Civilization and Its Discontents* (1930), defined the superego as "the internalized voice of the community," which is not so very different from the view of an earlier time that conscience is "God's voice speaking within man." Others have attempted to define conscience by saying that it is the repository or sum total of a person's values, standards, and beliefs. Although widely used in ordinary discourse and even in professional writings, these latter terms also suffer from ambiguity and vagueness. Moreover, all definitions of conscience relate to the problem of morality, which, within the past two decades, has undergone some major transformations. Until relatively recently, the prevailing ethic in our society was based on *moral absolutism* deriving mainly from the Judeo-Christian tradition. For a variety of reasons which need not be detailed here (Mowrer, 1961, 1973*c*), this tradition has been in the process of erosion for a century or more, a process which reached a climax in the Death-of-God ideology of the early 1960s (cf. Vahanian, 1961).

If God, in the sense of an anthropomorphic being with a special, benign concern for man, apart from all other living creatures, has indeed died, it is because the binding force of His Holy Word has, for vast segments of our population, been discarded or repudiated. Very few persons, even in the mainstream of the Judeo-Christian tradition, believe, for example, that the Ten Commandments were divinely revealed to Moses on Mount Sinai. Earlier documents, such as the Code of Hammurabi (1955? *B.C.*), suggest that the Mosaic Decalogue was the distillate of human experience which had long had social survival value and was attributed to a supernatural source in an effort to make it more authoritative, at a time when man understood his universe very poorly and was still struggling with the Herculean task of his own domestication. Modern theologians, painfully aware of this situation, have themselves often led the assault upon the mythological trappings of Judeo-Christian morality. A striking case in point is Bultmann's reference to the three-storied biblical interpretation of the universe (earth, heaven, and hell) as essentially mythological (1961); and Bonhoeffer (1948, 1954) has stressed the interpersonal rather than the metaphysical and mystical dimension of Christianity.

Thus, the collapse of moral absolutism, symbolized as it was by the purported "death" of God, undercut the basis for the inculcation by parents and the acceptance by children of anything as eternally fixed and unchanging as conscience. Beginning with the Industrial Revolution (about 1760), we have become increasingly industrialized and urbanized, with rapid and often confusing social changes which have raised all manner of doubts and questions with respect to traditional morality. Moreover, as anthropologists have often pointed out, consciences differ from society to society and even between subgroups within the same society. And since nothing so readily justifies our viewing others who do not share our own highest beliefs as inferior and even inhuman, moral absolutism, in both its social and subjective form, has been a pervasive cause of distrust, prejudice, persecution, exploitation, and even warfare.

In order to try to fill the ethical void created by the so-called Death of God, *i.e.*, the collapse of moral absolutism, Fletcher (1966)–to mention but one of several other theologians–introduced what he called *Situation Ethics,* which, during the late 1960s, had a brief vogue but is now hardly ever written about but has nevertheless been widely assimilated in our society in the form of the "do-your-own-thing" philosophy.

Having written at length on other occasions about the vagaries and deficiencies of Situation Ethics (Mowrer, 1973*c*; Mowrer and Vattano, 1975), I shall move on without further delay to an ethic based upon *Contract Psychology,* which has been a great boon to both the theory and practice in Integrity Groups and which offers everyone, we believe, a practical and surprisingly complete and dependable guide to everyday morality in a fluid and pluralistic society.

It was through the writings of Pratt and Tooley (1961, 1964, 1966; see also Shapiro, 1968; Kanfer and Phillips, 1972; and Parlour, 1975) that those of us associated with the evolution and development of Integrity Groups were first introduced to the concept of contracts, and from the outset we were captivated by it even though we did not immediately grasp its remarkable scope and power. The inception of our Groups dates back to 1957, and soon we were stressing the importance of three basic principles: Honesty, Responsibility, and Involvement (cf. Mainord, 1962); but it was not until later that we began to use the word *integrity* to explicitly characterize our groups and the major objectives we sought to achieve therein. At about this time we gradually began to see that Contract Psychology provides a uniquely useful conceptual framework for defining the nature and clearly identifying the goals of our Groups. By means of Contract Psychology, the term integrity can be quite precisely defined and operationalized as being *inversely proportional to the distance or discrepancy between the level of a person's promises (contracts) and the level of his performance (actual conduct).* Or, more simply, one can say that the better a person keeps his contracts (promises), the greater his Integrity.

We then found that we had equally concise ways of defining our three basic principles.

Honesty involves a person's disclosure of previously concealed contract violations, *i.e.,* promises not kept, obligations not discharged, agreements abrogated.

Responsibility, in the framework of Contract Psychology, is action designed to increase one's integrity, *i.e.,* reduce the discrepancy between what one has agreed to do, or not do, and what one has actually done. This task can be approached in any of three ways: (1) by making amends and reaffirming the violated contract or contracts; (2) by renegotiating the contract, in ways that may or may not involve restitution, but are mutually agreeable; or (3) by terminating the contract with whatever "settlement" seems equitable and just.

Involvement (concern) is manifested by efforts to help other persons become more honest and responsible, as defined above. In the context of Integrity Groups, this is often done through the sharing of one's own lapses from integrity and reporting on the benefits which have resulted from adopting a policy of honesty and responsibility.

Since one's identity is a function of one's integrity, the road back from identity crisis (Erikson's much more apt term than "neurosis," except as one may wish to use the latter term to denote actual neurophysiological disturbances) would seem to be clearly demarcated by the steps of Honesty, Responsibility, and Involvement. Involvement, in other words, is a form of repayment for the encouragement and support one has received from others in the pursuit of greater integrity and a clearer identity, which also has its own intrinsic rewards. In Alcoholics Anonymous there is a saying, "You can't keep it unless you give it away!" (Cf. the 12th Step of the A.A. Recovery Program.)

When a person enters an Integrity Group, that person's contracts are not inventoried to see if he or she has made the *right* ones. We are never quite sure what the right ones are. And although we may be willing to express an opinion as to the wisdom or practicality of this as opposed to that agreement or arrangement, it is always left up to each individual to make the final choice. But what we do ask a newcomer–and consider of the utmost importance–is how well he *keeps* his contracts and if, for some reason he defaults, does he acknowledge his errors and failures and quickly set things straight? Or does he cover his lapses in integrity with deception and do nothing more about them? If the latter course of action is followed, one's sense of community and the enjoyment of intimacy with others is blighted by apprehension lest one's guilty secrets be discovered; and every lie one tells to cover contract violation subtracts by that much from the definition and clarity of one's identity.

When, in the security of an Integrity Group, one develops enough trust and courage to acknowledge deceptions, one is not censored for having previously lied but is instead supported and reinforced for now *speaking the truth;* and then, in due course, a discussion ensues as to how one can most effectively rectify the past lapses from integrity which have, until now, been concealed by deception and, very probably, have caused the individual emotional discomfort. By the time a contract violator and deceiver has, in the presence of a supportive yet concerned group, decided upon a corrective course of action, this person is likely to report feeling much less "up tight" and sometimes even weeps from the relief he or she has already experienced and the hope of better days ahead.

Only a few other brief observations and our discussion will be finished. Obviously, active and successful participation in an Integrity Group involves a great deal more than can be delineated here, and the reader is referred to a manual entitled, *Integrity Groups; The Loss and Recovery of Community* (Mowrer and Vattano, 1975, Vol. I–a second volume is now in preparation). But it will already have become apparent how greatly Integrity Groups and classical psychoanalysis differ in the matter of "timing." Whereas the latter admonished analysands not to make or act upon any new decisions until the analysis was virtually completed and their "insight" hopefully improved, in Integrity Groups suffering persons, as soon as they understand the implications of contract psychology and see how to rectify mistakes they have made, are encouraged to do so, not precipitously but promptly, under the guidance of what Sullivan called "consensual validation," *i.e.*, after checking out plans for making amends, step by step, with one's Group and perhaps other trusted persons.

Thus, as a person is conscientiously working on amends with respect to past errors, he is also learning to be honest with respect to the future, *i.e.*, to disclose to others and solicit their "feedback" regarding actions which are still only intentions or plans. Thus, one learns to "take counsel"—and, in turn, to give it.

From a variety of sources (including neo-Freudian psychoanalysis), we are learning more and more about the importance of the therapeutic goal of increasing "ego strength" (Johnson *et al.*, 1968), but no very clear suggestions are emerging as to how this objective can be achieved. To paraphrase the biblical statement that one has to lose his life to find it, it has been our experience that the quickest way for one to acquire greater ego strength (in the sense of having control over one's life) is to give up something of one's automomy, independence, self-sufficiency. In coming into a small "intentional" group and admitting that one needs "help," one is appealing to what he believes to be a source of strength or power greater than lies within himself. And by becoming honest, not just with respect to behavior already committed but merely contemplated, one receives the support of his group in avoiding short-sighted, self-defeating actions and in choosing far-sighted, growth-producing behavior. This is the quintessence of ego strength, epitomized by the saying, "A person is never so strong as when admitting his or her weaknesses."

"Passive dependency" is a common clinical epithet. In Integrity Groups we do not encourage passivity, but we do stress dependency, as a means to strength, in the manner just indicated. In fact, our most often quoted slogan is: "You alone can do it, but you can't do it alone." Professional psychotherapists have long pushed their patients toward self-sufficiency, independence. One of the primary goals of classical psychoanalysis has been for the analysand to become "ego-syntonic," and contemporary writers have used "self-actualization," "autonomy," "self-reliance," and kindred words to denote the same objective. However indefinite and confused the movement known as Community Psychiatry may be, there seems to be almost universal agreement that isolating deviant persons in enormous, impersonal state-operated custodial institutions—or "tanks" as they have been aptly termed—is neither a logical nor effective way of rehabilitating already lonely, alienated persons, and that our last hope of real aid to such persons consists of keeping them as close as possible to such reference groups, support systems, or social networks as they may still have—or helping them find new ones. One of the great things about Alcoholics Anonymous is that it provides not only a Recovery Program but also a special Fellowship (dedicated to staying sober instead of getting intoxicated) which one can enjoy and benefit from indefinitely. Gradually we have come to realize the profound truth of the old German proverb, "Ein Mann Ist Kein Mann"—"One man is no man."

Yet, until recently, much professional effort was spent trying to "liberate" patients, reduce their social interdependence, and push them in the direction in which they had already gone too far. But we cannot censor the professionals alone; we have, for a long time, lived in a culture which stressed personal independence and "rugged individualism." What has become of such ancient admonitions as being one's brother's keeper and bearing one another's burdens?

We do not, of course, hold that *all* identity crises arise from the progressive erosion of a sense of who one is by habitual deviance and deception. One's identity can also be badly shaken by the death of a loved one or by loss of position, home, or other possessions (Goodall, 1972; Holmes, 1971). But such losses are usually public knowledge, and the factor of secrecy and festering guilt is not a part of the situation, which, in the course of time, and with support, normally mends itself. In our Groups we welcome the bereaved and bereft, to whom we give such understanding and assistance as we can; but it is with the bearer of long-hidden guilty secrets that, for reasons already given, we do our best work and achieve the most striking transformations.

Although we have occasionally used the word, "guilt," in the foregoing pages, we have deliberately delayed in attempting to define it. In terms of Contract Psychology, it should by now be clear that guilt or the condition of being guilty is the same as lack of integrity, *i.e.,* it involves violation of a contract, tacit or explicit, with or without concealment. We know, of course, that a person can *be* guilty by this definition without *feeling* guilty. Much therapeutic energy has been spent trying to get persons *not* to feel guilty, without much attention to the possibility that the capacity to feel guilt is a valuable personal asset and is socially indispensable. Suppose that for human beings guilt existed only objectively, as a court of law attempts to determine it, without any subjective counterpart, in the sense of regret, remorse, contrition, self-criticism, self-punishment. We obviously would all be sociopaths; and there would not be much point in having police and a judicial system, for the individuals involved in law enforcement would be as sociopathic and careless of character as the rest of us, and justice would be a very chancy thing. No, every society has learned that it must attempt to establish in all its members an inner tribunal, which criticizes and condemns until the individual socializes his guilt and makes restitution or takes whatever other consequences may be in store for him. Karl Menninger's *The Crime of Punishment* (1968), argues persuasively for restitution instead of legal retribution, and it seems likely that the former is far more likely than the latter to awaken and educate conscience. But now we see that in order for Contract Psychology (the rule of law) to work, it presupposed the phenomenon of conscience. However, we have already reviewed the weaknesses of such a concept, notably its apparent rigidity and unmodifiabiltiy. But Contract Psychology again comes to our aid, as will be indicated in the following, far from hypothetical, example.

As is well known, university students often "live together," without benefit of clergy, and this practice is generally accepted, or at least condoned, by both their peers and professors. But if one is in a small group with such students, one discovers that they often have a lot of guilt about this type of "arrangement" or "relationship." And if you ask them *why* they feel guilty, the common response it, "My mother and father always taught us children that it was wrong to have sex before you're married, and I just can't get over feeling this way even though I know, in my head, that it's all right."

The earlier discussion of so-called scrupulosity now comes to our assistance, and suggests that we look for something other than the "liberated" sex practices of students as the source of their difficulty. So we often ask unmarried "married" students if their *parents* know the existing situation. The probability is very high that the answer will be, "No," followed by elaborate rationalizations, including how much one's parents would be "hurt" if they knew the full truth. Since university students are very commonly still dependent upon their parents, financially, socially, and emotionally, it is obvious that a contract, tacit or explicit, is being violated; and we encourage students either to come into compliance with respect to the contracts they have with their parents regarding sex, *or* tell them "like it is" and take the consequences. In short, it is often not the sexual behavior of students, as such, that is producing their guilt, but the deception and cheating that this behavior involves with respect to parental expectations regarding truthfulness. Either compliance with or renegotiation of student–parent contracts in the area of sexual behavior brings quite satisfactory results in the majority of cases.

Valuable as it is to know that students' so-called "sexual hang ups" can often be happily, or at least honorably, resolved by recourse to Contract Psychology, an even larger advantage emerges here. We find that by changing one's *contracts,* one can seemingly change conscience, *i.e.,* the behavior about which one feels guilty. However, this change in conscience may be more apparent than real: The concern of conscience, all along, has not been with sex per se but with the loss of authenticity or integrity in the relation with parents. When one has faced the sex issue head-on with parents, *i.e.,* been fully honest with them, conscience may be said to be *satisfied,* but *not* changed. The ubiquity of interpersonal contracts seems to be a cultural universal, and so also does the need for subjective discomfort when individuals secretly violate, or cheat with respect to, these contracts. Hence the heading of this section, "Conscience or Contracts? Two Sides of the Same Coin."

EPILOGUE

Harry S. Truman was certainly not as striking and charismatic a figure as either his immediate predecessor or one of his successors in the presidency of the

United States; and I personally, was never as much interested in or drawn to him during his presidency, as I was to Roosevelt and Kennedy. Recently, largely by happenstance, I have read West's *Upstairs at the White House, The Kennedy Case* by Dallas and Ratcliffe, *All the President's Men* by Bernstein and Woodward, and *Plain Speaking, An Oral Biography of Harry S. Truman* by Merle Miller. And, at least for me, Truman emerges as in some ways the greatest, probably the happiest, and almost certainly the most truthful of our last several Chief Executives.

The following brief excerpts are taken from the Truman book, *Plain Speaking:*

> The pink complexion was white, and his eyes were tired and faded. He came up to me and said, "I'm very sorry I wasn't down there this morning" [for the making of a video film].
>
> There were sudden tears in my eyes, but I managed to say, "That's all right, Mr. President. You were ill.
>
> "I know," he said, "but I like to live up to my obligations" (p. 29).
>
> Judge Albert A. Ridge of the U. S. District Court in Kansas City and a veteran of Battery D [of which Truman was Captain during World War I] said, "Harry Truman grew up in a society in which a man's word was his bond. If a man's word could be trusted there was no place he couldn't go. Nobody around here ever doubted Harry Truman's word" (p. 45).
>
> Father Tierman gave the best summing up of Harry Truman I have ever encountered. He said, "Harry Truman had integrity and much more than normal intelligence, and there is no limit in a free society to what men with those attributes can attain" (p. 143).
>
> "But that's what I mean [Truman once remarked to Merle Miller]. That shows you that if you have a fundamentally honest background with the Senators, you can get things accomplished. The whole thing, our whole government works on trust. If you can't trust a fellow Senator or anybody else in our government, the whole thing breaks down" (p. 170).
>
> "What I think is deep in him," said Max Lowenthal who had discovered Truman's integrity on the railroad committee, "is a sense of the atmosphere of the American tradition. . . . That has something to do with the way he ran his investigations. It is an innate part of his personality to be fair and to know what is fair, and to exercise restraint when he possesses great power, particularly the power to investigate and detect, and the power to police" (p. 441).

Perhaps what's good for presidents is likely to be good for the rest of us, as well.

REFERENCES

Bergin, A. E. (1971). The evaluation of therapeutic outcomes. In *Handbook of Psychotherapy and Behavior Change* (A. E. Bergin and S. L. Garfield, Eds.). New York: Wiley & Sons.

Bernstein, C., and Woodward, B. (1975). *All the President's Men.* New York: Warner Communications Co.

Bonhoeffer, D. (1948). *The Cost of Discipleship.* New York: The Macmillan Co.

Bonhoeffer, D. (1954). *Life Together.* New York: Harper & Bros.

Bultmann, R., *et al.* (1961). *Kerygma and Myth: A Theological Debate* (H. Bartsch, Ed.). New York: Harpern.

Dallas, R., and Ratcliffe, J. (1973). *The Kennedy Case.* New York: Putnam.

Ellenberger, E. (1966). The pathogenic secret and its therapeutics. *J. Hist. Behav. Sci. 2,* 29–42.

Ellenberger, H. (1970). *The Discovery of the Unconscious.* New York: Basic Books, Inc.

Eysenck, H. J. (1952). The effects of psychotherapy: An evaluation. *J. Consult. Psychol. 16,* 319–324.

Fletcher, J. (1966). *Situation Ethics.* Philadelphia: Wesminister Press.

Freud, S. (1930). *Civilization and Its Discontents.* London: Hogarth Press.

Goodall, J. (1972). *In the Shadow of Man.* New York: Dell Publishing Co.

Harris, T. A. (1967). *I'm OK, You're OK.* New York: Avon Books.

Holmes, T. (1971). The hazards of change. *Time Magazine* (Behavior). March 1, p. 54.

Johnson, R. C., *et al.* (1968). Resistance to temptation, guilt following yielding, and psychopathology. *J. Consult. Clin. Psychol. 32,* 169–175.

Johnson, R. C., Dokecki, P., and Mowrer, O. H., Eds. (1972). *Conscience, Contract, and Social Reality* New York: Holt, Rinehart & Wintston.

Jurjevich, R. N. (1974). *The Hoax of Freudianism.* Philadelphia.: Dorrance & Co.

Kanfer, F. H., and Phillips, J. S. (1972). Contract psychology: An operational approach to the problem of conscience and self-control. In *Conscience, Contract, and Social Reality* (R. C. Johnson, P. Dokecki, and O. H. Mowrer, Eds.). New York: Holt, Rinehart, & Winston, pp. 417–427.

Mainord, W. A. (1962). A therapy. *Res. Bull. Ment. Health 5,* 85–92.

Menninger, K. (1968). *The Crime of Punishment.* New York: Viking Press.

Meyer, M. F. (1922). *The Psychology of the Other-One.* Columbia, Mo.: Missouri Book Co.

Miller, M. (1974). *Plain Speaking: An Oral Biography of Harry S. Truman,* New York: Berkley Publishing Co.

Mowrer, O. H. (1950). *Learning Theory and Personality Dynamics.* New York: Ronald Press Co.

Mowrer, O. H., Ed. (1953). *Psychotherapy–Theory and Research.* New York: Ronald Press Co.

Mowrer, O. H. (1961). *The Crisis in Psychiatry and Religion.* Princeton, N. J.: D. Van Nostrand Co.

Mowrer, O. H. (1964). *The New Group Therapy.* Princeton, N. J.: D. Van Nostrand Co.

Mowrer, O. H. (1966). Abnormal reactions or actions? An autobiographical answer. In *Introduction to Psychology: A Self-Selection Textbook* (J. A. Vernon, Ed.). Dubuque, Iowa: Wm. Brown Co.

Mowrer, O. H. (1968). New evidence concerning the nature of psychopathology. In *Studies in Psychotherapy and Behavior Change* (M. J. Feldman, Ed.). Buffalo, N. Y.: University of Buffalo Press, pp. 111–193.

Mowrer, O. H. (1971). Belated clinical recognition of the "pathogenic secret." Champaign, Ill.: University of Illinois, Department of Psychology (Mimeographed).

Mowrer O. H. (1973*a*). Stress, constitution, character, and Integrity Groups. *Psychotherapy: Theory, Research and Practice 10,* 243–270.

Mowrer, O. H. (1973*b*). My philosophy of psychotherapy. *J. Contemp. Psychother. 6,* 35–42.

Mowrer, O. H. (1973*c*). "Old wine in new bottles"–Blueprint for a humanistic ethic. In *Human Behavior in a Changing Society* (J. F. Adams, Ed.). Boston: Holbrook Press, Inc., pp. 241–267.

Mowrer, O. H. (1974). O. Hobart Mowrer. In *A History of Psychology in Autobiography* (G. Lindsey, Ed.). Englewood Cliffs, N. J.: Prentice Hall.

Mowrer, O. H. (1975). Needed: A candle, not a curse! (Review of R. H. Jurjevich's *The Hoax of Freudianism.) Contemp. Psychol.* (in press).

Mowrer, O. H., and Vattano, A. J. (1975). *Integrity Groups–The Loss and Recovery of Community.* Urbana, Ill.: Integrity Groups.

Parlour, R. R. (1975). Some behavior techniques in community psychiatry. *Am. J. Psychother. 29,* 79–91.

Pratt, S., and Tooley, J. (1961). Human systems actualization through "participative organization"–A strategy of contract-systems psychology (Mimeographed, pp. 104). Presented at a conference on Behavioral Modification as a Function of Social and Interpersonal Factors, V. E. Bixenstine, Chairman; Kent State University, April.

Pratt, S., and Tooley, J. (1964). Contract psychology and the actualizing transactional-field. *In. J. Soc. Psychiatry,* Congress Issue, 51–69.

Pratt, S., and Tooley, J. (1966). Human actualization teams: The perspective of contract psychology. *Am. J. Orthopsychiatry 36,* 881–895.

Sarason, S. B. (1974). *The psychological sense of community: Prospectus for a community psychology.* San Francisco: Jossey-Bass.

Shapiro, S. B. (1968). Some aspects of a theory of interpersonal constructs. *Psychol. Rep. 22,* 171–183.

Stafford, J. W. (1950). Psychology and moral problems. *Homiletic and Pastoral Rev. 51,* 118–124.

Vahanian, G. (1961). *The Death of God.* New York: Braziller.

West, J. B. (1974). *Upstairs at the White House.* New York: Warner Communications Co.

Discussion

RICHARD S. LAZARUS

The contributions to this section can be divided into two clear groupings: (1) a primarily psychophysiological group including Selye, Mason, and Frankenhaeuser, and (2) an exclusively psychological group including Atkinson and Mowrer. I shall have the most to say about the former authors, mainly because they reflect and confront each other with some important issues and controversies, while the latter two do not interdigitate at all.

I would like to begin by making an observation about chapters in the previous section of this volume that also applies to this group. It was said previously, correctly I believe, that one cannot understand human or animal adaptation very well without knowing something about the brain, the central organ of adaptation. The argument must also be turned around, however: Neither can you understand the functioning of the brain without sound knowledge at the behavioral or psychological level. These areas of research and thought represent two interdependent levels of analysis, two sides of the same coin, so to speak. If we have poor behavioral science, we will also have a poor understanding of neural mechanisms, and vice versa.

The chapters on the neurophysiology of adaptation seem to me figuratively to decerebrate man because they deal largely with lower brain centers and lower animal life. Attention seems to center in the midbrain and at peripheral systems such as the autonomic nervous system and the endocrine glands, mainly the adrenals at that. Knowledge of neural and hormonal control of emotion in general is as yet restricted mainly to the hypothalamus, and the reticular system

---

RICHARD S. LAZARUS · Department of Psychology, University of California, Berkeley, California.

and rarely or never extends to the limbic system and the cerebral cortex. Moreover, if the object of study consists of lower animals in the phylogenetic sense, this must leave out or understate the higher neural and behavioral functions characteristic of man and other primates.

I think we all know this, but it is important to say it again because in humans especially, the central psychological problem in stress and adaptation is to understand the signaling system by means of which a person differentiates between what is harmful and what is benign, and how he chooses to act in one way or another in his coping efforts. Besides, I think humans make many distinctions of importance to their welfare and adaptive functioning including, for example, harm that has already been done, threat of harm, challenge, and the potential for and occasions of positive outcomes. Moreover, we must seek information, plan and prepare, sustain long-term efforts, and relate our coping activities to the complex demands with which we must deal. This is not a matter of a primitive, first-level signaling system to mobilize adaptive energy and bodily defenses, but it has to do with evaluative perceptions, judgment, and thought. Presumably these functions reside to a large extent in the cortex of the brain, not in the reticular system or the hypothalamus.

Let us be reminded that, in order to survive, even so limited a creature as Tinbergen's fowl have to differentiate between the dangerous predator hawk silhouette and the benign goose silhouette, though presumably this evaluative perception and the adaptive emotional behavior that it releases is to a considerable degree wired into the brain of the organisms he studied. When we get to man, the most important sources of stress and adaptation involve much learning and complex mental processes using symbols. Much of psychological stress is anticipatory, or what we ordinarily speak of as "threat." In addition, successful adaptation under threat depends on vigilance, planning, taking preparatory action in the absence of the noxious agent but often in the presence of ambiguous cues, signals, or symbolic referents of some possible future harm. Among the most important and arousing of such harms and threats to humans is damage to something we loosely call the self which is also a motivational system (Hilgard, 1949). Such psychological processes in stress and adaptation have little or no counterpart in past and current research on neurophysiological mechanisms. We are, as yet, babes in the woods when it comes to spelling out this crucial kind of psychophysiology.

Having made this statement, I want now to move on to what I believe is an important debate that has begun to emerge between Selye and Mason on stress and hormonal responses, one which is very clearly generated in this symposium as a result of the report of data from Mason's laboratory. There is a polarity between Selye's highly generalistic position that all noxious stimuli or stressors produce the same pattern of nonspecific bodily defensive responses, and Mason's specificity position that the precise patterning of hormonal secretions depends on the nature of the stressor agent. In the face of Mason's evidence, I think we now have to state that Selye has overstated the generality position, and that this extreme stance clearly needs to be modified if not abandoned.

In thinking about this debate I am reminded of the oscillation of an earlier period between the classical neurophysiological concepts of mass action and localization of function in the brain. When technology had advanced to the point of making possible the study of the functions of limited groups of brain cells, it became increasingly clear that there was indeed a considerable degree of localization of function. Yet the brain surely also operates as an integrated system. Similarly, we have had highly general theories about intellectual functioning, for example, that of Spearman, and much more specific ones such as Thurstone. My hunch is that these are not alternative or mutually exclusive positions, but in some sense they are complementary in that anything can be analyzed in terms of both generality and specificity. That is, every process can be said to have both general and specific features.

In my view, and I think that of most informed psychophysiologists, Selye's General Adaptation Syndrome was a very important step forward in stress physiology, and we can be grateful for his insistance that there were general aspects to the body's response to all kinds of noxious agents. Selye also focused our attention on the adrenal cortex and its role in metabolic processes during sustained stress and adaptation. The notion of stages of reaction was surely a most significant one, and it parallels what we know at the psychological level too where we can observe stages in stress and coping, for example, in the progression of grief and in the changing patterns of coping from one period to the next, although we are still too vague about the details and variations in these. The notion that time is an important dimension in adaptive processes, physiological as well as psychological, has profound implications.

What I think is unfortunate is the territorial manner in which Selye insists on defining all stress as the nonspecific bodily response to noxious agents. Unless one speaks of stress à la Selye, or Selye stress as some do, this possessive definition tends to limit the concept to one level of analysis and to rule out all sorts of stress processes at different levels (for example, sociological stress, or strain if you like) and psychological stress. Most important, this position also restricts our analysis of the specific patterning of the bodily response, and now seems to fly in the face of a growing body of evidence that the reaction to different kinds of stressors is, indeed, quite different, especially if one goes beyond the adrenal cortex as Mason has and looks at a broad spectrum of endocrine reactions. I would like to convince Selye that it is no embarrassment to his great contribution to our understanding of the physiology of stress to recognize the importance of specific hormonal patterning linked to different stressors.

I am much impressed with Mason's modest statement describing his attempt to, in his words, "clean up his independent variables," meaning to eliminate the confounding between physical stressors such as heat, cold, exercise, and fasting, and the psychological significance of these in a sensate, evaluating animal such as man. He has also extended the analysis of endocrine patterning to several neurohumoral systems. His data strongly suggest that the pituitary–adrenal hormonal axis is especially sensitive to psychologically mediated stress, and that when psychological threat is eliminated the various physical stressors show different patterns of hormonal secretion, there being little or no adrenal cortical response. Indeed, the special role of psychological threat in the adrenal corticosteroid response has been suspected and proposed a number of times in the recent past, both by myself, Lazarus (1966, p. 398), and Mason (1971, pp. 328–329).

Mason now seems to have made the issue of generality and specificity an empirical one, capable of being tested and documented. His data appear to favor a high degree of specificity, with the adrenal cortical response of the GAS possibly being restricted to, or at least being particularly sensitive to, psychological threats. He does not find much in the way of common patterns of hormonal response to diverse physical stressors. We need, of course, more evidence, and perhaps future studies will, as I suspect, reveal both generality and specificity in the stress response and the extent and form of each.

Let me proceed now to some of Selye's ideas concerning the use of natural principles of cellular activity in social living, particularly those articulated in his recent book, *Stress without Distress* (1974), and also in his contribution to this volume, although they are not emphasized here. I do not find the attempt to derive social and ethical considerations about how people should live from the cellular level of analysis convincing. For one thing, cellular processes are no more "natural" than are social and psychological processes which social scientists are also trying to understand. For example, it is all very well to say that work commitment is both healthy and desirable—this certainly resonates with my personal outlook—but this misses the point that for large numbers of people, work is either boring or stressful, as Frankenhaeuser has observed in her research. While work may be an important or even essential source of joy in the lives of some of us, it is hardly so for those doomed to work on the assembly lines of a modern industrial society. Thus, while there is much appeal in what Selye has to say about social philosophy, it seems also out of touch with what social scientists observe about the modern social world. The tie between the laws of physiology and the laws of group and individual psychology seems rather tenuous: One does not provide close and functional parallels for the other.

Most important, however, is Selye's recent suggestion in *Stress without Distress* that some kinds of stress, such as that connected with striving or achieving, are good or desirable while others, such as failure, frustration, and hatred are bad. I am not altogether clear about what Selye is telling us, and I would like to pose this as a question involving two alternative possibilities. (1) Does Selye mean that some stressors are not injurious to the tissues because they do not produce the GAS and hence do not lead to the diseases of adaptation? If this is indeed what Selye means, then he is implicitly abandoning his extreme position of generality for a two-factor stress theory, namely, that one kind of stress is potentially damaging to tissues via the GAS, and another kind does not generate the GAS and hence is not damaging. Work and striving illustrate the latter; anger and failure the former. And if this is his meaning, then the next step must be taken of specifying the biochemical mechanisms underlying each. (2) Or does Selye mean that all stressors, both good ones and bad ones, produce the GAS, but the gains in morale and the social benefits of the good ones outweigh the bodily harm that these stressors might do? This alternative says, in effect, that it is worth tolerating or paying the price of diseases of adaptation in the interests of larger considerations of happiness and morality. Whether a stressor is good or bad in this second sense requires a sort of costs-benefit analysis which includes two elements, on the one hand biochemical and tissue damage (costs) and feeling good psychologically or contributing positively to the social system (benefits) on the other. Thus, if one wants to live to the fullest and in most joyous fashion, one must be exposed to certain kinds of stressors even though they produce bodily harm and aging via GAS-produced tissue wear and tear. If we are to understand what Selye is telling us, I think he must indicate whether the former or latter alternative is meant. Each has very different implications for the underlying and psychosocial and physiological stress mechanisms.

Let us turn now to Frankenhaeuser's empirically grounded chapter which constitutes a valuable effort to identify the psychological conditions of catecholamine section. Like Mason, her aim is to separate out, if possible, by both field and laboratory research, the many complex psychological factors contributing to stress reactions as indexed by adrenaline and nonadrenaline secretion. One of the most interesting and important of these variables is the subject's sense of control over environmental conditions. Not only does this variable tie into much recent research and speculation on the importance of the feeling of control over environmental conditions as a factor in the stress reaction (Averill, 1973), it also brings to mind research on control or mastery as a stable property of personality (Gal and Lazarus, 1975), (cf., Rotter, 1966).

Another variable of interest in Frankenhaeuser's research is sex-role identity as a determinant of stress reaction in achievement-centered settings. When the populations were divided on the basis of gender, a sociological variable rather than a psychological one, the problem of whether obtained differences in catecholamines are biological or psychological in basis remains unresolved, as Frankenhaeuser well recognizes. Future efforts along such lines might assess the sex role psychologically, as Bem (1974) has done with her scale of androgyny. In this scale, some men and women are much more male in role outlook, others more female, and still others inbetween, that is, neither predominantly male nor female. Such a psychological differentiation would help to sharpen the study of sex-role identity as a factor in hormonally assessed stress reactions to achievement threats.

Another significant finding in Frankenhaeuser's research concerns the stress impact of work overload and underload. In field research in Sweden she finds that rapidly paced and repetitive machine work, as well as highly monotonous and coerced work, *both* lead to elevated catecholamine secretion. In related research in the United States by Hulin and Blood (1968), however, not all workers reacted positively to efforts at job enlargement which were designed to reduce the sense of boredom and meaninglessness in work. Alienated urban workers looked to rewards outside of work itself, such as social contact, family relationships, income, etc., and were not attracted to efforts to make their work more meaningful. Such an American group was recently reported as reacting quite negatively on a visit to an auto plant in Sweden which was involved in an experiment on altering the assembly line work pattern so as to increase the individual's sense of active participation in the production process. It may be that Frankenhaeuser has overgeneralized her findings to all workers, many of whom no longer subscribe any longer to a strong work ethic. In effect, there may be cultural, subcultural, and individual variations in the extent to which underload is stressful. In any case, this provocative set of data on work underload and overload as a stressor needs to be explored further in other work contexts and populations to determine the extent of its generality and the mediating psychological factors.

The two chapters by Mowrer and Atkinson are more difficult to comment on than the others because, as I have stated, they do not interdigitate readily with each other and with the psychophysiological work on which I have been commenting. I personally found Mowrer's to be, in part, highly metatheoretical and difficult to evaluate. It pursues a theme for which Mowrer has long been well-known, namely, that Freud was wrong in emphasizing that debilitating guilt arises from imagined sins but is, instead, a product of the real failure of the individual to live up to his commitments; in short, objective guilt rather than imaginary guilt is centrally implicated in neurosis. Referring to studies I have not myself seen and which are only briefly alluded to by Mowrer, he reports that both psychopaths and healthy individuals are low in guilt and anxiety while neurotics are high. One of the difficulties I have is in not knowing how the various groups, psychopath, normal, and neurotic, were defined empirically, and how guilt and anxiety were measured. Moreover, nowadays categories such as neurotic or normal seem too broad and all encompassing to be convincing, when we know there are so many varieties of persons and developmental histories represented within such categories. Nor am I convinced that it is fruitful to see a single process, such as objective guilt, as the mainspring of all the diverse forms of neurotic disturbance, a criticism that also applies to Freud who made anxiety over sexuality the core basis of neurotic disturbance.

Nevertheless, Mowrer's rather personal, autobiographical account of how he came to the position he has, and his extension of these views to create a form of psychotherapy that emphasizes getting the individual to take social responsibility for his commitments is most interesting. This is surely a novel and distinctive kind of attack on the anomie and irresponsibility that seems so much a part of modern social living. Although it is difficult to evaluate this approach empirically from Mowrer's report, it seems to make a useful and humanistic contribution to our thinking and to the effort to help people find a more serviceable and ethical way of life.

Finally, Atkinson has given us a clear and interesting summary of a theoretical approach to achievement motivation which has been tremendously productive and useful over a long period. One of his contributions, in my view, has been to take value-expectancy theory which, in the hands of economists and other social scientists, has been exclusively normative (oriented to people in general), and to strengthen it with a personality-based emphasis on individual and group differences. In this he has been one of the pioneers in turning attention to psychological mediators of achievement striving and its inhibition.

My most serious quarrel with Atkinson is his tendency to assimilate cognitive concepts to motivational and emotional ones, while allowing such cognitive concepts to slip into his account almost unannounced as though they were still anathema in the extreme tradition of recent positivism. Thus, he uses cognitively oriented expressions in his account such as "lack of realism," impossible goals and tasks that are "seen" by the person as having important future implications, and so on. Yet these concepts do not appear in his formal theoretical model, which refers mainly to emotion and motivation constructs such as motivation to achieve, fear of failure and success, and the like. If cognitive concepts are needed, then why not use them explicitly and speak about how people view themselves and their environments? One of the major conceptual problems of psychology has long been and remains that of unscrambling the complex, three-way relationships between cognitive, emotional (or affective), and motivational processes. But Atkinson often confuses me about whether he is speaking about motivations, emotions, or cognitions.

It was a great pleasure for me to examine these chapters. Each of the authors is struggling in his own way to grapple with processes underlying stress reactions and adaptation at both the behavioral and physiological level. Mason (1975) has suggested elsewhere that the area of emotion and adptation has tended to bifurcate along psychological or behavioral and physiological lines, and to some extent I think this is true. Nevertheless, we can see in these contributions some real efforts to bring these levels together without one necessarily preempting the other. It is a very valuable feature of this conference that a real attempt has been made to bring together persons whose focus is primarily physiological with those whose concerns are largely psychological. In some cases both levels of analysis are represented in the same researcher. I am grateful for the challenging opportunity to comment on these splendid examples of high-quality research and thought, and to have an input in some of the areas of controversy and uncertainty which they reflect.

REFERENCES

Averill, J. R. (1973). Personal control over aversive stimuli and its relation to stress. *Psychol. Bull. 80*, 286–303.

Bem, S. L. (1974). The measurement of psychological androgyny. *J. Consult. Clin. Psychol. 42*, 155–162.

Gal, R., and Lazarus, R. S. (1975). The role of activity in anticipating and confronting stressful situations. *J. Human Stress 1*, 4–20.

Hilgard, E. R. (1949). Human motives and the concept of the self. *Am. Psychol. 4*, 374–382.

Hulin, C. L., and Blood, M. R. (1968). Job enlargement, individual differences, and worker responses. *Psychol. Bull. 69*, 41–55.

Lazarus, R. S. (1966). *Psychological Stress and the Coping Process.* New York: McGraw-Hill.

Mason, J. W. (1975). A historical view of the stress field. *J. Human Stress 1*, Nos. 1 and 2.

Mason, J. W. (1971). A re-evaluation of the concept of 'non-specificity' in stress theory. *J. Psychiatr. Res. 8*, 323–333.

Rotter, J. B. (1966). Generalized expectancies for internal versus external control of reinforcement. *Psychol. Monogr. 80* (Whole No. 609).

Selye, H. (1974). *Stress without Distress.* Philadelphia: J. B. Lippincott Co.

WORKSHOP II. Psychopathology of Adaptive Learning

Motivation, Anxiety, and Stress

Edited by STEWART WOLF

The early papers emphasized that both the occurrence and character of bodily disturbances associated with the threatening life experiences must be seen as the result of the interaction of several factors, demanding on the one hand and supportive on the other. Moreover the prevailing state of the individual as well as quantitative and timing factors themselves may be crucial to the outcome. So may the degree of novelty or familiarity of the experience. In the case of a repeated experience the important factor may be whether the individual is sensitized or habituated to it. As Selye put it, "The main thing is not what happens to you but how you take it." He further pointed out that intensely pleasant experiences, while often accompanied by pronounced autonomic and endocrine changes are rarely productive of symptoms of bodily disease. Dr. Isaac Marks recalled the well-known observation that individuals with clinically evident anxiety states are not especially likely to acquire psychosomatic illnesses.

There followed a long discussion of the concept of stress during which the utility of the term was brought into question. Selye's insistence on the nonspe-

STEWART WOLF · Professor of Medicine and Physiology; Director, The Marine Biomedical Institute, University of Texas Medical Branch, Galveston, Texas. (Workshop moderated by Stewart Wolf.)

cificity of stress was contrasted with the discrete patterned and quasipurposeful bodily responses that have been shown experimentally in humans to be elicited during situations of intense emotional meaning. Both Mason and Corson urged the measurement of multiple parameters in animal as well as human experiments and held that responses are patterned and not nonspecific. Mason further suggested that many of the reactions attributed to physical and chemical manipulations might occur as a consequence of the accompanying emotional suffering.

Dr. Mowrer introduced the concept of "overload" attributed to Miller and pointed out the similarity to the Holmes and Rahe work with Life Change Scores. This provoked the reminder that the "load" is in the mind of the "loadee." That is the demand for coping or adaptation depends not so much on the nature of the event as it does on its significance to the affected individual in the light of his motivations, aspirations, and goals. As Atkinson put it, "The desperate problem among the behavioral scientists who are interested in . . . the behavior of humans, is their almost infinite capacity for varied behavior." Thus one may not expect any measured bodily indicator to be yoked to a particular type of experience or even to a particular emotion or feeling state. Least of all should one expect a fixed correlation between specific physiological indicators and psychologic test procedures.

Repeatedly, the discussion returned to the concept of stress, elusive because each one had his own idiosyncratic definition. Selye even suggested the term, eustress for positive and beneficial effects of demands on an individual's adaptive mechanisms. Some pointed out a similarity of stress to arousal, others to homeostasis, and still others preferred to reserve the term for pathogenic or potentially pathogenic mechanisms. Most were unwilling to translate the term into measured quantitites of urinary catechols or corticosteroids. It was emphasized that there exist numerous neural and humoral mediators of responses and the point was made that one cannot consider them in an either-or context.

Lazarus maintained "Stress is not a process but a rubric for many processes that occur at a physiological, psychological, or social level, the language of which and the concepts of which change as you go from one level of analysis to another." He recognizes stress when demands are made which tax or exceed the resources of the system.

Mowrer urged more attention to Selye's moral philosophy, "Earn your neighbor's love" as a contribution to the understanding of the basic interdependence of human beings and the importance of human relationships to health.

We are gradually learning that what we call health and disease reflects a balance of particular psychologic and physiologic regulatory processes that operate through pathways that are enormously complex. The circuitry contains excitatory and inhibitory neurons that are subject to a variety of feedback and other influences at several levels of organization in the nervous system and in relation to locally elaborated or circulating humors. When the regulatory processes are comfortably balanced and adaptive, we speak of health. Disease occurs when there is too much or too little of some adaptive element. For example, a heart rate of 120 in a track man immediately after a 100-yard dash would not be considered abnormal, but in a bedridden or even sedentary patient the same pulse rate would be called tachycardia and would be considered a clear evidence of illness. Similarly the various manifestations of psychopathology such as suspiciousness, boisterous behavior, or unwillingness to speak may be healthy behaviors when they assist the individual in coping with his life experiences. Otherwise they are evidences of illness. Moreover, the classical grief reaction without an antecedent loss would be considered abnormal, but so would its absence following the death of a loved one. In the healthy subject, therefore, we speak of responses such as changes in blood pressure, body temperature, and a host of other indicators including emotional and cognitive changes as being "within normal limits." These same bodily processes become abnormal or pathological when, in relation to the exciting situation, they are excessive, insufficient, inappropriate, or unduly prolonged.

The quality and relevance of the exciting situation are of more concern than its quantity. A faint odor or taste, a fleeting sight or sound, a seemingly innocuous word or phrase may set off an untoward chain of events. Thus the effects of a symbolic stimulus on the integrative processes of the brain are much more difficult to evaluate than the stimulus itself. Similarly, the degree of response in the body's immune apparatus elicited by an invasion of microorganisms depends not so much on their quantity as it does on the nature of the organisms and the state of the host at the time, as well as any past exposure. Thus, although quantity is a factor in the equation, it does not alone determine the presence or absence or even the severity of an infection. The quantity of stimulus, while important, is not therefore crucial. With either microorganic or symbolic stimuli, one need only establish their relevance to the particular effects observed, before undertaking the exploration of the processes involved.

The concept that emerges from all this is that the tissue changes characteristic of disease are the result of basically normal bodily processes gone wrong, exaggerated, insufficient, or inappropriate in some way. Health is manifested by a behavior of bodily systems that achieves and maintains a comfortable interaction or relationship with the environment. Haldane stated this principle when he said that progress in medicine depends on understanding how the human organism adapts to changes in the environment. Thus, the healthy person increases his red blood count when living at altitude but not at sea level. Polycythemia developed at sea level spells disease. The bodily mechanisms required to increase the number of circulating red blood cells, however, are identical in health and disease. So it is with other bodily systems. For example, the difference between infection and mere exposure to microbes depends on a neat balance of activation and restraint of immunological and other defense mechanisms. Even in an extensive epidemic there is always a healthy segment of the population. Inadequate modulation of immunological function appears to be responsible for the exaggerated immune behavior of certain connective tissue diseases such as lupus erythematosis and rheumatoid arthritis. Disease, then, may reflect too much or too little of certain adaptive functions, resulting in essentially inappropriate physiological behavior.

While multicausality is now a widely accepted concept, it is still inadequate to the task of explaining the behavior of the human organism in health and disease. What is needed is a concept of interaction between the person and his social environment, one that takes into account the mechanisms that manage the adaptive behavior of bodily structures. Forces acting along several vectors and involving many structures in the brain as well as in the rest of the nervous system determine the outcome of any stimulus or group of stimuli. The frequent association of manifestations of physiological dysfunction with overt emotional disturbance has led to the widely accepted but confusing proposition that emotions are the cause of bodily reactions. On the contrary, bodily changes are not "caused" by emotions but derive from the individual's evaluation of his experience, consciously or unconsciously, and with or without overt emotional expression. The basic element is the fact that the autonomic excitatory and inhibitory pathways that regulate visceral function are subject to influence by neuronal circuits in the forebrain, neuronal interactions that subserve the interpretation of life experience.

Much evaluative function occurs in the human brain without conscious awareness, even during sleep. A mother may awaken to the faint cry of an infant but sleep through a much louder noise that lacks important meaning to her.

Unconscious mental activity may, indeed be much more precise that that undertaken during awareness. A fully alert man may miss by an hour or more in judging the time of day and yet discipline himself to awaken within a minute of a set time.

Important emotional conflicts linked to bodily disturbances or disease are likely to be shunted out of awareness for the general comfort of the individual. The information about such conflicts remains in the brain, however, perfectly capable of being recruited on appropriate stimulation and of entering the complex process of behavior. Thus threatening circumstances may evoke behavioral responses with or without awareness of the stimulus and with or without conscious fear, anxiety, or resentment. For example, a young man returning to his home town after years of living elsewhere, suddenly noted violent nausea as he turned to walk through a short alley. The sensation subsided but troubled him since he was aware of nothing that could have aroused the nausea. Only later, recounting the incident to his physician did he suddenly recall that his grandmother's house had been on the alley. He further remembered with intense emotion that as a child he had been repeatedly sent to stay with his grandmother when his father came home drunk and physically abused his mother.

As more and more has been learned about excitatory and inhibitory influences, about facilitatory and inhibitory regulation of synaptic transmission, the concept of the reflex nature of bodily regulation has given way to a concept of neural interaction in which virtually all parts of the nervous system are interconnected so that local perturbations may have widespread effects. Rich interconnections between somatic sensory, visceral sensory, and the effector neurons of all sorts have been discovered that link many zones of the central nervous system, including thalamus, hypothalamus, and limbic cortex with the frontal lobes. The extent of interrelatedness of all of these structures in the formulation of the behavior of people not only has led to the discarding of the too simplistic reflex concept of regulation but has made it clear that the somatic and visceral pathways are not two systems after all but a single system with different kinds of neuronal hookup in a state of continuous dynamic interaction. Sometimes the requirements for adaptation are conflicting, and sometimes the bodily response is either insufficient or exaggerated; hence an imbalance with the potential of tissue damage, disability, and even death.

It follows that investigations at the molecular, cellular, and tissue level that have contributed so much to the rapid progress of our understanding must now give way to a greater emphasis on studies at the organismal level, studies of the whole conscious behaving organism, preferably man. There will then evolve a clearer understanding of what we know now, that all parts of the organism are interdependent and that the adaptive behavior of the viscera, like that of skeletal muscles, becomes ultimately a matter of the needs, goals, and purposes of the individual.

Clinical Modification of Behavior

Sources of Stress in the Drive for Power

DAVID C. McCLELLAND

Physiologists, psychiatrists, and psychologists have long been interested in the impact of stress on the organism. Years ago Selye (1936) identified a general adaptation syndrome which characterized the way the body responded to any demands made upon it. Eyer (1975) has described the syndrome this way. "Stress . . . leads to psychological and physiological arousal. Arousal consists of a series of internal changes that prepare the body for 'flight or fight' . . . among the acute changes (over seconds, minutes) are: a rise in heart rate and blood pressure; changes in the distribution of blood, e.g., more to brain and muscle, less to skin and stomach; a release into the blood of energy producing compounds such as glucose and fatty acids . . . both the acute and restorative changes are initiated and coordinated by the brain through its control of the autonomic and endocrine systems. For example, during the acute response the changes in blood pressure, blood flow, and heart rate are mediated by activation of many parts of the sympathetic nervous system and corresponding suppression of activity in many parts of the parasympathetic system. During this period there is also heightened release of such hormones as cortisol, epinephrine, norepinephrine, growth hormone, thyroxine, etc., from endocrine organs. At the same time release of other hormones, for example, insulin and sex hormones, is suppressed. The net effect of these hormonal changes is an accelerated breakdown of carbohydrates, fats, and proteins to provide energy (catabolism) and a slowing of the body's synthetic processes (anabolism)." Repeated stress and

DAVID C. McCLELLAND · Harvard University, Cambridge, Massachusetts.

mobilization of the body in this way eventually predisposes the individual to all sorts of pathology, particularly cardiovascular diseases (cf. Levi, 1971). People suffering from essential hypertension and heart disease in turn are described frequently as having a particular behavioral style which suggests chronic sympathetic activation. They display "extremes of competitiveness, striving for achievement, aggressiveness (sometimes stringently repressed), haste, impatience, restlessness, hyperalertness, explosiveness of speech, tenseness of facial musculature, and feelings of being under the pressure of time and under the challenge of responsibility" (Jenkins, 1971). Psychiatrists studying essential hypertensives (individuals with chronic high blood pressure) conclude "that they live under a permanent life stress from which they are unable to free themselves, that they are sinking under the burden of responsibility . . . that they are unable to cope with their duties . . ." (see Brod, 1971, Jenkins, 1971). Similarly, Wolf (1971), notes that individuals susceptible to heart attack seem to be forever struggling very hard to live up to the demands of a situation which they cannot quite satisfy.

These personality characteristics of people suffering from stress and its associated pathologies caught my attention for they seemed to me to be very similar to the personality characteristics of individuals high in the need for Power whom Winter (1973) and I (see McClelland *et al.*, 1972) and others have been studying for the past 25 years. To explain the significance of this possible linkage between research findings in two quite different fields, it is necessary to explain just what we mean by the need for Power and in particular just how we go about measuring it. For, from the psychologist's point of view, descriptions of personality characteristics such as those listed above, are lacking in precision. They employ psychological terms too loosely, terms which overlap in meaning and are not clearly differentiated from each other conceptually or by way of the measures employed to assess them. For example, "competitiveness" and "striving for achievement" are not more or less the same thing as Jenkins implies. Our research has shown that it is important to distinguish between them, for they are aroused in different ways and have quite different effects on the way the person behaves. If we are to understand stress better, we must get a more precise fix on the psychological variables associated with it.

The technique used for developing a measure of the need for Power was to expose individuals to a variety of "power arousal" situations and to look for their effects on the most sensitive behavioral indicator available, namely on the fantasies the individuals produced under arousal as compared with more neutral or controlled conditions. Various types of power arousal were tried; student leaders were tested while waiting to hear whether they had been elected to an office they were seeking, others just after they had been given control over another student who was serving as a subject. And still others after they had been exposed to the stirring experience of a filmed presentation of President John F. Kennedy's inaugural address. In all cases imaginative stories written after these various types of power arousal were compared with stories written by comparable students under "neutral" conditions, the idea being that we were trying to compare their thoughts sampled under different conditions in much the same way as a physiologist might collect blood samples to detect the effect of a particular agent. Gradually we developed a coding system which differentiated quite precisely between the kinds of thoughts people have under normal conditions and under conditions of Power Arousal. The way in which this coding system was derived and elaborated very carefully over a period of years has been described by Winter (1973). It is a completely objective system in the sense that coders can be trained to agree almost perfectly in picking out power thoughts and distinguishing them say from achievement or other types of thoughts. It is valid in the sense that it differentiates very significantly between those whose power motivation is aroused or not aroused.

The key element in what we came to call the coding system for need for Power (*n* Power) is whether or not any thoughts are expressed which signify a desire "to have impact." Such thoughts include: arguing, agressive acts, attempts to persuade other people, trying to control another person, trying to impress somebody, or even helping someone if the help has not been requested. What interested us particularly was how people would behave who thought along these lines all the time rather than just in response to a power arousal situation. To find out we simply compared how people high in *n* Power under neutral testing conditions behaved as compared to those who scored low in *n* Power under the same testing conditions. A large number of such studies have been conducted, the results of which have been summarized in Winter (1973), McClelland *et al.* (1972), and McClelland (1975). In general men who score high in *n* Power tend to be more argumentative and aggressive; they engage more often in competitive sports; they are sexually more active; they accumulate prestige supplies like fancy clothes and cars, and they tend to join organizations and ally themselves with others who have influence. In other words, they seem to show many of the signs of greater aggressiveness and competitive striving that students of the stress syndrome have said characterize individuals who are susceptible to coronary heart disease and essential hypertension. Could it be *n* Power that makes life so much more stressful for such people?

Two findings of a physiological nature implicated *n* Power strongly in the stress syndrome. In a very extensive series of interrelated studies (McClelland *et al.*, 1972), we showed that high *n* Power, particularly if it is combined with low inhibition, is associated with heavier alcohol consumption. The relationship seems to be quite general in the sense that it exists not only for individuals whose power motivation and inhibition scores were obtained from their written imaginative stories, but also for cultures which score high in power orientation and low in inhibition in their folk tales. That is, such cultures consume significantly more alcohol than cultures with other orientations in their folk tales.

Furthermore, consumption of alcohol in experimental party settings increases power fantasies so that we were able to argue that what a person with high power motivation is seeking when he drinks is a kind of power "rush," a feeling of activation, excitement, or increased strength. Alcohol produces this power "rush" in two ways: first by providing a source of energy in the form of lipids which dissolve almost immediately through the stomach wall into the bloodstream and second by increasing epinephrine secretion (Myrsten *et al.*, 1971). Epinephrine in turn is the major agent in producing the general activation of the body associated with stress, e.g., increased heart rate and blood pressure and accelerated production of energy-producing compounds like glucose from the liver. To put the matter simply, it looks as if people with a certain kind of power motivation act by drinking alcohol (particularly distilled spirits) in such a way as to produce the physiological response normally associated with stress. In popular terms, they are seeking the physiological "rush" that also comes from stress, but under more benign conditions.

The other physiological finding is that among a group of men aged 25–50 from all walks of life, those with high *n* Power reported that they had more difficulty sleeping at night ($r = 0.28$, $N = 85$, $p < 0.05$). The effect was even more marked among those who were high in *n* Power and also high in inhibition ($r = 0.41$, $p < 0.05$) as contrasted with low inhibition ($r = 0.18$, n.s.). Excess epinephrine secretion seems implicated, for it is known that people suffering from insomnia show a high level of epinephrine production (see Myager, 1971). It stands to reason that people in a constant state of physiological arousal associated with a high level of epinephrine secretion would have difficulty relaxing enough to go to sleep. Could it be that men high in *n* Power develop a chronically high level of epinephrine production that makes it difficult for them to sleep?

But as Wolf remarks in another connection (1971) the major difficulty in relating psychosocial or psychological variables to stress-related pathologies like heart disease is "not a matter of leads or likelihoods; the problem is to martial firm evidence." Can we establish a more direct link between *n* Power and epinephrine secretion? Fortunately, Steele (1973) has conducted an experimental study which goes far toward providing firm evidence of such a linkage. He was interested in the relationship of physiological activation to various motivational states as measured in the usual way through content analysis of thought samples. His main hypothesis, derived from Schachter and Singer's theory (1962), was that physiological arousal would be elaborated into or associated with either a power or achievement motivational state depending on the cognitive context in which the arousal took place. So he designed four experimental conditions. In the first or Power Arousal condition he attempted to arouse power motivation by having the subjects listen to a tape of famous inspirational speeches. In imaginative stories written after exposure to the tape, subjects scored significantly higher on *n* Power than subjects writing stories in any of the other conditions. In the second or Power Control condition, the subjects listened to a series of tape-recorded travel descriptions. In the third or Achievement Arousal condition the subjects were asked by a very serious test administrator to perform a number of tasks which were described as measures of intelligence, critical abilities, and intellectual alertness. The same tasks were administered in the fourth or Achievement Control condition except that the subjects were told by an informally dressed undergraduate that their performance was not critical, that their cooperation was needed only in order to ascertain if there were any problems in the format of the test.

Steele's main interest was in how changes in physiological activation among these different conditions related to motive scores. He employed two different measures of activation, the first being Thayer's adjective check list in which the subject is asked to describe how he feels "at this moment" in terms of such adjectives as sleepy, intense, wide awake, or quiet. The second measure was of catecholamines excreted in the urine, following the procedure developed and used extensively by Frankenhaeuser and associates (cf. 1973). Urine samples were collected on three occasions, the first on a day previous to that of the actual experiment to get them used to the procedure, the second on the day of the experiment an hour and half after arriving in the laboratory to get a base line measure, and the third at the end of the experiment. The crucial question is how increases in epinephrine in the urine from the second to the third measure related to motive scores obtained from stories written after various experimental conditions.

The main results are summarized in Table 1. As expected epinephrine production increased more in the Power Arousal condition than in the other conditions, but the most important finding involves the association of increased sympathetic activity with motive content. In the Power Arousal condition, individuals whose epinephrine output increased more had significantly higher *n* Power scores at the end of the session. Yet their *n* Achievement scores did not similarly increase. Physiological, emotional arousal was not associated with an increase in *all* motive scores. The reverse was true in the Achievement Arousal condition. There individuals whose epinephrine output increased more did not display higher *n* Power scores, but if anything tended to have higher *n* Achievement scores, although the last correlation is not significant. What is not altogether clear is why the same results seem to hold to a lesser degree of significance for the two control conditions. However, in the main, Schachter's hypothesis is confirmed: Under conditions of power arousal physiological activation (epinephrine production) is associated with power motivation and not achievement motivation whereas the reverse tends to be true under conditions of achievement arousal. The cognitive elements in the situation—whether they have to do with power or achievement—determine how the physiological activation will be elabo-

Table 1. Relation of Epinephrine Increase to Motivational Content After Different Types of Experimental Arousal

Condition	N	Mean gains in epinephrine secretion (ng/min)	Correlations of epinephrine change with motive scores	
			n Power	n Achievement
Power arousal (stirring tapes)	16	+1.51	0.71[a]	0.04
Power control (travel tapes)	14	+ .27	0.38	0.06
Achievement arousal (ego-involving tasks)	19	+ .35	−0.02	0.25
Achievement control (relaxing tasks)	14	+ .47	0.10	0.36

[a]After Steele (1973). $p < 0.05$.

rated in the motivational thought patterns of the individual. Furthermore, epinephrine output is much more significantly associated with power arousal than it is with achievement arousal. One might suspect that achievement motivation arousal may not be associated with increased epinephrine secretion at all: The low and insignificant correlation shown in Table 1 might result, if it is to be taken seriously at all, from the fact that achievement arousal also contains some elements of power motivation arousal. Telling someone that you are testing his intelligence may arouse his power motivation as well as his achievement motivation.

It is reasonable to infer further that a person with a strong dispositional need for Power (a trait measure) might have a chronically elevated level of epinephrine production. Steele obtained data which can be used to check this hypothesis. In most of the conditions the measure of *n* Power obtained at the end of the session cannot be regarded as a measure of dispositional *n* Power because it was obviously influenced by the experimental conditions that preceded it. However, the *n* Power score obtained in the Achievement Control condition may be considered a measure of dispositional *n* Power for several reasons. Nothing was done in this condition that should have aroused power motivation, so as to contaminate the measure with temporary states of motive arousal. In fact, the average *n* Power score of the subjects was very close to what it was among a comparable group of students who took the test under neutral conditions. Furthermore increases in epinephrine output were not associated with higher *n* Power scores, as in the Power Control condition, which would suggest arousal effects were operating. And only in this condition did the *n* Power score correlate even at a moderate level with an alternative measure of dispositional *n* Power obtained from an activities index. So if we may assume that the *n* Power scores in this condition represent a *dispositional n* Power, what is the relationship of these scores to the *resting level* of epinephrine production? The base level output of epinephrine is obtained from the second urine sample and it correlates 0.62 ($p < 0.05$) with the *n* Power scores for subjects in this condition. Furthermore, its correlation with the *n* Achievement score in this condition is only 0.05, strongly supporting the inference that it is only people high in the need for Power (not those high in any motive trait) who tend to be in a state of chronic physiological activation, to judge by their resting level output of epinephrine. Furthermore, the *n* Power score is often divided into two components, representing the kind of power motivation that is associated with low inhibition (and high alcohol consumption) which is labeled *p* (for personal) Power, or high inhibition which is labeled *s* (for socialized) Power. Among the subjects in the Achievement Control condition *s* Power scores correlated 0.64, $p < 0.05$ with resting level of epinephrine production and *p* Power scores only 0.20, (p = *n.s.*) with the same measure. In other words those with a *controlled* need for Power have a high chronic level of epinephrine output, not those with an impulsive power need. Perhaps the latter express their power drive more readily in action which "uses up" the epinephrine so that it does not accumulate in the bloodstream and the urine. Such results are certainly consistent with our discovery that people high in *n* Power and in inhibition have difficulty sleeping. For it is just those people who have a higher resting state of epinephrine input which ought to interfere with their sleeping.

It is time to organize these bits of evidence into some overall scheme linking power stresses to pathologies of the circulatory system. Figure 1 outlines the supposed sequence of events. Power stresses increase epinephrine output,[1] as any number of experimental studies has demonstrated (cf. Frankenhaeuser, 1973). Increased output of epinephrine in a power context is associated with a higher *n* Power score—a state measure. If this sequence of events is repeated over and over again, the individual should develop a higher dispositional need for Power—a trait measure—which if the person is also high in inhibition, is associated with chronically elevated epinephrine production which makes the person susceptible to such circulatory disorders as essential hypertension and heart attacks. Another element in the picture is sketched in at the upper right-hand corner of Fig. 1. Power stress should have a greater effect on individuals who are already high in the dispositional need for Power. In Steele's study this hypothesis could be checked only indirectly since he had no direct measure of a dispositional need for Power in the Power Arousal condition. Instead, he used an index based on activities the subjects reported that they engaged in, which other

[1] They also cause the release of other stress hormones, like cortisol, from the adrenal cortex, but to simplify the discussion we will confine our discussion to sympathetic arousal and epinephrine. In any case the effect of the other stress hormones released is also to mobilize the body for action.

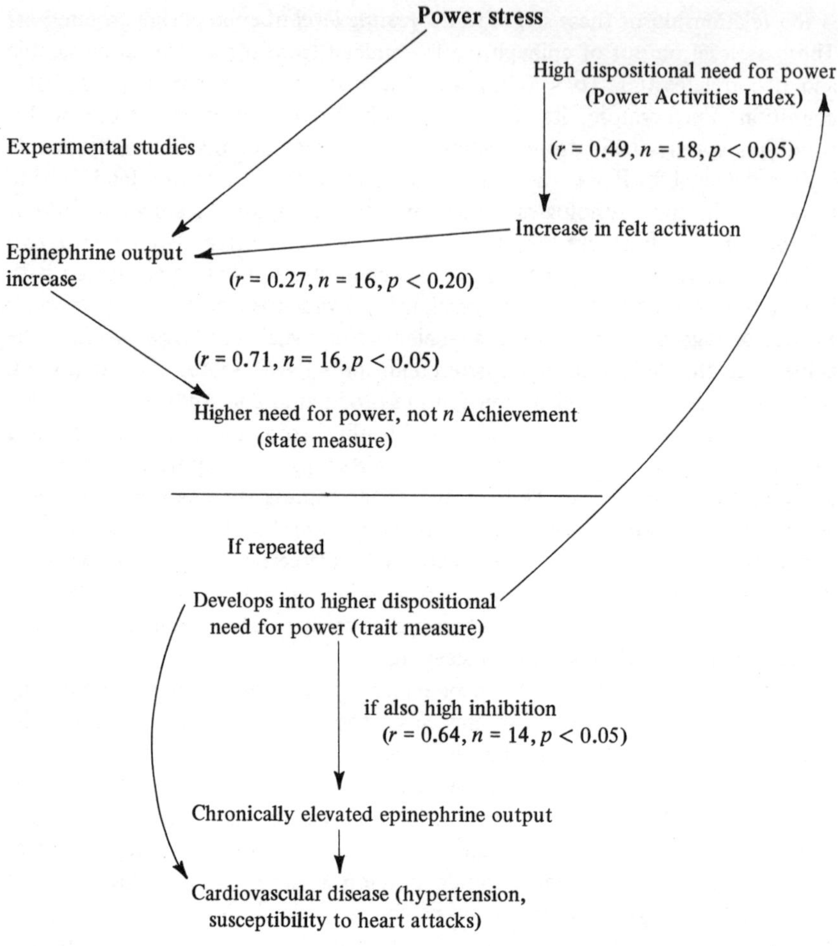

Fig. 1. Relation between power stresses, power motivation, physiological activation, and susceptibility to hypertension and heart disease.

studies have shown are highly correlated with dispositional *n* Power. People high on the power activities index checked more adjectives indicating they felt more energized or activated in the Power Arousal condition than subjects lower on the power activities index but for some reason they did not display a significantly greater increase in epinephrine–perhaps because everyone's epinephrine output was at a higher level under conditions of power stress. At any rate felt activation *is* related to epinephrine output increase and there is ample theoretical and empirical support for the belief that people high in dispositional *n* Power are more sensitive to power stresses and more activated by them. Furthermore, it is important to note that the reported increase in felt activation for subjects high in dispositional *n* Power occurred *only* in the Power Arousal condition. It occurred in none of the other conditions including the Achievement Arousal condition. That is, it is not *any stresses* which makes power-oriented individuals feel more aroused: It is only power stresses.

It is at this point that the research on motivation may be able to introduce more precision into the conclusions drawn from the very extensive studies of stressors which increase adrenalin output. For what Steele's data suggest is that it is a more limited type of stress that increases adrenalin output than most scholars in the field now believe to be the case. At first glance, it appears that almost any kind of psychological stress increases epinephrine output: I.Q. testing, sexual films (in males), electric shock, vigilance tasks, physical work, admission to a hospital, airplane flying, noise, embarrassing interrogations, and even viewing a pleasant movie (see Raab, 1971; Frankenhaeuser, 1973). It is small wonder that Frankenhaeuser concludes "the empirical data clearly show that any event which is perceived as emotionally arousing . . . will generally be accompanied by increased adrenalin output" (1973, p. 13). In view of this variety is it possible to argue that it is primarily, and perhaps only, power stressors that give rise to increased epinephrine output? To begin with, it is clear that most of the stressors just listed do involve a demand for increased power from the subject whether it be to be alert and vigilant so as to perform a task better, to withstand an electric shock or prolonged physical work, or to cope with the threat of failure on an intelligence test or the physical danger involved in parachute jumping or flying. But some of the emotional stimuli seem not to involve power stress at all—such as the sexual film, or a pleasant film evoking laughter, or a pleasant game. On closer examination, however, even these instances have strong elements of power arousal in them. Sexual stimulation in males is closely associated with power motivation (see Winter, 1973). The pleasant film used was *Charley's Aunt*, a slapstick comedy in which there are many aggressive cues, and the pleasant game employed was Bingo which is certainly competitive, particularly when the players could actually win some money. Looked at from this point of view, it appears that earlier investigators of stress almost unconsciously chose power stress situations as most likely to increase adrenalin production.

But the critical question then is: What emotionally arousing situations do *not* increase adrenalin output? The evidence cited from Steele's experiment in Table 1 suggests that a pure achievement arousal situation does not significantly increase adrenalin output, but it is difficult to separate completely the two types of arousal. Almost any achievement arousal involves elements of power stress as well. Far more convincing is the evidence that affiliation arousal leads to a *decrease* in epinephrine output rather than an increase. Frankenhaeuser reports that "individuals with depressive tendencies showed a relatively weaker adrenalin reaction when required to perform under disturbing external conditions" (1971, p. 33). And Myager reports "grief resulted in a decline of the adrenalin content. During crying there was a sharp decline of the catecholamine content in blood, substantiating the view that crying can be considered a protective mechanism against emotional 'overstress' " (1971, p. 258). In other words, it seems highly unlikely that the emotional arousal which comes from grief or loss of a loved one would be associated with adrenalin increase. What has happened is that the term "stress," as it has been operationalized in the laboratory, has come to mean "power stress." While not many studies have been done which have attempted to introduce stress in other motivational systems, it seems clear that not all emotionally arousing states increase epinephrine output.

SEX DIFFERENCES IN POWER STRESS

Another strong reason for believing that power demands are especially involved in the stress syndrome is to be found in the data on sex differences in epinephrine production. Frankenhaeuser and other investigators have been somewhat puzzled by the fact that the kinds of stress like intelligence testing which regularly increase adrenaline excretion in males do not do so in females (1973, p. 19). This result has been found often enough for Frankenhaeuser to conclude tentatively that "adrenal-medullary activity is a less sensitive indicator of behavioral arousal in females than in males" (1973, p. 21). Why should this be so? The interesting fact is that the evidence reviewed above and summarized in Fig. 1 was all collected on males and where similar data are available on females, the same relationships do not hold. For example, the dispositional *n* Power score among women is not associated with difficulty in sleeping ($r = 0.03$, $N = 115$, $p = n.s.$) nor is it associated with heavier drinking (see Wilsnack, 1974) nor are females who are high in *n* Power more aggressive, argumentative, or competitive so far as their activities are concerned. Finally, the cardiovascular pathologies, such as heart attacks and hypertension, are much less common in women than they are in men. It has long been noted that women live longer than men on the average and as Eyer concludes (1975), "the sex differential in the total death rate has widened with development because women have benefited more from public health and medical measures reducing infectious disease while suffering less rise of coronary heart disease, hypertension, ulcers, suicide, cirrhosis and other stress related causes. These data clearly imply that there has been a disproportionately large rise of social stress on men in modern development."

It begins to look as if the chain of events outlined in Fig. 1 linking power stress to cardiovascular disease holds only for men: The chain of cause and effect is somehow broken for women. But where and why? Two types of explanations readily present themselves. In the first place, women may simply be constructed differently from men, as Frankenhaeuser suggests. According to this explanation, power demands on men and women are the same, but the female adrenal-medullary system is simply less reactive: They do not respond with increased adrenalin output the way men do. Evolution may have built in some protective biological mechanism which prevents the development of the chronic high levels of adrenalin secretion that seem to be responsible for circulatory pathologies in men.

The other explanation is psychosocial, rather than biological. It assumes that both role requirements and the nature of his physique make it more likely that more demands for strong, powerful action will be placed on the male of the species. If in fact the male is more subject to power stresses than the female, then the sequence outlined in Fig. 1 is more likely to develop a strong dispositional *n* Power in men with all the accompanying pathological dangers as outlined. According to this line of reasoning, men may not start out in life more physiologically reactive to stress than women, but since they are regularly subjected to more *power stressors*, they may be more apt to develop the dispositional *n* Power which in turn makes them more sensitive to further power stresses. If this is so, then men should on the average have higher *n* Power scores than comparable women do—which in fact is the case, although the difference in scores cannot be interpreted unambiguously, because one cannot be certain that the pictures used to elicit stories have exactly the same power cue value for both men and women (see McClelland, 1975). But even if men do not have a higher average *n* Power score, few would disagree that their social role will often demand strong, assertive responses from them which in turn lead to the somatic mobilization of energy governed by the adrenal-medullary system. Thus, according to this argument, males are more likely to die of heart attacks because they are more subject to *power* stresses, not because they are more subject than women to all kinds of stress. So a distinct advantage of the hypothesis that it is primarily power stressors that increase adrenalin output is that it explains the sex differences in mortality due to heart attacks and the like.

But can we link the series of events outlined in Fig. 1 more directly to various cardiovascular pathologies? While most researchers are still somewhat cautious in drawing conclusions about human functioning, evidence from experimental studies with animals show a very clearcut connection between chronic stress and heart disease and high blood pressure. For example, Lapin and Cherkovich (1971) have demonstrated how prolonged immobilization or interruption of diurnal rhythms in monkeys produced neurosis which in time resulted in chronic high blood pressure, coronary insufficiency, and eventually myocardial infarctions. Selye (1971) has reported an extensive series of studies showing how stress hormones can produce cardiovascular diseases. Can we establish a connection between motivational dysfunctions and such pathologies more directly? Unfortunately the definitive study has not yet been done which determines whether individuals with cardiovascular disease are high in *n* Power and high in inhibition. However, there is a lot of indirect evidence that such a linkage should occur. For example, the differential susceptibility of men and women to cardiovascular disease is indirect evidence suggesting that the difference may be due to the fact that power stress and power motivation are less strong among women on the average than among men. Even more persuasive is a line of reasoning which grows out of the recent series of studies on the pathological effects of stress-inducing life changes. Rahe (1972) has developed a scale for measuring degree of life stress based on assigning numerical values to any life change which has occurred say in the last 6 months of an individual's life. Losing one's job receives a high score on the scale while being absent from work because of illness receives a much lower score. Most of the stresses, though not all, can be classified as power stresses in our terms, in the sense that they involve threats to the person's power to cope—such as loss of a job, illness, financial setbacks, trouble on the job, etc. Further work would have to be done to separate out power stresses from, say, affiliation stresses if our hypothesis is to be carefully checked that it is primarily power stresses that lead to cardiovascular difficulties. But at any rate with Rahe's very general measure of life stress, various studies have demonstrated that the greater the individual life change score, the more susceptible he is to accidents and diseases of various types. Of particular importance here is the finding by Theorell *et al.* (1972) that the mean weekly level of life changes is positively correlated with a person's mean 24-hr level of epinephrine secretion. Lundberg *et al.* (1973) report further that patients suffering from heart disease have higher levels of recent life changes than normal. In other words, in humans as in animals, the greater the recent stress, the greater the production of a stress hormone like epinephrine, and the greater likelihood of cardiovascular disease.

What is more, Boyatzis and Dailey (1975) have found in fairly large samples of adult men and women that the life change unit score (measuring stress) for the past six months is significantly correlated with reports of the number of antisocial aggressive acts committed and the frequency of drunkenness, both in men and women. What is important about this finding for our present argument is that *n* Power has also been empirically linked to these same two effects—namely aggressive antisocial acts and heavier drinking (McClelland *et al.*, 1972). Figure 2 summarizes the general line of argument which seems to follow from these interconnected findings. Both life changes and the need for Power have been associated with increased production of epinephrine, a stress hormone. They are both also connected with more aggressive acts and heavier drinking. Stress hormones and life changes have both been connected with cardiovascular disease. All that is missing in the argument is empirical evidence that *n* Power is

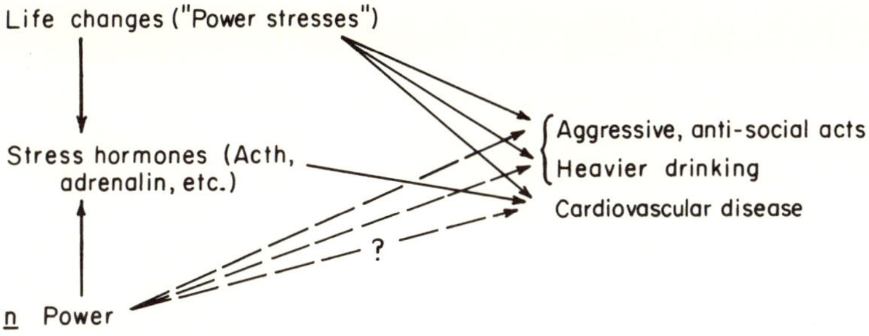

Fig. 2. Relation of life changes and the need for Power to pathology in humans.

also connected with heart disease. It seems reasonable to assume that it will turn out to be.

National statistics do implicate the need for Power in cardiovascular disease. It has been argued by Eyer (1975) and others that increasing death rates from heart disease are due to the increased stresses of modern life. For example, he notes that death rates from heart disease have increased from 68/100,000 in 1860 to 562/100,000 in 1960 in Massachusetts. One cannot, of course, attribute such an increase, as he recognizes, entirely to such social factors as stress because death rates from other causes like the infectious diseases of infancy or tuberculosis have dropped sharply. It may be that people who would have died in 1860 of tuberculosis or some infant disease live long enough in 1960 to die of a heart attack. Despite the difficulty in interpreting the meaning of disease specific death rates, it seemed worth examining them among different countries which have roughly the same exposure to modern medicine. It would obviously be absurd to compare the death rates from heart disease in say India and Sweden, since in India, as in Massachusetts in 1860, deaths from the kinds of diseases which have now disappeared in Sweden would make it much less likely that an Indian would survive long enough to develop heart disease. On the other hand, it seems reasonable to compare death rates from heart disease in say Sweden, the United States, and Japan, since such countries have at least in recent years employed methods of modern preventive and curative medicine rather widely. The statistics on disease specific death rates published in the United Nations demographic yearbook are limited anyway to those countries which have a well enough developed system of modern medical practice to collect the data. So it seems reasonable to compare death rates from heart disease and essential hypertension among these countries according to whether the countries are oriented strongly or not toward power and inhibition. Elsewhere (1961, 1975) I have explained how these motivational indexes were obtained, by coding popular literature (i.e., in children's textbooks) in the countries, and how these indexes can be used to predict what happens subsequently in the same countries. Generally speaking, I have found it necessary, both on theoretical and empirical grounds, to allow some time lag before assessing indicators of changes in collecting behavior. In the present instance I assumed that levels of power motivation and concern for inhibition around 1950 ought to predict death rates from heart disease and hypertension 15–20 years later, after the individuals exposed to these orientations had had time to develop and die of the diseases involved. It is not clear what the level of various motivational variables in popular literature represents. It may reflect average levels of motives in individuals or it may reflect the pressures these individuals feel from the general orientations of the culture, which they would pick up from reading newspapers or other popular literature. But in either case the prediction would be the same: Death rates from heart disease and hypertension should be higher after a time in countries where power motivation is high and is combined with a high concern for control of action or inhibition.

Table 2 assembles the data in a form that permits testing this hypothesis. Death rates were included for such countries as Taiwan and Chile with some hesitation because it is difficult to be sure that death reports and public health practices are comparable to what they are elsewhere, but in the end every country save one was included in the table on which both death rates and motivational data were available. Only Mexico was omitted because it had a much higher proportion of deaths unclassified than the other countries and because the figures reported seemed way out of line with those reported for any other country, being much lower. This strongly suggested either that Mexicans were still dying predominantly of diseases controlled in other countries or that methods of reporting were very different. The results for the remaining countries strikingly confirmed both hypotheses, even though the statistics are obviously not as standardized as one could wish for. The countries in the upper right-hand quadrant, above the world average in need for Power and for control, have a higher average death rate from heart disease than the countries in the other quadrants although the average is almost as high for the countries high in need for Power and low in the need for control. Generally speaking the countries low in the need for Power have lower death rates from heart disease, but the effect is most marked for those low in need for Power and high in need for control. In fact, if one compares the rank orders of death rates from heart disease of the countries in the high high quadrant from the low high quadrant, the difference is nearly significant by the Mann Whitney U-Test. What this finding suggests is that to avoid heart attacks it is best to have a low need for Power and to be careful and lead a controlled life. On the other hand what is worst for producing heart attacks is a strong drive for power combined with a very high control mechanism which bottles up the anger, assertiveness, and physiological activation mobilized by the need for Power. This picture, of course, confirms our

Table 2. Death Rates per 100,000 Inhabitants in Various Countries from Heart Disease (A) and Hypertension (B) Around 1968 as a Function of Need for Power and Control (1950)[a]

	Need for E Control					
	Low			High		
		A	B		A	B
High need for power	Italy	246	33.0	West Germany	252	25.9
	U. So. Africa	248	24.8	Switzerland	247	23.6
	Belgium	320	23.9	England	367	22.8
	Finland	330	19.6	U.S.	339	22.8
	New Zealand	306	13.6	Argentina	156	22.8
	Canada	255	13.6	Denmark	363	12.5
	Spain	132	12.8			
	Taiwan	38.2	8.7			
	Average	277.4	18.8	Average	287.3	21.7
Low need for power	Hungary	302	33.0	Austria	326	28.4
	Sweden	350	21.2	Uruguay	189	27.2
	Portugal	135	18.8	Ireland	374	25.2
	Norway	290	16.9	Poland	140	20.0
	Australia	328	15.5	Japan	76	18.2
	France	200	11.9	Bulgaria	176	16.0
	Netherlands	228	11.5	Greece	131	12.5
	Chile	92	8.9	Israel	194	11.2
	Average	240.6	17.2	Average	200.7	19.8

[a]Mexico omitted; very high ratio not classified. Mann Whitney U-tests, heart disease, high high $>$ low high, $p = 0.07$; hypertension, high high $>$ low low, $p = 0.04$.

earlier findings showing that high *n* Power and high Inhibition is related to insomnia and high resting level production of epinephrine.

A slightly different picture emerges for hypertension. Once again the "high high" group of countries has the highest average death rate from hypertension, but the control variable seems to have a greater impact on the incidence of hypertension than the power variable. Note that both groups of countries low on the control variable show a lower average evidence of deaths from hypertension. Hence those countries low both in the need for Power and for control rank significantly lower in death rates from hypertension than countries high on both variables. Only Hungary is an outstanding exception. In other words, too much concern for control or too much inhibition seems likely to contribute to high blood pressure, even though it may protect the people with low power motivation from heart attacks because it contributes to leading a more regular life.

The findings make good sense in terms of what is known about the neural mechanisms contributing to the development of heart disease. Regestein and Schwartz (1975) have developed a model arguing that it is the simultaneous activation of the sympathetic and parasympathetic innervation of the heart that is particularly likely to bring on heart attacks. It is as if one branch of the autonomic nervous system (the sympathetic) is acting to increase the speed and rate of contraction of the heart, while the other branch (the parasympathetic) is operating to check this increased activity. Since we know that the need for Power is related to increased sympathetic activity, we need only further assume that the inhibition variable is likewise associated with greater parasympathetic activity to have a very similar model for explaining susceptibility to heart attacks.

CAUSES AND CURE FOR THE MOTIVATIONAL DISPOSITIONS LEADING TO CARDIOVASCULAR DISEASE

A physiologist or a doctor might be willing to grant for the sake of argument that psychological variables like the needs for Power and for control may be involved in cardiovascular disease, but wonder whether it is really necessary to take psychological variables into account. The advantage of doing so lies in our ability to find out more about the disease process from what affects the psychological variables. What, for instance, raises or lowers the need for Power? At the level of experimental arousal we have noted that power demands elevate the need for Power. Let us be a bit more specific about what those demands involve: In each case the individual was under some kind of pressure to act strongly on his own as a leader or as an experimenter. So we might reasonably infer as a beginning hypothesis that conditions which continuously call for strong independent assertive action might lead to the development of a dispositional need for Power. Such conditions are very similar to those stressors described as critical for producing prolonged physiological arousal, with the exception that it is usually further assumed that stress is even greater if a stronger response is called for than the individual can make. There is no evidence existing at the moment that need for Power is elevated even further the less able a person is to make a strong response which is called for. It seems more likely that what serves as the equivalent of this type of frustration at the psychological level is a need for control which inhibits internally the expression of the assertive response. To induce neurosis and eventually heart disease in a monkey, he is both enraged and restrained (Lapin and Cherkovich, 1971). The equivalent at the personality level would appear to be a strong disposition to act assertively which is simultaneously checked by an inner desire for control and restraint.

What reduces the need for Power? Here findings on the low incidence of heart disease among Italian residents of Roseto, Pennsylvania, are very suggestive. As reported by Wolf (1971), extensive studies of the population of this town over a 12-year period demonstrated a remarkably low death rate from myocardial infarctions as compared with comparable communities in the neighborhood, or even as compared with death rates of Rosetans from the same families who had migrated to nearby cities. Furthermore the low death rate from heart disease in Roseto occurred despite the prevalence of such risk factors in the population as heavy consumption of animal fats, cigarette smoking, and little muscular exercise. The conclusion seems inescapable that some psychosocial factor must be responsible for the lower death rate from heart disease, since other explanations seemed ruled out. Socially the Rosetans were different. "The study revealed that, unlike most American towns, Roseto is cohesive and mutually supportive, with strong family and community ties. . . . The family was found to be the focus of life. Children and teenagers related primarily to siblings and cousins of all ages rather than to peer groups, such as cub scouts, little league baseball teams, teenage clubs and gangs, as is customary in other American communities. . . . Problems are solved by family conclaves in which each person takes responsibility and may make some sacrifice. . . . The overall effect is one of mutual support and understanding and unfailing sustenance in time of trouble" (Wolf, 1971, pp. 328–329). Conditions of life seem to be the opposite of what would lead to the development of a high need for Power. Heavy demands are not made on the individual for strong independent forceful action; rather he is part of a supportive network which acts as a whole. So it seems reasonable to infer that the need for Power would be low among individuals in Roseto. Some support for the inference comes from the fact that Italians living under similar conditions in Italy tend to have a lower need for Power (McClelland, 1961).

At any rate, we now have a hypothesis: Chronic demands for individual assertiveness should raise the average level of need for Power and social cohesiveness, group action, and support for the individual should lower it.

Two sources of data exist which can be used to test the hypothesis. One consists of the information which was collected on a number of small preliterate societies for use in the study of economic achievement (McClelland, 1961) and drinking patterns (McClelland *et al.*, 1972). The need for Power was assessed by coding folk tales from these cultures. The second source consists of a sample of nations whose children's stories were scored for *n* Power around 1950. In both samples, the *n* Power measure was entered into a correlation matrix which also contained scores on a number of other structural and behavioral variables relating to characteristics of the cultures or nations involved. Table 3 reports every correlation of *n* Power with a structural variable in either matrix which reached approximately the 10% level of significance. The pattern of correlations confirms the hypothesis to a surprising degree considering the miscellaneous

Table 3. **Correlations of *n* Power with Collective Characteristics**

A.	Cross-cultural sample	(N = 36)
	Colder climate	0.47[d]
	Percent dependence on hunting	0.27[a]
	Percent dependence on agriculture	–0.27[a]
	Lineal descent	–0.44[d]
B.	Cross-national sample[a]	(N = 38–40) *n* Power measured around 1950
	Internal war (1955–1960)	0.27[b]
	Gains in electric power use (1937–1954)	0.34[c]
	Percent union membership (1966)	–.38[c]

[a]From Southwood (1969).
[b]$p < 0.10$.
[c]$p < 0.05$.
[d]$p < 0.01$.

nature of the variables included. Among preliterate societies *n* Power is associated with a colder climate and with greater dependence on hunting as a means of subsistence. Both hunting and struggling to keep warm involve a demand for individual assertiveness which in turn stimulates the production of adrenalin and other stress hormones. On the other hand cultures low in *n* Power tend to be those which are more dependent on agriculture and which reckon kinship and descent in lineal terms (either patrilineal or matrilineal). Both institutional characteristics reflect more organization and cohesiveness in the culture. Dependence on agriculture requires and is associated with greater interdependence and so is a lineal descent pattern. Figuring descent in nonlineal terms means that the individual is less apt to be part of any cohesive kin group or lineage. The correlations support the hypothesis that social cohesiveness hinders the development of *n* Power; the demand for individual assertiveness promotes it.

The pattern of significant or near significant correlations for the cross-national sample can be interpreted in a similar way. A high need for Power among these nations is associated with greater internal violence and with a greater gain in electric power available for use in the period prior to the assessment of the motive levels. Much internal violence indicates that the individuals in these countries are assertive and aggressive, though it is not possible to figure out from this correlation alone whether the aggression is the cause or result of high *n* Power. But why should rapid gains in the use of electric power seem to foster the growth of the need for Power? Actually the measure of electricity output was employed in another study as a way of estimating more accurately the rate of growth of the economy as a whole (McClelland, 1961). So what this correlation suggests is that more rapidly developing technological societies tend to demand more individual assertiveness or at least to encourage the individuals to believe that they are more powerful. For what technology does is make an individual think he is more powerful by placing more power at his disposal. With electric power, he can turn night into day, level mountains, travel rapidly over great distances, heat his dwelling against the cold, even wash and dry his clothes and beat his eggs. Many observers have pointed to the growth of technology as a major reason for man's greater belief in his own powers to dominate nature and control everything that happens to him Apparently this same growth in technology, as reflected in gains in electric power consumption, also increases the collective power orientation of a country. Thus we have an explanation for the greater stress which many of these same observers assume is characteristic of modern technological society. It is the result of a greater *n* Power which, as we have shown in Fig. 1, tends to accentuate the impact of power demands on physiological arousal.

On the other hand, the only variable in the matrix associated with lower *n* Power is percent of the working force which is unionized. When I first noted this correlation some years ago, I could not understand it. However, in the light of the present hypothesis, it makes good sense because unionization promotes exactly the kind of social cohesiveness which we have been arguing tends to lower the need for *individual* assertiveness and the need for Power. It is a different kind of solidarity than that which characterizes well-organized lineages or the Italian community of Roseto, but in all these instances the group seems to take a greater responsibility for responding to challenges, leaving the individual less under pressure to be assertive on his own.

Eyer (1975) has argued that it is the increased competitiveness of the modern technological capitalist society which increases stress and death rates from related diseases, particularly among certain age groups. He interprets the association between economic depression and increased rates of mental hospital admissions established by Brenner (1973) in the same way. During economic downturns, unemployment rises and men must compete harder than ever with each other in order to get work. This in turn creates greater stress and its associated pathologies, including mental disease. The argument if interpreted in terms of the need for Power has a testable implication. Economic downturns which cause a substantial rise in unemployment should be followed soon after by a rise in *n* Power. Elsewhere I have reported (1975) the estimates of *n* Power levels obtained from popular literature for each decade in the history of the United States from 1780 to 1970. So it is only necessary to find a time series which will serve to estimate rates of unemployment to test the hypothesis. Easterlin (1968) has assembled a number of such series. Unfortunately none of the relevant ones goes back as early as 1780. But beginning in 1820 he reports the gross rate of alien immigration per thousand total population per year. These figures are not direct estimates of unemployment, but they are the next best thing to it. For in bad times, the flow of immigration almost stopped, while in good times it increased greatly. This same index could not be used over the entire time period because in the twentieth century laws were passed controlling immigration.

However, direct estimates of unemployment rates are available from 1890 on. So employment opportunities were estimated from immigration rates from 1820 to 1890 and from unemployment rates from 1890 through the 1960s. Table 4 has been assembled in a way which permits a test of the association between economic downturns and *n* Power. It was assumed that competition for jobs or relatively high unemployment should *precede* a high need for Power. That is, if times were bad that would be reflected in fewer people immigrating to America, or in higher rates of unemployment in the first half decade in the early period later on. Bad times should lead to a rise in *n* Power. Somewhat arbitrarily it was decided that a steadily high rate of unemployment (at least 5.5%) was crucial in creating the mood of competition leading to high *n* Power. This should create more of a mood of struggle than a time when unemployment was high, as in 1940, but opportunity was definitely improving year by year as men were drawn off into the Army so that unemployment was down from almost 15% in 1940 to 2% by 1943. So job opportunities in the first half of each decade are classified

Table 4. Relation of Employment Opportunities to Relative Strength of the Need for Power in U.S. History[a]

		Need for Power in that decade is		
Epoch	Employment opportunities	less than	greater than	need for Affiliation
1820–24	Poorer[b]		+	
1830–34	Poorer		+	
1840–44	Poorer		+	
1850–54	Better	+		
1860–64	Poorer[c]	–		
1870–74	Better	+		
1880–84	Better	+		
1890–94	Poorer[d]		+	
1900–04	Better	+		
1910–14	?[e]			
1920–24	Better	+		
1930–34	Poorer		+	
1940–44	Better	+		
1950–54	Poorer		+	

[a]+ = Confirmation of hypothesis. – = Nonconfirmation of hypothesis. Association of competition for jobs with predominance of power motivation is unlikely to be due to chance ($p < 0.01$).

[b]Estimates based on whether alien immigration for the 5-year period is lower or higher than averages for 5-year periods from 1820 to 1913 (Easterlin, 1968).

[c]This estimate may be incorrect because fewer people migrated to the U.S. at this time because of Civil War.

[d]From here on estimates based on whether unemployment rate is 5.5% or more in four out of five years, or less. (Easterlin, 1968, *Historical Statistics of the United States, etc.).*

[e]Motivational data unavailable for this decade.

as better or worse for the period to which each time series applies. Then it is noted on the right-hand side of Table 4 whether *n* Power in popular literature published throughout that decade was higher or lower than the need for Affiliation. This index was used because, while the association between economic downturns and *n* Power existed as predicted, it was not significant, and as I have reported extensively elsewhere (1975), the *n* Power-*n* Affiliation ratio is a more sensitive indicator of the tendency toward aggressive competitiveness. At any rate, there is a clear and significant relationship between relatively poor economic opportunities in the early part of a decade and a subsequently higher *n* Power than *n* Affiliation. The relative importance of *n* Power is correctly predicted 13 out of 14 times from economic conditions in the first half of the decade. What is more, a significant relationship does not exist between unemployment rates in the last half of a decade and predominance of *n* Power, strongly suggesting that it is economic competitiveness that leads to a rise in *n* Power rather than a high need for Power which leads to economic competitiveness and unemployment. Furthermore Veroff *et al.* (1960) have reported that men from families whose incomes were in the lowest category (under $2000 annually in the late 1950s) had the highest *n* Power scores; strongly suggesting once more that unemployment or *economic* competitiveness for jobs is one of the factors leading to high *n* Power in men. And Eyer (1975) has noted that the death rates from various stress-related diseases are highest among the economically disadvantaged and blacks, presumably, if our argument is correct, because pressure to find work is even greater when it is hard to find and these continuing demands for individual assertiveness elevate *n* Power.

When I started assembling data for this paper, I had no idea that so much evidence could be found which implicates the need for Power in the stress syndrome. The final link in the chain of evidence has yet to be forged. We need the results of studies now under way on motivation levels among patients with cardiovascular disease, and on the connection of high levels of *n* Power and Inhibition in early life with later development of essential hypertension. But let us assume for the sake of the argument that our case is complete and speculate a bit on what it all means. One implication is immediately obvious. America and many of the other advanced industrial nations have regularly sacrificed community to what Bakan (1966) calls agency. Americans believe that the individual is more important than the group, that his freedom to act on his own and realize his potential is the supreme value of our way of life. We further believe that man is more powerful than nature, that doing is more important than being, and that the future is more important than the present or the past. See Kluckhohn and Strodtbeck (1961). What our argument above implies is that we are paying a tremendous price for making these values supreme. They make us more susceptible to competitive stress which in turn takes its toll in cardiovascular and other pathologies. As Eyer puts it "the most human solution, and in the long run the only real one, is to halt the social disruption and create relaxed community. Work should no longer be a high pressure activity kept going by the threat of a variety of social punishments. People should not be socialized to put themselves under chronic stress in order to produce. The ideals of competitive material achievement must be replaced by ideals of cooperative mutual development of social relationships" (1975). But how are we to accomplish this end? America has repeatedly tried to develop communal, cooperative living arrangements usually after a period of high *n* Power, as in the 1830s, the 1940s, and again now in the 1970s. But such experiments have not lasted long because they are not based on a very clear understanding of the psychological variables involved. Idealists interested in more cooperativeness generally attempt to reject both *n* Power and *n* Achievement altogether in their passionate concern for decreasing competitiveness and increasing affiliative ties. But *n* Achievement need not be rejected because as it is understood and measured by psychologists it does not necessarily lead to a competitive drive for material achievement. Rather it represents a concern only for doing something well, or better than, or more efficiently than it has been done before. It can express itself just as well in learning to play the guitar better as it can in building a better mousetrap. It is an essential part of economic propserity because a concern for efficiency seems to be necessary to build enough surplus to feed everyone well (see McClelland, 1961). Furthermore, as I have shown elsewhere the need for Power is not in itself destructive, for it expresses itself in various ways at different stages of maturity. At its earliest stage, it is perhaps best represented as mother or father or hero worship, as wanting to be strong by being around someone who will make you feel strong. In the next stage it is best represented by a desire to control oneself and order one's life. In the third stage the desire for Power turns outward into assertiveness and the desire to control others. It is this stage which most people identify with the "real" need for Power, but ideally it should be only a passing phase which one grows out of as he grows older. The difficulty is that the American value system has tended to idealize this stage, particularly for the male who must be assertive and independent in his job, in his loves, and in all his affairs, if he is to have any self-respect. As a consequence, he particularly suffers the penalties of death at an earlier age from cardiovascular disease. But there is a still higher stage of development for the power motive. The need for Power can express itself in helping others and at the highest stage in a kind of egoless service of the type idealized in many Eastern religions. For such a person it is precisely the higher values of the group–his family, his community or his religion–which are more important to him than his selfish desires for impact. Ideally the *self* disappears either as the source of power or as an object to be promoted. There is neither the space nor the need to go into the subtleties of this theory here, since I have done so extensively elsewhere (1975). The point that needs emphasis is only that *n* Power itself can undergo development and therefore we need not attempt to escape our stress-related dilemmas in the same old, ineffectual way by trying to reject the need for Power or deny that anyone should have it. Instead I take hope from those young people who both as practitioners and researchers are exploring in depth the possibility of dissipating the stresses and strains of the ego-related need for Power through meditation and other spiritual disciplines. Wallace and Benson (1972) have argued that meditation works by counteracting the physiological stress syndrome which has been the key concern of this paper, and of course meditation as a psychic discipline is precisely the means recommended by Eastern sages for transcending the self, for getting beyond the ego-oriented power drive characteristic of Stage III, into the egoless, "flow through" *n* Power condition, characteristic of the fourth or highest stage. Thus science in the end may provide us with the technological breakthroughs which will help us cope more effectively with stress in the modern world and develop our power motives to the stage in which transcendent values guide us more than our egotistic concerns for assertiveness.

REFERENCES

Bakan, D. (1966). *The Duality of Human Existence.* Chicago: Rand McNally.

Boyatzis, R., and Dailey, C. (1975). The effects of stress-inducing life changes. Boston, Mass.: McBer and Co.

Brenner, M. H. (1973). *Mental Illness and the Economy.* Cambridge: Harvard University Press.

Brod, J. (1971). The influence of higher nervous processes induced by psychosocial environment on the development of essential hypertention. In *Society, Stress and Disease* (L. Levi, Ed.). London: Oxford University Press, pp. 312–323.

Easterlin, R. A. (1968). *Population, Labor Force and Long Swings in Economic Growth: the American Experience.* New York: Columbia University Press.

Eyer, J. (1975). Stress-Related Mortality and Social Organization. Philadelphia: Department of Biology, University of Pennsylvania.

Frankenhaeuser, M. (1973). Experimental Approaches to the Study of Catecholamines and Emotion. Stockholm: Psychological Laboratories, University of Stockholm.

Frankenhaeuser, M. (1971). Experimental approaches to the study of human behaviour as related to neuroendocrine functions. In *Society, Stress and Disease* (L. Levi, Ed.). London: Oxford University Press, pp. 22–35.

Friedman, M. (1969). *Pathogenesis of Coronary Artery Disease.* New York: McGraw-Hill.

Jenkins, C. D. (1971). Psychologic and social precursors of coronary disease. *N. Engl. J. Med. 284*, 244–255, 307–317.

Kluckhohn, F., and Strodtbeck, F. L. (1961). *Variations in Value Orientations.* Evanston, Ill.: Row Peterson.

Lapin, B. A., and Cherkovich, G. M. (1971). Environmental changes causing the development of neuroses and corticovisceral pathology in monkeys. In *Society, Stress and Disease* (L. Levi, Ed.). London: Oxford University Press, pp. 266–279.

Levi, L. (Ed.) (1971). *Society, Stress and Disease.* London: Oxford University Press.

Lundberg, U., Theorell, T., and Lind, E. (1973). Life Changes and Myocardial Infarction. Stockholm: Psychological Laboratories, University of Stockholm.

McClelland, D. C. (1961). *The Achieving Society.* New York: Van Nostrand.

McClelland, D. C. (1975). *Power: The Inner Experience.* New York: Irvington, Halsted-Wiley.

McClelland, D. C., Davis, W. N., Kalin, R., and Wanner, E. (1972). *The Drinking Man.* New York, Free Press.

Myager, V. (1971). Psychic trauma and cortical-diencephalic interrelationships. In *Society, Stress and Disease* (L. Levi, Ed.). London: Oxford University Press, pp. 258–260.

Myrsten, A.-L., Post, B., and Frankenhaeuser, M. (1971). Catecholamine output during and after acute alcoholic intoxication. *Percept. Mot. Skills 33,* 652–654.

Rahe, R. H. (1972). Subjects' recent life changes and their near future illness susceptibility. *Adv. Psychosomatic Med. 8,* 2–19.

Raab, W. (1971). Cardiotaxic biochemical effects of emotional-environmental stressors–fundamentals of psychocardiology. In *Society, Stress and Disease* (L. Levi, Ed.). London: Oxford University Press, pp. 331–337.

Regestein, Q., and Schwartz, G. (1975). A psychophysiological model of sudden cardiac death. Unpublished paper, Department of Psychology and Social Relations, Harvard University, Cambridge, Mass.

Schacter, S., and Singer, J. E. (1962). Cognitive, social, and physiological determinants of emotional state. *Psychol. Rev. 69*, 379–399.

Selye, H. (1971). The evolution of the stress concept–stress and cardiovascular disease. In *Society, Stress and Disease* (L. Levi, Ed.). London: Oxford University Press, pp. 299–310.

Selye, H. (1936). A syndrome produced by diverse nocuous agents. *Nature 138*, 32.

Southwood, K. E. Some sources of political disorder: a cross-national analysis. Champaign, Ill.: Unpublished doctoral dissertation, University of Illinois.

Steele, R. S. (1973). The physiological concomitants of psychogenic motive arousal in college males. Ph.D. thesis, Harvard University, Cambridge, Mass.

Theorell, T., Lind, E., Fröberg, J., Karlsson, C.-G., and Levi L. (1972). A longitudinal study of 21 subjects with coronary heart disease: Life changes, catecholamine secretion, and related biomedical reactions. *Psychosomatic Medicine 34,* 505–516.

Vander, A. J., Sherman, J. H., and Luciano, D. S. (1970). *Human Physiology: The Mechanisms of Body Function.* New York: McGraw-Hill.

Veroff, J., Atkinson, J. W., Feld, S. C., and Gurin, G. (1960). The use of thematic apperception to assess motivation in a nationwide interview study. *Psychol. Monogr. 74*, 12 (whole number 499).

Wallace, R. K., and Benson, H. (1972). The physiology of meditation. *Sci. Am. 226*, 84–90.

Wilsnack, S. C. (1974). The effects of social drinking on women's fantasy. *J. Personal. 42*, 43–61.

Winter, D. G. (1973). *The Power Motive.* New York: Free Press.

Wolf, S. (1971). Psychosocial forces in myocardial infarction and sudden death. In *Society, Stress and Disease* (L. Levi, Ed.). London: Oxford University Press, pp. 324–330.

Advances in the Healing of Psychopathology

Exposure Treatment

ISAAC MARKS

The experimental tradition is not new to psychology or medicine, but it is a recent arrival in psychological treatments. One of its fruits is behavioral psychotherapy, which can relieve several kinds of formerly untreatable psychopathology. Examples include anxiety syndromes like phobic and obsessive-compulsive disorders, and some sexual and social problems.

The value of behavioral psychotherapy is that it not only relieves certain forms of distress but also lends itself to an experimental approach which can refine the efficacy of our treatments and our understanding of psychopathological mechanisms. The ingredients of treatment can be systematically varied so that those aspects which are crucial for effecting improvement can be detected, and redundant elements may then be omitted.

This paper will outline advances in the treatment of anxiety syndromes by behavioral psychotherapy and will trace the search for essential determinants of outcome. We will first glance at evidence that exposure treatments have an effect, and then examine the importance of exposure as a therapeutic ingredient. Detailed evidence will be drawn mainly from work carried out at the Maudsley Hospital with colleagues and students, but outside data will be cited where appropriate. Data will be cited mainly from phobic and obsessive-compulsive

ISAAC MARKS · Reader in Experimental Psychopathology, Institute of Psychiatry, London, S.E.5., England.

disorders. This is not the place to discuss the reasons for separating phobias from obsessions despite their similarities. The approach to their treatment employs a common principle. This principle can be called exposure. It states that relief of phobias and compulsions requires continued contact of the sufferer with those situations which evoke his discomfort until it subsides. The exposure principle predicts some clinical conditions necessary for successful treatment. Clinicians need to search for those cues which trigger phobias or rituals, and confront the patient with the cues concerned. The principle of exposure does not explain why improvement ensues under those conditions.

RESULTS OF EXPOSURE IN PHOBIC AND OBSESSIVE-COMPULSIVE DISORDERS

Numerous controlled trials in phobic and obsessive-compulsive disorders have found that exposure treatments produce significant relief in patients with phobias or compulsions up to the latest 2–4-year followups which are available These results can be obtained with real-life exposure in phobics after 1–20 sessions and in obsessive-compulsive disorders after 5–30 sessions (reviewed by Marks, 1975). Clinical experience indicates that the more complex problems with wider ramifications take longer to help. Chronicity, however, is not important and even long-standing problems can be quickly relieved.

An early form of exposure treatment was desensitization in fantasy, in which the patient is relaxed and asked to repeatedly imagine himself gradually approaching the object which causes him fear. The phobic images are only visualized for a few seconds at a time, and the subject is asked to relax between images. Figure 1. shows results in phobic disorders treated in an early series of trials by Gelder and myself (Marks, 1971). Desensitization in fantasy reduced phobias more than a comparison group which had dynamic psychotherapy. Patients who did not improve with group analytic psychotherapy were then given desensitization in fantasy and these then improved in their phobias. The superiority of desensitization was gradually eroded by other patients slowly catching up over the years, but desensitization subjects improved earlier and with less treatment, and maintained their improvement until 4 years followup.

These results were replicated by Gillan and Rachman (1974) in a similar group of phobic patients. Desensitization in fantasy was found to be superior to dynamic psychotherapy for the relief of phobias, and improvement continued to 6 months follow-up. The amount of improvement with each treatment was very comparable to that obtained in the earlier investigation. In many other studies in volunteers, desensitization in fantasy achieved superior fear reduction to control groups which experienced no phobic imagery (reviewed by Marks, 1969, 1975).

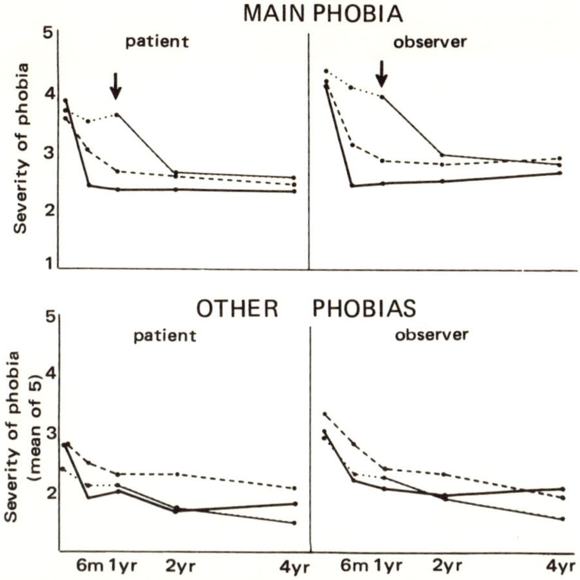

Fig. 1. Improvement in phobias continues to 4 years follow-up. Comparison is between desensitization in fantasy and dynamic psychotherapy. Dynamic psychotherapy was given either individually or in groups. Patients whose phobias failed to improve after 18 months (72 sessions) of group psychotherapy were then given desensitization in fantasy, upon which their phobias improved. (——) Desensitization only (*n* = 15); (– – –) psychotherapy only (*n* = 18); (· · · ——) psychotherapy followed by desensitization at ↓ (*n* = 7). Data from Gelder *et al.* (1967) and Gelder and Marks (1968) as reported by Marks. [(1971). *Brit. J. Psychiat. 118,* 683–688.]

More recently emphasis has been on exposure *in vivo,* which gives quicker results by bringing the patient into contact with his discomforting situation in real life without relaxation exercises. A partially controlled experiment found that exposure *in vivo* improves obsessive-compulsive rituals significantly more than does relaxation treatment (Fig. 2, Marks *et al.,* 1975). In Fig. 2 all patients are depicted as having relaxation treatment first, which would leave the question open whether we are simply seeing an order effect, i.e., that the second treatment is best. In fact, this is not so. In a subexperiment patients who had exposure *in vivo* without preceding relaxation did just as well as those whose exposure followed relaxation (Fig. 3). Improvement with exposure *in vivo* thus occurred whether it was given as the first or as the second treatment block (Rachman *et al.,* 1973). Improvement continued to 2 years followup (Fig. 4).

Exposure treatments come in many forms. The patient may be brought into contact with his distressing situation in fantasy or in real life, for shorter or longer periods, with or without modeling, in which the therapist first demonstrates to the patient what to do, with or without praise (operant conditioning

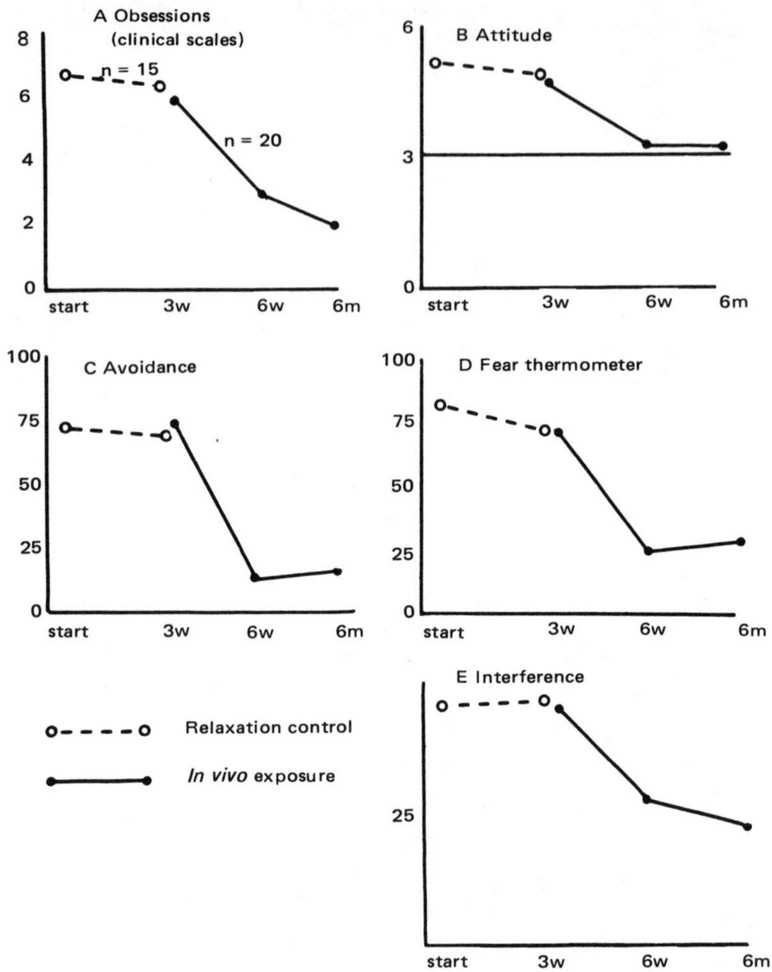

Fig. 2. Changes in five measures of obsessive-compulsive rituals until 6 months follow-up (n = 20). Clinical scales are mean of patients' and assessors' ratings. Attitude ratings are evaluative scales of semantic differential completed by patients. Interference score is from Leyton Obsessional Inventory. [Taken from Rachman *et al.* (1973). *Behav. Res. Ther. 11,* 463–471.]

or shaping) and with the patient either relaxed or very anxious during the period of exposure. There is a natural tendency to seek for a single explanatory principle behind any treatment, though reality is usually more complex than our constructs. A therapeutic element more pervasive than most is that of exposure to a noxious stimulus until the organism gets used to it. Alternative terms for "gets used to it" are habituates, extinguishes, or adapts, each of these terms

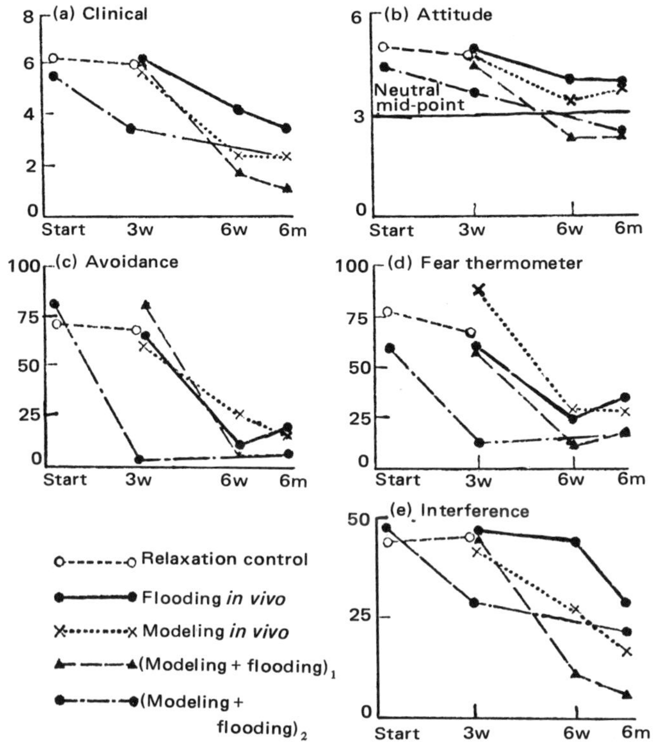

Fig. 3. Exposure *in vivo* is not potentiated by preceding relaxation. Comparison of exposure *in vivo* without preceding relaxation (labeled as modeling + flooding$_2$) with three other groups which had relaxation beforehand. Flooding *in vivo* had fast exposure without modeling, modeling *in vivo* had slow exposure with modeling, modeling + flooding had fast exposure with modeling) (each group $n = 5$). [Taken from Rachman *et al.* (1973). *Behav. Res. Ther. 11,* 463–471.]

having overlapping implications. The noxious stimulus may be a troublesome fantasy, a feeling of uncertainty about the future, an airplane journey, or sexual intercourse. It is not our brief to discuss how such stimuli come to be perceived as noxious in the first place, why one man's meat becomes another man's poison. However, once situations repeatedly produce discomfort, as in phobias and obsessions, then sufferers usually lose this discomfort by agreeing to remain exposed to the situations until they feel better.

Exposure is a similar concept to extinction, which holds that repeated applications of a CS without a US leads to elimination of a CR. Extinction simply means that a response ceases to occur, without explaining why that is so. The same applies to exposure, which simply holds that given enough time in

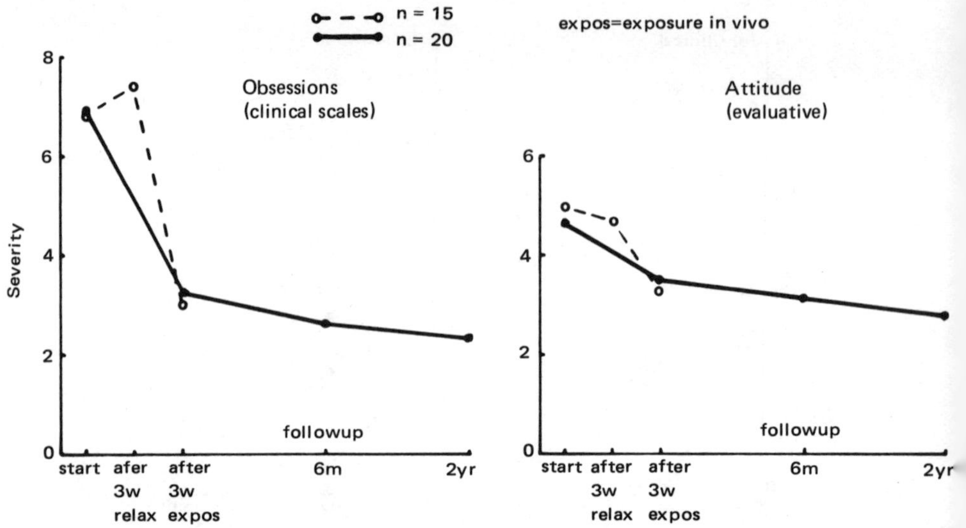

Fig. 4. Improvement in obsessive-compulsive rituals continues to 2 years follow-up (n = 20). Self-ratings. [Taken from Marks *et al.* (1975). *Brit. J. Psychiat. 127,* 349–364.]

contact with the provoking situation a phobic or an obsessive ceases to respond with avoidance, distress, or rituals. Just as animals sometimes fail to extinguish and instead show paradoxical enhancement of avoidance responses, so do humans rarely sensitize rather than habituate to exposure, but this is fortunately most uncommon under clinical conditions. We fail to understand as yet the conditions which produce sensitization (or paradoxical enhancement) rather than habituation (or extinction). The question is why exposure to a trauma sometimes produces phobias while at other times it cures them. Which sets of conditions predict a traumatic or a curative outcome have still to be delineated, though systematic research has begun to chart some dimensions of this problem.

AROUSAL LEVEL DURING EXPOSURE

Low-Anxiety. Several hypotheses have stressed the importance of the level of arousal (anxiety) during exposure. Professor Wolpe stated that relaxation or other procedures were necessary to "reciprocally inhibit" anxiety during contact with the phobic stimulus so that improvement could follow (Wolpe, 1958). Lader and Mathews (1968) carried this idea a stage further by hypothesizing that habituation was maximized during states of low arousal. Many experimenters since then have found that the outcome to desensitization in fantasy is not impaired by omitting relaxation; the evidence points strongly to relaxation being a redundant element in the therapeutic package (Marks, 1975).

One disconfirmation of the hypothesis of reciprocal inhibition came from Benjamin *et al.* (1972) in a study of chronic phobic patients. These patients imagined phobic images up a hierarchy while they were either relaxed or in a neutral affective state. The hypothesis of reciprocal inhibition predicts that relaxed subjects would show less anxiety to phobic images during treatment, and have a superior outcome for fear reduction. Our experimental manipulation was successful in producing two significantly differentiable treatment conditions. During treatment sessions relaxed patients had significantly less skin conductance activity between phobic images than patients who had not been relaxed, i.e., they were less aroused. However, contrary to prediction from a reciprocal inhibition or maximal habituation model, arousal between images during treatment did not correlate with anxiety decrement either during or after treatment. Patients lost their fear steadily at the same speed during and after treatment sessions. During sessions subjective anxiety and heart rate diminished at the same rate whether the phobic images were visualized in a state of relaxation (desensitization) or neutral affect (exposure) (Fig. 5). After the end of each treatment condition the reduction in phobias was similar.

This result was replicated by Gillan and Rachman (1974), who found comparable outcomes in phobic patients after treatment by desensitization in fantasy with and without accompanying training in muscular relaxation. We can conclude that training in muscular relaxation exercises is a redundant element of exposure treatments and can be omitted without impairing results.

So far we have dealt with relaxation accompanying exposure treatment. Relaxation by itself *without* exposure is also unhelpful (Rachman *et al.*, 1973, Fig. 1). In that design compulsive rituals did not improve after 15 sessions of muscular relaxation but did reduce significantly after 15 sessions of exposure *in vivo*. This was replicated in a subsequent study with Roper *et al.* (1975). In fact, so inert is relaxation for phobias, compulsions, and even sexual deviations that it can safely be used as a placebo control to contrast with more active treatments under investigation (e.g., Rooth and Marks, 1974).

High-Anxiety. A view opposite to that of reciprocal inhibition is the notion of implosion (Stampfl, 1967). This holds that for improvement to occur anxiety has to be maximally aroused during exposure until the patient is so exhausted that he cannot experience any more emotion. This is conceived as an extinction process. The evidence was based on uncontrolled clinical experience, and can be paraphrased from Stampfl for one obsessive-compulsive handwasher: "He who has lived in a cesspool for a few days (in his mind) will not worry later about a bit of dirt on his hands." In this procedure the patient is asked to worry about the worst possible things which might happen to him for up to several hours at a

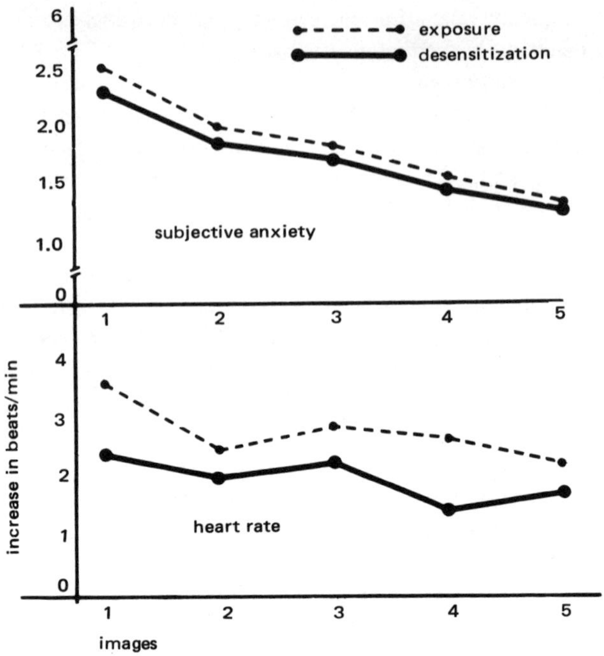

Fig. 5. Relaxation does not speed up fear reduction. Improvement in subjective anxiety and heart rate during phobic images presented on five successive occasions (n = 8). Exposure consisted of desensitization in fantasy. [Taken from Benjamin *et al.* (1972). *Psychol. Med.* *2*, 381–390.]

time, after which he feels better about his troubles. This is implosion or flooding in fantasy. A similar procedure could be done for real-life exposure to the troublesome situations which provoke phobias or compulsions.

Evidence for the value of fantasy implosion or flooding is conflicting in volunteers, but review of the literature indicates that two variables influence outcome. Where flooding in fantasy is given by a tape-recorded narrative from the therapist results tended to be poor, and outcome was better where sessions were longer (Marks, 1975).

In phobic patients flooding in fantasy is partially therapeutic (Watson and Marks, 1971) and has similar results to desensitization in fantasy (Gelder *et al.*, 1973; Mathews *et al.*, 1974). The latter study was carried out in Oxford. In it fantasy exposure was continuous during flooding but intermittent during desensitization, and an *in vivo* phase which followed fantasy treatment might have blurred differences between the two fantasy procedures.

Flooding *in vivo* is therapeutic in phobic and obsessive-compulsive patients. The question is whether high arousal is an essential part of the treatment. A direct test was undertaken of the thesis that high arousal facilitates improvement during *in vivo* exposure (Hafner and Marks, 1976). Chronic agoraphobic patients were exposed continuously for 3 hours a day over 4 days to their real phobic situations, e.g., they were asked to shop in crowded supermarkets or to ride in subway trains until they felt better. In a high anxiety condition the therapist commented how bad they looked and mentioned all the catastrophes which might befall them In a lower anxiety condition the therapist was reassuring, though he couldn't eliminate all anxiety.

The experimental manipulation produced two significantly differentiable treatment conditions, with patients experiencing significantly more discomfort during exposure in the high than in the lower-anxiety condition. However, this produced no difference to outcome on any measure (Fig. 6). Lower-anxiety patients improved at the same speed and to the same extent as did high-anxiety patients. This experiment thus disconfirmed the idea that high anxiety was facilitatory for improvement during exposure.

Further evidence that high arousal was not especially helpful came from a second controlled experiment by Hafner and Marks (1976). Chronic agoraphobics were exposed as groups to their real phobic situations in a double blind balanced design. During exposure some patients had their anxiety damped down by small doses of diazepam, while others had a placebo. Patients in two diazepam conditions who had less discomfort during exposure improved at the same rate as those on placebo who reported more panics (Fig. 7).

Results from these studies of phobics are in line with findings from three earlier experiments (Marks *et al.*, 1971; Watson and Marks, 1971; Stern and Marks, 1973) that anxiety *during* exposure does not predict subsequent outcome of phobias.

In brief, phobias and obsessions improve with exposure treatments, but it

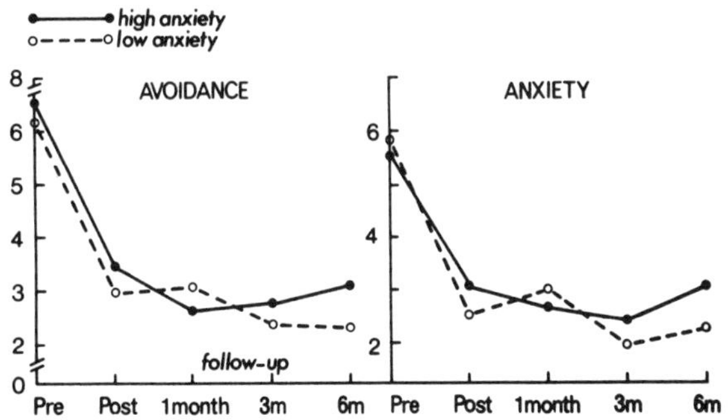

Fig. 6. High anxiety during exposure *in vivo* of agoraphobics does not enhance fear reduction (high and low anxiety each n = 6). Patients were exposed individually for 3 hr an afternoon over 4 days. [Taken from Hafner and Marks (1976). *Psychol. Med. 6,* 71–88.]

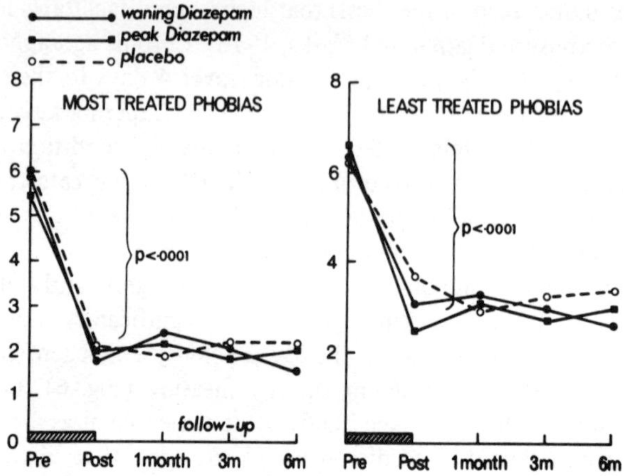

Fig. 7. Group exposure *in vivo* had a similar outcome in agoraphobics whether their anxiety was high with placebo or damped down with diazepam (each condition n = 13–14). [Taken from Hafner and Marks (1976) *Psychol. Med. 6*, 71–88.]

matters not whether patients are relaxed, neutral, or anxious during such exposure. *Duration of exposure* seems a more important variable. In animal experiments longer durations of response prevention or of CS confrontation hasten extinction procedures. Though it is difficult to define the CS or response prevention in humans, longer exposure seems better in volunteers (reviewed by Marks, 1975) and in patients.

In a Latin square design Stern and Marks (1973) exposed chronic agoraphobic patients for long or short periods in fantasy and *in vivo* (Fig. 8). Exposure *in vivo* gave significantly better results when carried out for two continuous hours than in 4 interrupted half hours over the same afternoon (Fig. 9). This effect was true for reduction in heart rate during treatment sessions and also for decrease in phobias at the end of treatment. This experiment was in chronic patients, and the optimum time of exposure might well be less for those whose phobias are of more recent duration. Duration of exposure is important presumably because it gives certain unidentified processes more time to work while exposure is going on, e.g., it might be giving people time to develop self-regulatory strategies to control their own emotions or to reach critical levels of habituation which may be necessary for lasting change to occur. The latter is implied in the question "Is it best to end on a good note?" We have no answer as yet.

The duration of exposure required for improvement may also depend upon genetic variables. In rats Sartory and Eysenck (1976) found that genetically reactive Maudsley strains needed longer durations of CS confrontation (3 to 10 min rather than 0 or 1 min) for extinction to occur of a step-down avoidance response which had been acquired to shock 2 days previously. With Maudsley

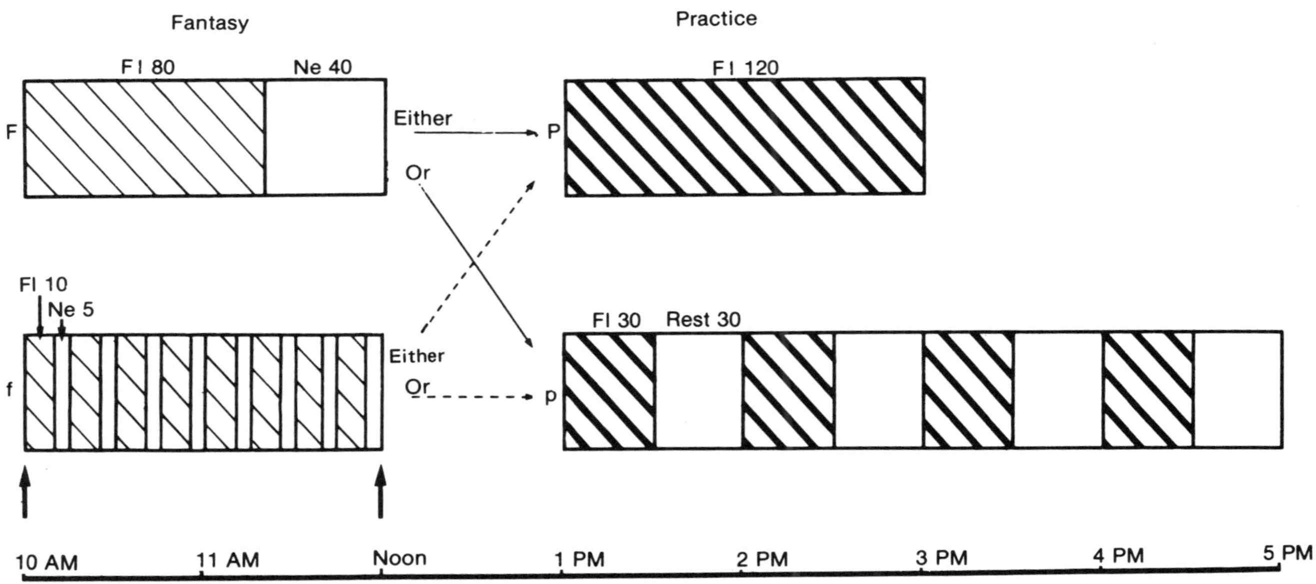

Fig. 8. Experimental design which shows long to be better than short durations of exposure *in vivo* (n = 16). Fantasy exposure was by tape-recorded voice and always preceded exposure *in vivo*. *Fl* = flooding, *Ne* = neutral imagery, *F* = long fantasy, *f* = short fantasy, *P* = long practice, *p* = short practice. [Taken from Stern and Marks (1973). *Arch. Gen. Psychiat. 28,* 270–276. Copyright 1973–1976, American Medical Association.]

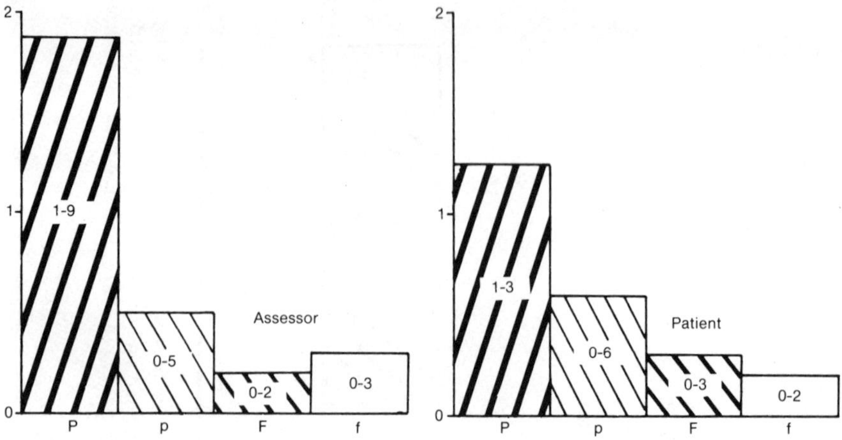

Fig. 9. Long exposure (practice) *in vivo* reduces phobias significantly more than short exposure (n = 16). Fantasy exposure has little effect, perhaps because it was given by tape-recorded voice. [Taken from Stern and Marks (1973). *Arch. Gen. Psychiat. 28,* 270–276.]

nonreactive strains duration of CS exposure was less important. The genetics of anxiety syndromes are by no means clear in man, but will eventually have to be examined in this respect.

FANTASY VERSUS LIVE EXPOSURE

Another important variable is whether exposure is to the real or merely to the fantasied situation. Strong evidence is accumulating in volunteers (Lopiccolo, 1969; Sherman, 1972) and in agoraphobic patients (Stern and Marks, 1973) that phobias are reduced much more efficiently by live than by fantasy exposure. A classic experiment by Bandura (1969) which was originally used to illustrate the value of modeling is in fact a stronger argument for *in vivo* over fantasy exposure rather than for modeling.

Figure 10 shows results in Bandura's four experimental groups of snake phobic volunteers. A group which had no treatment did not improve. Two other groups improved partially with exposure in fantasy—one of these had individual desensitization in fantasy without a model, the other watched a film of a model gradually approaching a snake (so-called "symbolic modeling," but fantasy exposure is clearly present). A fourth group watched a live model gradually approach and play with a real snake, and were then encouraged to handle the snake themselves. (so-called "live participant modeling," but *in vivo* exposure is present). As both the live modeling and the symbolic modeling groups had modeling, the superiority of "live modeling" cannot be attributed to modeling. A more likely explanation is the presence of *in vivo* exposure in the live modeling group.

Exposure as the Active Ingredient in Many Procedures. Nearly all claims for the value of modeling for fear reduction contain the critical component of exposure, i.e., of approach to the discomforting situation. Rachman *et al.* (1973) found no value in adding modeling to exposure *in vivo* for obsessive-compulsives. The critical aspect for fear reduction is thus not modeling per se, the act of observing a therapist doing something. Rather, the important point is what is actually observed. If observation brings the patient into contact with the phobic or obsessive cue then he improves, and if it doesn't he does not lose his phobia or ritual. In a test of this point Roper *et al.* (1975) found that observation of a therapist merely carrying out relaxing exercises did not reduce rituals, but observation of the therapist contaminating herself was followed by a significant decrease in the patient's compulsions.

The same argument applies to operant conditioning procedures. Wherever these have been shown to reduce fear they have been inextricably mixed with exposure procedures. Where contingent reward has been omitted but the amount of exposure was kept constant then phobics continued to improve at the same rate (e.g., Emmelkamp and Ultee, 1974).

It is thus clear that what at first sight seem to be widely different forms of fear reduction, e.g., desensitization in fantasy, flooding *in vivo,* exposure in real life, modeling, or operant conditioning, can all be subsumed under the same rubric of exposure. The different names simply describe variations on the theme of exposure. The current treatment of choice for obsessive-compulsive and for phobic disorders seems to be prolonged live exposure of the patient to the noxious stimulus with his consent. Evidence for the importance of consent comes only from clinical anecdotes. Experimental data on this point are naturally hard to get.

Response prevention may be yet another example of a treatment which finally works through exposure. In compulsive ritualizers active response prevention during exposure may be unnecessary except as a way of prolonging exposure. This idea was supported in a pilot experiment by Lipsedge (1974). He found that compulsive handwashers who were allowed to wash but were contaminated continually throughout the washing ritual improved as much as those who voluntarily refrained from washing. Exposure was to their feeling of contamination after they touched what they thought was "dirty." This experiment still requires replication.

The principle of exposure to the stressful situation is not only important for phobias and obsessions but also for current treatments of social deficits and

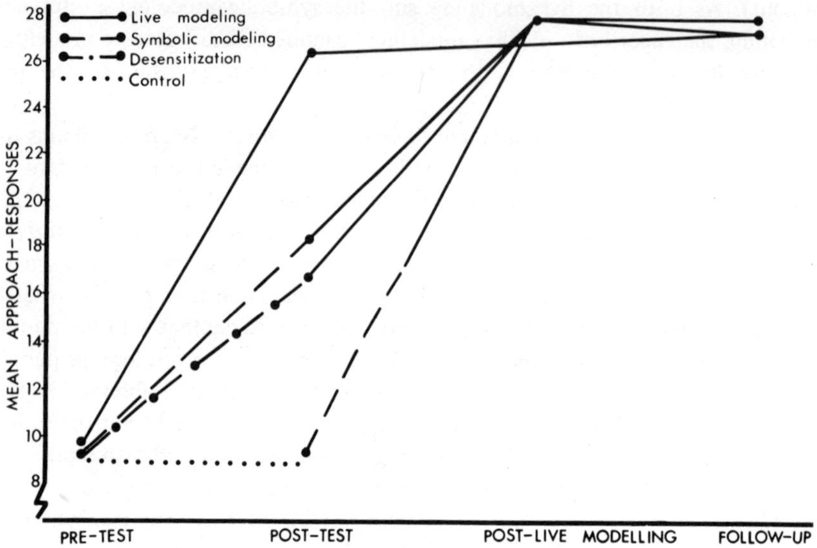

Fig. 10. Exposure *in vivo* has greater effect than exposure in fantasy. Live modeling had exposure *in vivo* with modeling, symbolic modeling had exposure in fantasy with modeling, desensitization was in fantasy without modeling, control had no treatment. Snake phobic volunteers. [From Bandura (1969). *Principles of Behavior Modification.* New York: Holt, Rinehart & Winston.]

sexual problems, though here it is bound up with training in interpersonal and other skills. In social skills training and in sexual retraining programs, of which Masters and Johnson pioneered one form, patients are required to rehearse appropriate behavior repeatedly in their problem situations until they lose their discomfort and acquire the requisite fluency of performance. Controlled evidence for the value of this approach still comes mainly from volunteers, though trials in patients are nearing completion (Crowe, 1975).

UNEVEN PROGRESS DURING EXPOSURE

During exposure we commonly observe uneven rates of improvement in different components of psychopathology. An emotional sequence is like a rope with many strands interwoven, only some of which may be traced completely through from start to end, others of which only appear at certain points along the path. Treatment of particular strands may lead to change in those strands while leaving others unaffected, and exposure treatments only acquire clinical utility when the effect eventually generalizes to embrace most of the maladaptive problem.

An experiment of feedback of heartrate will illustrate the point. Specific phobic patients were exposed to their real phobic situations (live animals) either with or without feedback of heart rate (Nunes and Marks, 1975). In a balanced controlled design ten patients were exposed individually for 2 continuous hours to their phobic stimulus. During this exposure, for 2 half-hour periods this was omitted. The order of the feedback and no-feedback conditions was balanced. Figure 11 illustrates the process in one patient.

Results during the session were instructive. Heart rate dropped significantly during periods of exposure with feedback compared to periods of exposure without such feedback (Fig. 12). However, despite the significant lowering of heart rate this did not hasten improvement in subjective anxiety, which improved equally throughout the exposure period regardless of feedback (Fig. 13). It is possible that had feedback gone on for longer this effect might have

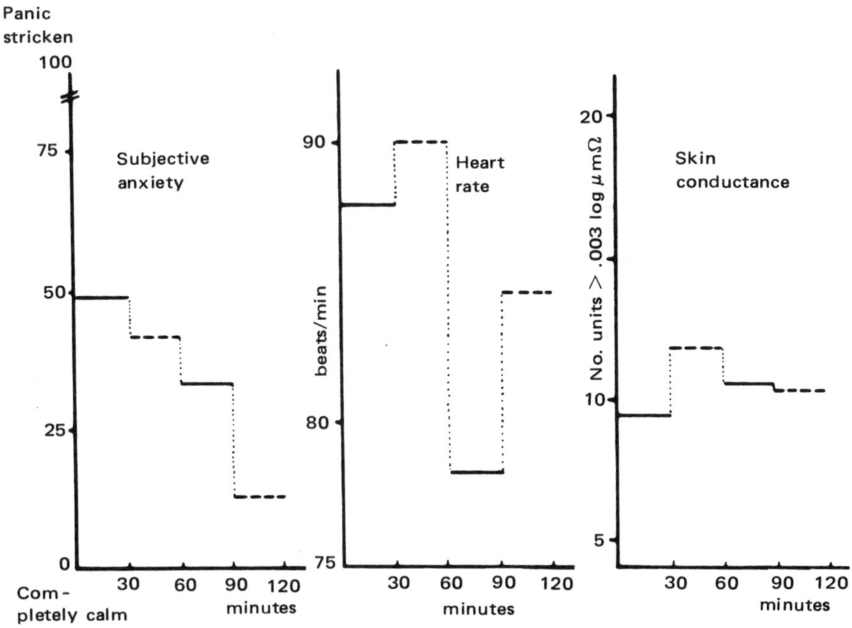

Fig. 11. Design of heart rate feedback in one patient with specific animal phobia. Two periods of exposure with feedback (– – –) were followed by two without feedback (· · ·). (from Nunes and Marks (1975). *Arch. Gen. Psychiat. 2,* 933–936. Copyright 1973–1976, American Medical Association.)

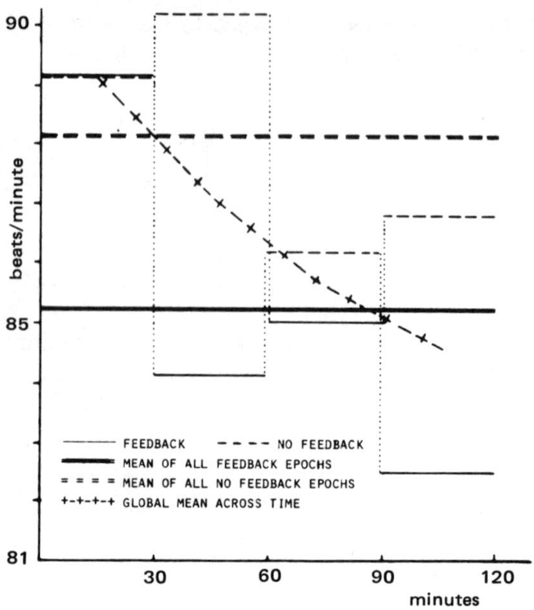

Fig. 12. Heart rate is significantly lowered during exposure with feedback of heart rate. Pooled results in ten patients. [From Nunes and Marks (1975). *Arch. Gen. Psychiat. 2,* 933–936. Copyright 1973–1976, American Medical Association.]

eventually generalized to subjective anxiety, or that subjective habituation would have occurred anyway in time. Treatments of this kind tend to produce improvement in avoidance behavior and physiological responses first, with attitude changes only occurring from hours to several weeks later, a phenomenon which might be called "cognitive lag."

Perhaps the differential rate of change in different strands of psychopathological emotions reflects a simple rule that whichever strand happens to be treated first improves first. Exposure treatments tend to concentrate on approach behavior rather than on subjective states. Perhaps treatments directed first at attitude rather than at avoidance might reverse the order of change.

SHORTCOMINGS OF THE EXPOSURE MODEL

The exposure hypothesis of fear reduction leaves several important facts unexplained. (1) A few subjects (fortunately a small minority) do not habituate during exposure despite their fulfillment of all criteria for predicted success, i.e.,

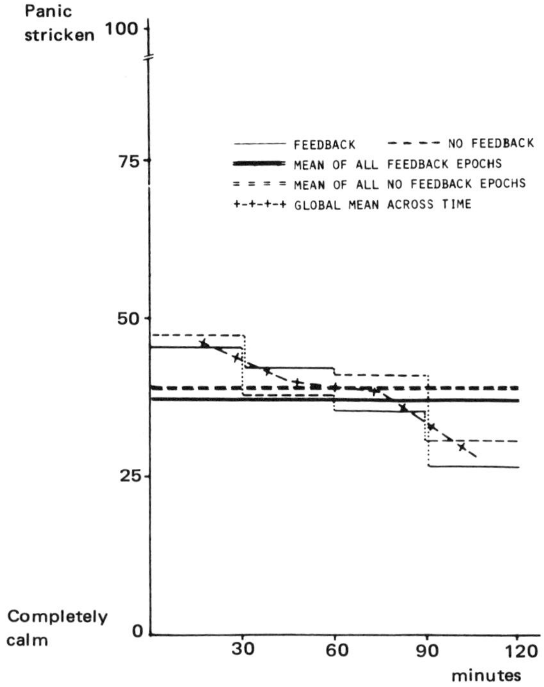

Fig. 13. Despite lowered heart rate, accompanying subjective anxiety does not improve at a faster rate during the session. Pooled results in ten patients. [From Nunes and Marks (1975). *Arch. Gen. Psychiat. 2,* 933–936. Copyright 1973–1976, American Medical Assosiation.]

adequate motivation, absence of serious depression, no attempts to escape in fantasy or reality during exposure, and adequate duration of treatment. This is not the problem of relapse after improvement, but rather the failure of any improvement to occur at all. In these patients some crucial factor for success is missing, the nature of which is hard to determine. (2) Some phobic, obsessive-compulsive and other forms of anxiety improve without any exposure merely by the taking of antidepressant drugs. (3) Without any exposure to the specific phobic stimulus anxiety sometimes remits after the abreaction of intense emotion, e.g., fear which is irrelevant to the phobic stimulus (Watson and Marks, 1971), anger (Marks, 1965), or other feelings. In an experiment which compared the effect of fantansied fear which was either relevant or irrelevant to the phobic stimulus, both types of fear proved to be equally (though only partially) therapeutic. Internal evidence suggested that the two procedures produced their equal effects through different mechanisms. One might then argue that relevant fear acted as a specific extinction procedure, and irrelevant fear as an abreactive process. The two mechanisms need not be mutually exclusive and could in fact interact synergistically if the patient faced both relevant and irrelevant fear cues during the treatment sesssion.

A tempting explanation for the (limited) value of irrelevant fear might be that this teaches the patient a form of coping. This is a widening of the exposure hypothesis, and leads to important theoretical and practical issues, even if evidence from Watson and Marks (1971) does not support this idea for reasons we have no time to discuss now.

A widened exposure hypothesis of coping states that a patient benefits from exposure to irrelevant fear as he would to unpleasant emotions in general. It predicts that the experience teaches the patient general emotional control, which also helps him over his specific phobia of obsession. This explanation is testable, but is not altogether convincing. Many patients have excellent general emotional control despite intense phobias or obsessions. The concept may perhaps hold better for patients with widespread emotional problems.

An important unknown is whether improvement can follow not only negative but also strong positive emotions. Whether the valency of the emotion is negative or positive may not be crucial and sexual or religious ecstasy might well be therapeutic on occasion; the design of experiments to test this notion leaves much to the imagination. Clearly orgasm by itself is not much help as many patients with intense anxieties have fully normal sex lives without this improving their obsessions, phobias, or social deficits. If intensely positive emotions turn out to be as therapeutic as negative ones, then the hypothesis could not be sustained that abreaction acts as a device for the teaching of coping with unpleasant emotions. One might then broaden the hypothesis into one of learning to modulate any intense emotion.

The concept of coping with unpleasant emotions is related to that of stress immunization. This raises the possibility of preventing disorders by appropriate procedures in childhood and subsequently. The idea was not new in ancient Sparta and amounts to the teaching of stoicism. Experiments in children (Surwit, 1973) and students (Meichenbaum, 1974) indicate this can be done for specific situations like visits to the dentist or pain in the arm from experimental ischemia. The question is how far stress immunization can be generalized, when it should be applied, in what way, for how long, and at what ages. We are only at the start of a long road of research into adaptive behavior.

There has been increasing study recently of self-regulation and coping in anxiety syndromes. An immense range of potential coping devices exists for different people. At various times patients may find it useful to breathe deeply, to relax, or to tell themselves to pull themselves together, or that they are dying of a heart attack, or that it will be tough but they can pull through the difficult treatment process. A coping process might be defined as one which alters the frequency of behavior. Circular reasoning has to be avoided, e.g., "procedure X works, therefore it is a coping device." To avert this pitfall one needs to predict in advance which will constitute coping responses for a given population.

In conclusion, recent advances in behavioral psychotherapy have produced successful treatments of psychopathology. Those which help anxiety syndromes generally contain the critical component of exposure of the patient to the distressing situation, but only a few parameters are known which decide that the outcome of exposure will be therapeutic. The exposure hypothesis unifies the fact that desensitization, implosion, flooding, modeling, shaping, response prevention, and related procedures are all helpful. However, the hypothesis raises many questions about its relationship to stress immunization, coping, and self-regulatory mechanisms. It is a sign of progress in this sector that as soon as we have answered one set of questions, the search moves onto fresh ground which has to be mapped out.

REFERENCES

Bandura, A. (1969). *Principles of Behavior Modification.* New York: Holt, Rinehart & Winston.

Benjamin, S., Marks, I. M., and Huson, J. (1972). Active muscular relaxation in desensitisation of phobic patients. *Psychol. Med. 2,* 381–390.

Crowe, M. (1976). Behavioral treatments. In *Recent Advances in Clinical Psychiatry* (K. G. Grossman, Ed.). Edinburgh: Churchill, Livingston.

Emmelkamp, P. M. G., and Ultee, K. A. (1974). A comparison of "successive approximation" and "self-observation in the treatment of agoraphobia. *Behavior Therapy* (in press).

Gelder, M. G., Bancroft, J. H. J., Gath, D. H., Johnston, D. W., Mathews, A. M., and Shaw, P. M. (1973). Specific and non-specific factors in behavior therapy. *Brit. J. Psychiat. 123,* 445–462.

Gillan, P., and Rachman, S. (1974). An experimental investigation of behaviour therapy in phobic patients. *Brit. J. Psychiat. 124,* 392–401.

Hafner, J., and Marks, I. M. (1976). Exposure in vivo of agoraphobics: contributions of diazepam, group exposure, and anxiety evocation. *Psychol. Med. 6,* 71–88.

Lader, M. H., and Mathews, A. M. (1968). A physiological model of phobic anxiety and desensitisation. *Behav. Res. & Ther. 6,* 411–418.

Lipsedge, M. S. (1974). Therapeutic approaches to compulsive neurosis. Unpublished M. Phil. dissertation, University of London.

Lopiccolo, J. (1969). Effective components of systematic desensitization. Unpublished doctoral dissertation, Yale University.

Marks, I. M. (1965). *Patterns of Meaning in Psychiatric Patients.* Maudsley Monograph No. 13. Oxford University Press.

Marks, I. M. (1969). *Fears and Phobias.* New York: Academic Press.

Marks, I. M. (1971). Phobic disorders 4 years after treatment. *Brit. J. Psychiat. 118,* 683–688.

Marks, I. M. (1975). Behavioral treatments of phobic and obsessive-compulsive disorders: a critical appraisal. In *Progress in Behavior Modification* (R. Herson, Ed.). New York: Academic Press.

Marks, I. M., Hodgson, R., and Rachman, S. (1975). Treatment of chronic obsessive–compulsive neurosis by in vivo exposure. A two year followup and issues in treatment. *Brit. J. Psychiat. 127,* 349–64.

Marks, I. M., Boulougouris, J., and Marset, P. (1971). Flooding versus desensitisation in the treatment of phobic patients: a crossover study. *Brit. J. Psychiat. 119,* 353–375.

Mathews, A. M., Johnston, D. W., Shaw, P. M., and Gelder, M. G. (1974). Process variables and the prediction of outcome in behaviour therapy. *Brit. J. Psychiat. 125,* 256–264.

Meichenbaum, D. (1974). *Cognitive Behavior Modification.* University Programs Modular Studies, General Learning Press.

Nunes, J., and Marks, I. M. (1975). Feedback of true heart rate during exposure in vivo. *Arch. Gen. Psychiat. 32,* 933–936.

Rachman, S., Hodgson, R., and Marks, I. M. (1973). The treatment of obsessive–compulsive neurotics by modeling and flooding in vivo. *Behav. Res. & Ther. 11,* 463–471.

Rooth, G., and Marks, I. M. (1974). Persistent exhibitionism: short-term response to aversion, self-regulation and relaxation treatments. *Arch. Sex. Behav. 3,* 227–247.

Roper, G., Marks, I. M., and Rachman, S. (1975). Passive and participant modeling in exposure treatment of obsessive–compulsive neurotics. *Behav. Res. Ther. 13,* 271–279.

Sartory, G., and Eysenck, H. J. (1976). Strain differentials in the acquisition and extinction of fear responses in rats. *Psychol. Rept.* (in press).

Sherman, A. R. (1972). Real life exposure as a primary therapeutic factor in desensitization treatment of fear. *J. Abnorm. Psychol. 79,* 19–28.

Stampfl, T. G. (1967). Implosive therapy: The theory, the subhuman analogue, the strategy and the technique: Part 1. The theory. In *Behavior Modification Techniques in the Treatment of Emotional Disorders.* Battle Creek, Mich.: V. A. Publication, pp. 22–37.

Stern, R. S., and Marks, I. M. (1973). Contract therapy in obsessive-compulsive neurosis with marital discord. *Brit. J. Psychiat. 123,* 681–684.

Surwit, R. (1973). McGill University, unpublished Ph.D. dissertation.

Watson, J. P., and Marks, I. M. (1971). Relevant and irrelevant fear in flooding–a crossover study of phobic patients. *Behavior Therapy 2,* 275–293.

Wolpe, J. (1958). *Psychotherapy by Reciprocal Inhibition.* Stanford: Stanford University Press.

How Laboratory-Derived Principles of Learning Have Conquered the Neuroses

JOSEPH WOLPE

INTRODUCTION

Neuroses are a very widespread cause of long-term human suffering, which is often severe, occasionally to the point where suicide is attempted. The ages have witnessed many theories of neurosis and countless prescriptions for treatment; but only in the past quarter century have laboratory studies identified their basis, making possible the development of rational methods of therapy, as the following pages will describe.

In broad terms, a neurosis is a habitual susceptibility to emotional disturbance, usually anxiety, in specific situations in which such disturbance is inappropriate. It has been estimated that one person in every ten suffers at some time in his life from a severe and disabling neurosis; and less severe neuroses are far more common. A survey of university students (Willoughby, 1932) revealed substantial degrees of neuroticism in 50% of them. In addition to the personal suffering they cause, neuroses impose a serious financial strain on the entire community. In the United Kingdom, the annual cost in lost productivity was computed in 1959 to be in excess of $800,000,000.

The most common action by a medical practitioner who diagnoses a neurosis is

JOSEPH WOLPE · Temple University School of Medicine and Eastern Pennsylvania Psychiatric Institute, Philadelphia, Pennsylvania to provide palliative treatment, usually in the form of reassurance and sympathy combined with tranquilizers, stimulants, or other drugs. Palliation in fact is often achieved; but since the effects of drugs soon fade, the need for treatment arises again and again, except under special circumstances which will in due course be stated. When the neurosis persists, the practitioner may refer the patient to a psychiatrist to attack the problem "at its roots." In most cases, the psychiatrist is psychoanalytically oriented. Although psychoanalysis continues to be widely regarded as the *crème de la crème* of methods, the truth is that psychoanalytic theory has no scientifically acceptable support (Wohlgemuth, 1923; Salter, 1952; Wolpe and Rachman, 1960; Wolpe, 1961); and psychoanalytic practice has an undistinguished record of results (Eysenck, 1952; Wolpe, 1964). An individual's psychoanalysis frequently continues for four or more years, and may last as long as twenty years, without overcoming the neurosis. In a study conducted by the American Psychoanalytic Association (Brody, 1962; Masserman, 1963) of cases regarded as *completely analyzed*, only 60% were rated as recovered or much improved after an average of 3 or 4 sessions per week for 3 or 4 years, i.e., 600–700 sessions. A certain number of lasting recoveries do follow psychoanalysis, but no more than follow other conventional treatments of neurosis. When neurotic patients unburden themselves to a sympathetic therapist of *any* persuasion, substantial and lasting beneficial change occurs in at least 40% of them. If half of a therapist's patients do well, he naturally attributes the benefit to the methods he uses. This human tendency has helped to maintain faith in psychoanalytic methods, a faith that has been a barrier to the general acceptance of the new and more effective modes of treatment that will be described below.

THE NATURE OF NEUROSIS

Careful interrogation of patients diagnosed as neurotic usually reveals inappropriate anxiety response habits as the central problem.[1] The anxiety is sometimes combined with depression, and frequently produces secondary functional disturbances, which can be of many kinds, such as stuttering, frigidity, impotence, obsessions or compulsions, or psychosomatic reactions like asthma. In many cases, the secondary manifestations overshadow the anxiety, whose cardinal role, then, becomes clear only after probing. For example, a patient with psychogenic asthma may not volunteer anything about anxiety; but investi-

[1] We define anxiety (or fear) as the individual organism's characteristic pattern of autonomic responses to noxious stimulation. This constellation is readily conditionable, so that, as life goes on, conditioned anxiety is more commonly elicited than unconditioned anxiety.

gation will usually reveal anxiety to be the precursor of the asthmatic attack. Similarly, investigation of stutterers almost invariably reveals that the patient can speak fluently to members of his family or close friends, and that the stutter appears only when he speaks to people who evoke anxiety in him.

As will be seen, *neurotic anxiety response habits are learned habits.* Overcoming them means procuring their unlearning; and this is no less necessary in cases of stuttering or asthma than when anxiety is the presenting complaint; for when the anxiety habit is removed, so are its consequences. However, removing it is easier said than done. Neurotic anxiety habits are remarkably persistent. It was work on the neuroses of animals that provided us with reliable ways of dislodging them.

EXPERIMENTAL NEUROSES AND THEIR TREATMENT

One can produce a neurosis in a higher animal if one repeatedly arouses anxiety in him in a particular situation. That situation progressively acquires, through conditioning, an autonomous ability to elicit anxiety, eventually at a very high level. The original, unconditioned anxiety is aroused either by subjecting the animal to a severe conflict (which may take the form of simultaneous powerful impulses to eat and not eat); or by administering noxious stimulation, usually electric shock. In experiments that I conducted many years ago, I employed the latter agent on cats (Wolpe, 1952, 1958). The animal would be placed in a small experimental cage with a grid on its floor that could be electrified. When the animal stopped exploring, I sounded a buzzer and followed it by a high-voltage low-amperage electrical stimulus 2 sec in duration, administered through the grid—a stimulus that was very disturbing but not capable of producing tissue damage. The shock put the animal into a state of great agitation, causing him to run about the cage, clamber onto its sides, or crouch in a corner. It also caused muscular tenseness, and autonomic responses that were predominantly sympathetic—hair erect, pupils widely dilated, rapid respiration. At the cessation of the first shock, these responses subsided. After subsequent shocks, which were administered at irregular intervals of a few minutes, they subsided less and less. Finally, when about a dozen had been administered, the animal was almost as disturbed between shocks as he had originally been only when shocked. This was the criterion level of disturbance for our experimental neuroses; and *thereafter no further shocks were given.*

In all respects in which comparison is possible, experimental and clinical neuroses are similar. There are plainly at least nine points of similarity between animal and human neuroses (Wolpe, 1967); but only three need be referred to here. To begin with, both are extremely resistant to extinction. The neurotic animal is tranquil in his living cage, but he always lastingly reacts with fear in the experimental cage, with an intensity that does not decrease whether he is put back in the cage day after day or kept away for weeks or months–like our human patients, who are disturbed by particular situations indefinitely, even after many exposures without adverse effects other than the anxiety itself.

A second common feature of experimental and clinical neuroses is generalization. The animal is most anxious in the experimental cage, but also to a marked degree anywhere in the experimental room where he was shocked. Generalization is manifested in the fact that he is also anxious in other rooms according to how much they resemble the experimental room. Generalization also characterizes human neuroses. A person who has a fear of heights, for example, is more afraid on the fifth floor than on the second floor, and so on.

A third feature common to animal and human neuroses is unadaptiveness. My experimental animals showed this in several ways, one of which was particularly striking. If a cat who had been starved for a period of 24 hr was placed in the experimental cage, and if pellets of fresh meat were then dropped near him, he would not touch that food no matter how long it was there in his presence. It is hard to think of anything more convincingly maladaptive than the failure of a hungry animal to eat food that is easily available. There is a human parallel in anorexia nervosa. More frequently, however, anxiety impedes other human functions. The ability to work, for example, may be interfered with if there is anxiety at being watched while working, or at the presence of authority figures; and anxiety in sexual situations is the usual cause of impotence or frigidity.

The method by which I succeeded in overcoming the experimental neuroses made use of two of the observations mentioned above. I reasoned that if an animal is prevented by anxiety from eating, there must be, behind that anxiety, a stronger excitation than that underlying his drive to eat. It seemed reasonable to suppose that if the relative strengths of the two drives could be reversed, then instead of the anxiety inhibiting the eating, eating would occur and anxiety be inhibited. It was possible systematically to vary the relative strengths of the drives by offering the animal meat in the rooms to which anxiety was generalized, going in descending order of their resemblance to the experimental room. A room was always found where the animal would eat. There, he ate successive pellets of meat more and more readily as the strength of evoked anxiety declined. We moved up the sequence of rooms, one by one, as feeding eliminated the anxiety from each. Eventually, the anxiety-evoking potential of the experimental room and the experimental cage was eliminated.

These therapeutic experiments suggested that reciprocal inhibition might afford a therapeutic principle. It was postulated that *if in relation to a stimulus to anxiety a response is evoked that is incompatible with anxiety, the anxiety will be to some extent inhibited and, in consequence, the anxiety habit weakened.* This principle has lent itself to implementation in a variety of ways.

CLINICAL APPLICATIONS OF THE RECIPROCAL INHIBITION PRINCIPLE

It was natural first to consider feeding when the question arose of applying the reciprocal inhibition principle to human subjects. The therapeutic use of feeding had in fact been reported as early as 1924—by Jones at the University of California; but her work aroused no interest at that time. One of her reports (Jones, 1924) dealt with the now famous case of Peter, aged 2 years and 10 months, who was afraid of a rabbit and other furry animals. He was put in a chair at one end of a long room and given food that he liked; and then the rabbit was brought in a wire cage as close as possible without interfering with his eating. The rabbit was brought closer by degrees without the feeding being interrupted in successive sessions over a period of weeks. Eventually Peter could handle the animal and enjoy playing with it.

Feeding has not been found to be of much value in treating adult neuroses, presumably because it loses much of its emotional impact with maturation. Fortunately, a variety of other counteranxiety responses are available, and have been successfully used in adult neuroses. Notable examples (Wolpe, 1958, 1973) are the emotional responses that go with deep muscle relaxation; sexual responses; the appropriate expression of anger, affection, and other feelings; and a variety of directly suggested emotional responses.

To utilize any of these responses it is first necessary for the therapist to identify unequivocally the stimuli that trigger his patient's neurotic anxiety responses. This crucial operation, which is called *behavior analysis*, must be undertaken with care. It entails a detailed history of the onset and development of each neurotic response and an exploration of the stimulus factors that currently control it; and a background history of the patient's early life, his educational and social experiences, and his love life. Only with an accurate picture of the stimulus patterns relevant to the patient's neurotic anxiety responses can a rational therapeutic strategy be formulated.

Having determined the categories of stimuli that trigger neurotic responses, the next step, following the model of the treatment of experimental neuroses, is to arrange for anxiety-countering responses to compete with the anxiety responses. This requires that we have at our disposal stimuli that can evoke weak levels of anxiety. We compile a list of stimuli in each area of neurotic disturbance, and arrange them in order of their anxiety-evoking potential. Such a list is called a *hierarchy*. The "weakest" stimulus will be the first to be introduced in treatment, the others following rank order, as weaker items lose their power to evoke anxiety. This process of gradual breakdown of an anxiety response habit is known as *desensitization.*

According to the requirements of the case, we may use either real stimuli or their imagined equivalents. For example, if a person is fearful of crowds, we may employ the counter-anxiety effects of deep muscle relaxation (Jacobson, 1938) to inhibit the anxiety evoked by exposure to real crowd situations, at first of low anxiety-evoking power. The patient will "switch on" the relaxation of muscles not in active use for the sake of these anxiety-inhibiting effects.[2]

Systematic Densensitization Employing Imagined Scenes

Using muscle relaxation as the source of inhibition of anxiety, the method most commonly practiced is "standard" systematic desensitization. The patient, who has been trained in relaxation by a modification of Jacobson's (1938) technique, relaxes in the consulting room and then imagines, with closed eyes, the feared situations in hierarchical order. In the instance of a fear of crowds, the "weakest" crowd situation is presented repeatedly to his imagination, a few seconds at a time, until it no longer evokes any anxiety. Then the next "stronger" situation is introduced. Eventually the patient can imagine the maximum crowd situation without anxiety. Desensitization with imagined scenes has the advantage of convenience in those subjects (who make up about 85%) who can imagine realistically. It has the further advantage of giving the therapist precise control of the stimulus situations to which the patient is exposed. And what is most important is the fact that the patient becomes free from anxiety in all the corresponding real situations, no less than if real situations had been involved in the treatment.

The example of a fear of crowds is a simple one; and even then the relevant stimulus material has to be extracted, for the patient does not provide it in advance, laid out on a platter. Many cases are much more complex. We have already noted those whose presenting complaints do not include mention of anxiety, but relate only to functional disabilities or psychosomatic symptoms. But even when the immediate complaint is of anxiety, a skillful behavior analysis is usually needed to establish the actual stimulus-response relations.

Cases of agoraphobia afford a good example. It is extremely unsafe to make the assumption that the treatment of agoraphobia necessarily involves desensitizing to a hierarchy based on distance. A behavior analysis may show that what the patient really fears is attacks of pain in the chest; despite medical assurances of a health heart; and the farther away from possible help he is, the more anxious he becomes. Desensitization here must be to his own chest pain, and not

[2] Numerous studies have shown that deep muscle relaxation produces autonomic effects opposite to those of anxiety. (e.g., Jacobson, 1938; Paul, 1969*a*; Van Egeren *et al.*, 1971) and that they are able to weaken anxiety habits (Paul, 1969*b*; Wolpe and Flood, 1970).

to distance. Another basis of agoraphobia is fear of people. If contact with strangers elicits much anxiety, a person may be afraid of going away from home to places where he is sure to encounter strangers. It is then necessary to decondition the interpersonal anxiety. The commonest context of agoraphobia is involvement in an unsatisfactory relationship. Usually, the patient is an unhappily married woman very low in self-sufficiency, whose phobia has appeared after several years of marriage. In contrast to a woman with normal self-sufficiency, who, having tried unsuccessfully to improve her marital situation, would take steps toward divorce, the woman low in self-sufficiency remains frozen in her situation, because the idea of being on her own is extremely frightening; and the fear generalizes to the physical situation of being out alone. In such a case, it is obvious that interpersonal reconditioning is needed rather than desensitization to spatial separation.

As mentioned above, other anxiety inhibiting agents can take the place of relaxation in a desensitization program. One that is used especially for patients who are not able to relax depends upon external inhibition (Pavlov, 1927). The patient, sitting in a comfortable chair with eyes closed, is asked to imagine the "weakest" scene from a hierarchy and to raise the index finger of his right hand as soon as the image is clear. Two or three intrusive but not painful galvanic stimuli are then delivered to his left forearm in quick succession. As soon as possible, he again visualizes and raises his finger, and is again shocked. Scenes can be repeated as many as eight times per minute. As a rule, after 10 to 30 presentations of a scene, the anxiety response to it is completely eliminated; and then the therapist introduces the next scene in the hierarchy, and so on.

Another method of anxiety-inhibition depends on the evocation of other emotional responses, generally of a pleasant character, by direct suggestion (Rubin, 1972) or by the use of imagery pleasing to the individual, such as skiing or lying on the beach (Lazarus and Abramovitz, 1962). Transcendental meditation (Wallace, 1970; Boudreau, 1972) and biofeedback (e.g., Budzinski *et al.*, 1970) are additional means of inhibiting anxiety.

Special mention must be made of the inhibition of anxiety by tranquilizing drugs, since this is a method that has great potential public health importance because of its economy. For some years, barbiturates have been used as relaxation-adjuvants in systematic desensitization (e.g., Brady, 1966; Friedman, 1966; Reed, 1966). An experiment by Miller *et al.* (1957) has pointed the way to the possible use of tranquilizing drugs as deconditioning agents in the life situation. Animals made fearful in an experimental situation rapidly overcame their fear if they were exposed to that situation repeatedly while injected with chlorpromazine, but not after injections of saline. A considerable number of individual patients have been systematically treated by exposing them to disturbing real-life situations only when they are effectively tranquilized by such drugs as Librium (e.g., Miller, 1967; Wolpe, 1973). In a high proportion of these patients, if pains are taken never to expose them to these situations unless anxiety has been controlled by the drug, it is found that in a matter of weeks diminishing doses of the drug can control it; until finally the patient has no more anxiety in the situation even without any drug. This is a method that deserves thorough exploration, because it takes up very little of the therapist's time.

The Therapeutic Use of Assertive Training

The term "assertive" covers behavior involving the expression of practically all feeings other than anxiety, on the assumption–strongly supported by experience as well as by formal studies (e.g., Arnold, 1945, 1960; Simonov, 1967; McFall and Lillesand, 1971)–that such expression tends to inhibit anxiety responses. The patient has difficulty in expressing himself where it is reasonable and right to do so because he is inhibited by the anxieties elicited in him by other persons. He may, for example, when unfairly imposed upon by others, be prevented by these anxieties from protesting, even though he feels injured and annoyed. The therapist encourages the outward expression of his just annoyance and coaches him in this, often with the help of role-playing in the office ("behavior rehearsal"). When other kinds of emotional expression like affection, admiration, and revulsion are inhibited by fear, their expression is likewise taught and encouraged.

Frequent repetition of emotionally expressive behaviors leads to the weakening and elimination of the constricting anxieties. Success is evidently due to a summation between the patient's spontaneous reaction to a situation (e.g., his resentment) and the coaxing of the therapist toward action in the same direction. A total action potential is thus produced that is great enough to overcome the inhibitory effects of the anxiety and reciprocally to inhibit the latter. In the course of the same operations, another learning process also goes on–the operant conditioning of the motor acts of assertion, which are reinforced by such rewarding consequences as the attainment of reasonable goals previously out of reach.

The Use of Sexual Responses

Sexual responses afford a convenient way of inhibiting anxiety that is conditioned to sexual situations and that interferes with sexual performance. They have had their chief application in cases of impotence, especially premature ejaculation.

In all its phases prior to ejaculation, the male sexual response is essentially a parasympathetic function. Anxiety impairs male sexual performance because sympathetic arousal inhibits parasympathetic function, with the result that penile erection is weakened; or else the excessive sympathetic activation prematurely precipitates ejaculation, a sympathetic activity. In using the sexual response for therapeutic purposes, it is necessary at all times to ensure that it is "stronger" than the anxiety against which it is pitted. Hence, a preliminary step is to determine at what point in the sexual approach the man *begins* to feel anxious. If it is when lying next to his wife in bed in the nude, he is at first allowed to do that and no more (of course with the collaboration of his wife). It is generally found that after two or three repetitions of this situation the man no longer feels anxiety in it. He is then permitted to go on to the next stage—such as fondling her breasts; and to this he is now confined until all anxiety subsides. Subsequent steps will include approximating the penis to the vulva, increasing degrees of intromission without movement, and finally increasing movement. In an average of about 8 weeks, about two-thirds of patients can be expected to achieve normal sexual function, and another one-sixth to function well enough to satisfy their wives.

The Use of the Patient's Emotional Responses to the Therapist

As indicated in the introduction, all traditional forms of psychotherapy (i.e., therapies other than behavior therapy) produce lasting beneficial changes in about 40% of patients. The incidence of such changes seems to bear no relation to the techniques employed (Landis, 1937; Wilder, 1945). A reasonable supposition, still to be rigorously tested, is that the basis of the changes is that the therapeutic interview arouses in some patients emotional responses that inhibit the anxiety responses to verbal stimuli that crop up during the interviews. These effects are fortuitous in that the therapist neither regulates the stimuli that are introduced, nor deliberately elicits the counter-anxiety emotions. Therapeutic effects of this nature of course also occur in the sessions conducted by behavior therapists; but the behavior therapist often also takes additional advantage of the "antianxiety" emotions that the interview situation may evoke, by using them to inhibit the anxiety of stimuli he systematically introduces from hierarchies. Especially in patients who cannot learn deep relaxation, the behavior therapist may employ these "spontaneous" emotions as a basis for desensitization.

More often, behavior therapists exploit these interpersonal emotions in the context of real stimuli, to bring about "*desensitization in vivo*." For example, in the case of a lawyer whose practice had been restricted by a fear of making mistakes in the gaze of other people, a two-dimensional hierarchy was constructed, ranking mistakes according to their "seriousness," and audiences according to their threat. Then I arranged for him to make a speech in a small lecture room before a "mild" audience and to include a minor mistake. A little anxiety was aroused by this; and then the same speech was repeated four times, in the course of which the anxiety aroused by the mistake declined to zero. Increasingly "serious" mistakes were serially introduced, and then, later, more threatening audiences. Progressive diminution of anxiety in real public-speaking situations resulted from these maneuvers.

Flooding

A method that has recently aroused considerable interest is "flooding." It consists of treating neurotic anxiety by exposing the patient to relatively strong anxiety-evoking stimuli—in contrast to what is done in desensitization. A straightforward account of it was given by Malleson (1959), but it has also been carried out with various trappings under such labels as "paradoxical intention" (Frankl, 1960) and "implosive therapy" (Stampfl and Levis, 1968). A man who has a phobia for spiders may at once be asked to imagine a spider a few feet away and to continue the image until there is clear diminution of the anxiety it elicits—which often happens when the image is persisted with—for a time that varies from a few minutes to half an hour or longer. It is usually more satisfactory to present the stimulus *in vivo*, because it is vital to ensure continuity, as indicated by Rachman's (1966) finding that even 200 min of exposure to strong stimuli were without therapeutic effects when the exposure was broken up into 2-min periods.

Although flooding is much less pleasant than desensitization, and probably on the whole less effective (e.g., Willis and Edwards, 1969; De Moor, 1970; Mealiea and Nawas, 1971), it is certainly an important addition to our therapeutic armamentarium. In those obsessive-compulsive patients whose lives are ruled by fear of contamination, prolonged exposure to degrees of contamination that evoke moderate disturbance leads to increasing and finally complete tolerance of the contamining agent, usually in a matter of weeks—a significant breakthrough in what used to be a most refractory syndrome (Meyer, 1966; Hodgson *et al.*, 1972).

The successful effects of flooding have been regarded by some (e.g., Marks, 1972) as constituting an argument against the operation of reciprocal inhibition in general, on the reasoning that the high levels of anxiety appear to decline without the intervention of a competing response. Even if this were true of flooding, it would not be possible to conclude that it was also true of desensitization. Actually the fact that the anxiety level drops makes it clear that *some* inhibitory process must be at work. A few years ago the possibility was suggested (Wolpe, 1969) of transmarginal inhibition (Pavlov, 1927; Gray, 1964). Recently, however, there have been increasing grounds for thinking that flooding also, after all, depends on reciprocal inhibition. Experiments performed by Gath and Gelder (1971) have strongly suggested that flooding and desensitization are on a continuum; and observations in our department, as well as those of Marks (1972) indicate that flooding is relatively ineffective in the absence of a therapist. It is interesting to note the parallel between this and the observation of Grinker and Spiegel (1945) that abreactions related to war neurosis were never therapeutic when they took place outside a formal therapeutic setting.

Reciprocal Inhibition in the Overcoming of Unadaptive Habits of Thought

Although conditioned anxiety is the core of most neuroses, there is also frequently a need to deal with unadaptive learned habits of other kinds. The most important of these are misconceptions—beliefs that do not conform with reality. Some of these are serious sources of anxiety. The erroneous thought of an imminent heart attack can be an antecedent of anxiety; and so can entering an elevator if one has the thought that one may suffocate. Some severe neuroses have developed when bizarre sensations have been taken as evidence of being on the brink of insanity. In all these cases, the anxiety would be appropriate if the belief were true: It is appropriate to feel anxious if one is really threatened by suffocation. The therapeutic problem here is to remove the false belief. This involves bringing correct information to bear on the stimulus situation that arouses the fear—pointing out, for example, the presence of air vents in the elevator. I have elsewhere reported (Wolpe, 1973, pp. 57–79) an example of rapid recovery from a severe neurosis of 10 years' duration achieved by correcting the patient's belief that her failure to achieve coital orgasms was due to constitutional inadequacy, by demonstrating to her that situational factors were responsible.

Although displacing one idea by another is a very common happening, it too is an instance of the operation of reciprocal inhibition. It has long been known, on the basis of experiments on pairs of nonsense syllables, that a bond A-B is broken by repeated presentation of the sequence A-C. Osgood (1948) was the first to note that the evocation of C is accompanied by the reciprocal inhibition of B, since the two responses are incompatible.

HOW FAR HAVE THE NEUROSES BEEN CONQUERED?

In the introduction to this paper, reference was made to the fact that a wide range of psychotherapeutic transactions all produce marked benefit in about 40% of neurotic patients. That fairly uniform success rate indicates that there is a change agent common to them all and that their distinctive methods have very little, if any, effective role. The common agent would appear to be the anxiety-inhibiting emotional responses that the interview situation may evoke. Anyone who claims that his particular therapeutic procedures have additional efficacy must demonstrate a success rate significantly higher than the 40% baseline.

It would be expected on the basis of the evidence of the learned character of neurotic behavior, that the relearning methods embodied in behavior therapy would procure for their practitioners results far exceeding the baseline. This seems indeed to be the case. Studies of the effects of behavior therapy on unselected patients in clinical practice show that when the procedures are carried out by skilled practitioners on the basis of adequate behavior analyses, the recovery rate is usually in the region of 80–90% (see Table 1).

In my own series (Wolpe, 1958) of 210 patients, 89% were either recovered or at least 80% improved, as jointly estimated by therapist and patient, in a mean of about 30 sessions. The criteria were those suggested by Knight (1941)—symptomatic improvement, increased productiveness, improved adjustment and pleasure in sex, and ability to handle ordinary psychological conflicts and reasonable reality stresses. Hussain (1964) reported that in 99 out of 105 patients he achieved complete or almost complete removal of symptoms by methods based on the reciprocal inhibition principle. A series of 27 patients treated by Hain *et al.* (1966) is of particular interest because the therapists considered themselves "avowed eclectics" rather than behavior therapists, and because they limited their behavioral interventions largely to systematic desensitization; and yet 70% of their patients showed marked improvement in a mean of 19 sessions. The figures of Latimer and Groves are from a preliminary study (1975) of results obtained by trainees in behavior therapy in our department in cases which have afforded a reasonable trial of the reciprocal inhibition techniques that seemed applicable to them

A little-mentioned fact about behavior therapy is that there is *a clear relation*

Table 1. Outcome Studies of Behavior Therapy in Unselected Neuroses (See Text)

Series	N (total)	Apparently cured or much improved	
		Number	Percentage
Wolpe (1958)	210	188	89.5
Hussain (1964)	105	99	94.2
Hain *et al.* (1966)	27	19	70.0
Latimer and Groves (1975)	21	17	81.0

in time between specific interventions and therapeutic change in target behaviors.

Ever since behavior therapy came into being, psychoanalysts have prognosticated that those patients who benefit from it would be liable to relapse and symptom substitution because the "nuclear conflicts" that psychoanalytic theory postulates are not resolved. The truth is that when behavior therapy is skillfully conducted, it is *very uncommon* for relapse or symptom substitution to occur. For example, in 45 cases who had been followed up for periods ranging from 2 to 7 years (Wolpe, 1958), only one presumptive relapse could be found. I have at present followed up 24 cases for periods ranging from 20 to 25 years with no relapses or symptom substitutions, and with a general trend toward increasing emotional freedom, functional competence, and the waning of any residual symptoms. The adequacy of the behavior analyses and the skill with which the behavioral methods are used are, however, of cardinal importance. Incomplete or misplaced treatment can leave the patient with a liability to anxiety-evocation that may be a basis for reconditioning.

Up to this point we have been evaluating the efficacy of combinations of behavioral methods tailored to individual neurotic constellations. What is of greater scientific interest is to evaluate the effects of particular procedures. Systematic desensitization, the most widely used of all behavior therapy techniques, has been the subject of a large number of controlled studies. One of the most significant of these is by Paul (1966) who investigated its effects on subjects with severe fears of public speaking, employing exclusively the therapeutic services of psychoanalytically oriented therapists. Densensitization was compared with the therapists' accustomed insight therapy and with a procedure called "attention-placebo" that involved suggestion and support. Each patient had five therapeutic sessions. In the outcome, systematic desensitization was found significantly superior on cognitive, physiological, and performance measures. On standard clinical criteria, 86% of the patients treated by desensitization were much improved, compared with 20% for the insight group and none for the attention-placebo group (Table 2).

Subsequently, Paul (1968) reviewed the outcome data of systematic desensitization based on the work of more than 90 different therapists with nearly 1000 different clients. Of 55 uncontrolled studies, 46 showed positive correlations between desensitization and favorable change, 6 were ambiguous in results or reporting, and only 3 were negative. Of 8 studies that involved designs out of which case-effect relationships could be evaluated—Lang and Lazovik (1963), Lang (1964), Lang (1965), Lang *et al.* (1965), Paul (1966, 1968), Moore (1965), and Davison (1968)—all found solid evidence for the specific effectiveness of systematic desensitization. Paul comments that "systematic desensitization is the first psychotherapeutic procedure in history to withstand rigorous evaluation."

Table 2. Percentage Breakdown of Cases in Traditional "Improvement" Categories from Stress-Condition Data[a]

Treatment	N	Unimproved	Slightly improved	Improved	Much improved
Desensitization	15	–	–	14%	86%
Insight	15	7%	46%	27%	20%
Attention–placebo	15	20%	33%	47%	–
Treatment–control	29	55%	28%	17%	–

[a]From Paul, G. L. (1966). *Insight versus Desensitization in Psychotherapy*. Stanford, California: Stanford University Press.

The evidence is clear that therapeutic interventions based upon experimentally established principles have greatly enhanced the treatment of neuroses. About 10% of them still defy our best efforts, either because we fail to ascertain the crucial stimulus-response relations or because none or our available methods can effectively combat the unwanted responses. But a vast amount of research is going on whose results are mainly channeled into four journals entirely devoted to behavior therapy. The fruits of this research are progressively cutting down the residue of unsuccessfully treated cases.

REFERENCES

Arnold, M. B. (1945). The physiological differentiation of emotional states. *Psychol. Rev. 52*: 35.

Arnold, M. B. (1960). *Emotion and Personality, Vol. 1.* New York: Columbia University Press.

Boudreau, L. (1972). Transcendental meditation and yoga as reciprocal inhibitors. *J. Behav. Ther. Exp. Psychiat. 3*: 97.

Brady, J. P. (1966). Brevital-relaxation treatment of frigidity. *Behav. Res. Ther. 4*: 71.

Brody, M. W. (1962). Prognosis and results of psychoanalysis. In *Psychosomatic Medicine* (J. H. Nodine and J. H. Moyer, eds.). Philadelphia: Lea and Febiger.

Budzinski, T., Stoyva, J., and Adler, C. (1970). Feedback-induced muscle relaxation: Application to tension headache. *J. Behav. Ther. Exp. Psychiat. 1*: 205.

Davison, G. C. (1968). Systematic desensitization as a counterconditioning process. *J. Abnorm. Psychol. 73*: 91.

DeMoor, W. (1970). Systematic desensitization versus prolonged high intensity stimulation (flooding). *J. Behav. Ther. Exp. Psychiat. I*: 45.

Eysenck, H. J. (1952). The effects of psychotherapy: An evaluation. *J. Consult. Psychol. 16*: 319.

Frankl, V. (1960). Paradoxical intention: A logotherapeutic technique. *Am. J. Psychother. 14*: 520.

Friedman, D. E. (1966). A new technique for the systematic desensitization of phobic symptoms. *Behav. Res. Ther. 4*: 139.

Gath, D., and Gelder, M. G. A. (1971). Treatment of phobias–desensitization versus flooding. Paper delivered to Dept. Psychiatry, Temple Univ. Medical School, November 8, 1971.

Gray, J. A. (1964). *Pavlov's Typology.* Oxford: Pergamon Press.

Grinker, R. R., and Spiegel, J. P. (1945). *War Neuroses.* Philadelphia: Blakiston.

Hain, J. D., Butcher, R. H. G., and Stevenson, I. (1966). Systematic desensitization therapy: An analysis of results in twenty-seven patients. *Brit. J. Psychol. 112*: 295.

Hodgson, R., Rachman, S., and Marks, I. M. (1972). The treatment of chronic obsessive-compulsive neuroses: Follow-up and further findings. *Behav. Res. Ther. 10*: 181.

Hussain, A. (1964). Behavior therapy using hypnosis. In *The Conditioning Therapies* (J. Wolpe, A. Salter, and L. J. Reyna, Eds.). New York: Holt, Rinehart & Winston.

Jacobson, E. (1938). *Progressive Relaxation.* Chicago: Univ. of Chicago Press.

Jones, M. C. (1924). A laboratory study of fear. The case of Peter. *J. Genet. Psychol. 31*: 308.

Knight, R. P. (1941). Evaluation of the results of psychoanalytic therapy. *Am. J. Psychiat. 98*: 434.

Landis, C. (1937). A statistical evaluation of psychotherapeutic methods. In *Concepts and Problems of Psychotherapy* (L. Hinsie, Ed.). New York: Columbia Univ. Press.

Lang, P. J. (1964). Experimental studies of desensitization psychotherapy. In *The Conditioning Therapies* (J. Wolpe, A. Salter, and L. J. Reyna, Eds.). New York: Holt, Rinehart & Winston.

Lang, P. J. (1965). Psychotherapy, pseudotherapy, and behavior therapy. Paper presented at the Annual Meeting of the Midwestern Psychological Association, May, 1965.

Lang, P. J., and Lazovik, A. D. (1963). Experimental desensitization of a phobia. *J. Abnorm. Soc. Psychol. 66*: 519.

Lang, P. J., Lazovik, A. D., and Reynolds, D. J. (1965). Desensitization, suggestibility, and pseudotherapy. *J. Abnorm. Psychol. 70*: 395.

Latimer, P., and Groves, G. (1975). Study of clinic population and therapy outcome–preliminary report. Unpublished manuscript.

Lazarus, A. A., and Abramovitz, A. (1962). The use of "emotive imagery" in the treatment of children's phobias. *J. Ment. Sci. 108*: 191.

McFall, R. M., and Lillesand, D. B. (1971). Behaviour rehearsal with modeling and coaching in assertion training. *J. Abnorm. Psychol. 77*: 313.

Malleson, N. (1959). Panic and phobia. *Lancet 1*: 225.

Marks, I. M. (1972). Flooding (implosion) and allied treatments. In *Behavior Modification Principles and Clinical Applications* (W. S. Agras, Ed.). Boston: Little, Brown & Co.

Masserman, J. H. (1963). Ethology, comparative biodynamics and psychoanalytic research. In *Theories of the Mind* (J. Scher, Ed.). New York: The Free Press.

Mealiea, W. L., Jr., and Nawas, M. M. (1971). The comparative effectiveness of systematic desensitization and implosive therapy in the treatment of snake phobia. *J. Behav. Ther. Exp. Psychiat. 2*: 85.

Meyer, V. (1966). Modifications of expectations in cases with obsessional rituals. *Behav. Res. Ther. 4*: 273.

Miller, G. E. (1967). Personal communication.

Miller, R. E., Murphy, J. V., and Mirsky, I. A. (1957). Persistent effects of chlorpromazine on extinction of an avoidance response. *Arch. Neurol. Psychiat. 78*: 526.

Moore, N. (1965). Behavior therapy in bronchial asthma: A controlled study. *J. Psychosom. Res. 9*: 257.

Osgood, C. E. (1948). An investigation into the causes of retroactive inhibition. *J. Exp. Psychol. 38*: 132.

Paul, G. L. (1966). *Insight Versus Desensitization in Psychotherapy.* Stanford: Stanford Univ. Press.

Paul, G. L. (1968), Two-year follow-up of systematic desensitization in therapy groups. *J. Abnorm. Psychol. 73*: 119.

Paul, G. L. (1968*a*). Physiological effects of relaxation training and hypnotic suggestion. *J. Abnorm. Psychol. 74*: 425.

Paul, G. L. (1969*b*). Outcome of systematic desensitization II. In *Behavior Therapy Appraisal and Status* (C. M. Franks, Ed.). New York: McGraw-Hill Book Co., p. 105.

Pavlov, I. P. (1927). *Conditioned Reflexes.* Trans. by G. V. Anrep. New York: Liveright.

Rachman, S. (1966). Studies in desensitization II: flooding; Studies in desensitization III: speed of generalization. *Behav. Res. Ther. 4*: 1.

Reed, J. L. (1966). Comments on the use of methohexitone sodium as a means of inducing relaxation. *Behav. Res. Ther. 4*: 323.

Rubin, M. (1972). Verbally suggested responses as reciprocal inhibition of anxiety. *J. Behav. Ther. Exp. Psychiat. 3*: 273.

Salter, A. (1952). *The Case Against Psychoanalysis.* New York: Holt, Rinehart & Winston.

Simonov, P. V. (1967). Studies of emotional behavior of humans and animals by Soviet physiologists. Paper read at Conference on Experimental Approaches to the Study of Behavior, New York.

Stampfl, T. G., and Levis, D. J. (1968). Implosive therapy, a behavioral therapy. *Behav. Res. Ther. 6*: 31.

Van Egeren, L. F., Feather, B. W., and Hein, P. L. (1971). Desensitization of phobias: Some psychophysiological propositions. *Psychophysiology 8*: 213.

Wallace, R. K. (1970). Physiological effects of transcendental meditation. *Science 167*: 1751.

Wilder, J. (1945). Facts and figures on psychotherapy. *J. Clin. Psychopath. 7*: 311.

Willis, R. W., and Edwards, J. A. (1969). A study of the comparative effectiveness of the systematic desensitization and implosive therapy. *Behav. Res. Ther. 7*: 387.

Willoughby, R. R. (1932). Some properties of the Thurstone Personality Schedule and a suggested revision. *J. Soc. Psychol. 3*: 401.

Wohlgemuth, A. (1923). *A Critical Examination of Psychoanalysis.* London: Allen and Unwin.

Wolpe, J. (1952). Experimental neurosis as learned behavior, *Brit. J. Psychol. 43*: 243.a

Wolpe, J. (1958). *Psychotherapy by Reciprocal Inhibition.* Stanford: Stanford Univ. Press.

Wolpe, J. (1961). The prognosis in unpsychoanalyzed recovery from neurosis. *Am. J. Psychiat. 118*: 35.a

Wolpe, J. (1964). The comparative clinical status of conditioning therapies and psychoanalysis. In *The Conditioning Therapies* (J. Wolpe, A. Salter, and L. J. Reyna, Eds.). New York: Holt, Rinehart & Winston.

Wolpe, J. (1967). Parallels between animal and human neuroses. In *Comparative Psychopathology–Animal and Human* (J. Zubin and H. Hunt, Eds.). New York: Grune & Stratton.

Wolpe, J. (1969). *The Practice of Behavior Therapy*, First ed. New York: Pergamon Press Inc.

Wolpe, J. (1973). *The Practice of Behavior Therapy*, Second ed. New York: Pergamon Press Inc.

Wolpe, J., and Flood, J. (1970). The effect of relaxation on the galvanic skin response to repeated phobic stimuli in ascending order. *J. Behav. Ther. Exp. Psychiat. 1*: 195.

Wolpe, J., and Rachman, S. (1960). Psychoanalytic evidence: A critique based on Freud's case of little Hans. *J. Nerv. Ment. Dis. 131*: 135.

Recurrent Dilemmas in Behavioral Therapy

HOWARD F. HUNT

For a variety of reasons that are easy to understand, the literature on behavioral therapy tends to underemphasize problems and difficulties. My comments here arise out of my own experience and my joint consulting activities with Drs. Aronoff, Lagos, Luhrmann, Olds, Singh, and, particularly, Albert over the past 8–9 years. These comments will touch on the most common recurrent issues encountered in applying a behavioral approach to psychotherapeutic treatment, and on some tentative solutions. The enumeration only notes the major points, and does not pretend to be exhaustive, of course.[1]

Most of my consulting and supervision has dealt with regressed psychotic or borderline patients who have failed to improve or have regressed further under the normal therapeutic regimes of my home institution. By the time of referral, the responsible therapist often has become desperate, the staff has polarized, and the patient is on his way out. Often the therapist has become bogged down in a power struggle with the patient and staff, or a victim of rescue fantasies and/or countertransferential resentment. Or I may be asked to help with a refractory, irritating ward management problem. What has seemed possible or feasible to do usually has been attempted already, including direct application of garden-variety behavioral methods, and has produced whatever effects it could. I am usually the last stop.

[1] For a more comprehensive discussion, see Hunt (1975).

HOWARD F. HUNT · N.Y. State Psychiatric Institute and Columbia University

In most instances, we find ourselves looking into "systems problems," contexts that extend well beyond the patient and his behavior and intrapsychic life and into the ward and the family. Fortunately, substantial gains often may be achieved by tipping the psychosocial balance of the total system toward responsiveness in support of healthy, prosocial behavior, and away from the usual balance in support of pathology. One does not need total control–1984 in 1975–to produce significant improvement. Indeed, attempts to gain such control often are counterproductive in practice, given contemporary concerns for the rights of patients and conditions of staff employment. But one does need robust, relatively foolproof procedures that work well enough often enough to deserve the confidence of patients and staff. And the aim, of course, is not to have the therapy be behavioral, but rather to have it be effective. In these junctures, however, behavioral approaches often produce positive change after other approaches have failed.

DOES THE BEHAVIORAL APPROACH HELP BECAUSE IT IS "SCIENTIFIC"?

In my own clinical experience, I have not found behavioral approaches to psychotherapy to be as simple, straightforward, "scientific," and unambiguous as behavioristic protagonists sometimes have implied.

The logical problems in clinical application differ somewhat from those of the laboratory. In the laboratory, we work from theories or hypotheses that suggest how behavior, as a dependent variable, should be affected by manipulations of independent variables–causative or antecedent variables–that we can control. If things do not work out, we can go back to the drawing board, develop new hypotheses, and rerun the whole enterprise with new, "naïve" subjects substantially equivalent to the ones already studied. Here, the context of justification or confirmation flows naturally out of the context of discovery or invention (Reichenbach, 1938). Our theories tell us what should happen, and we can find out if it does, exploring the problem repeatedly under supposedly comparable conditions until the data satisfy the conventions of scientific proof.

In contrast, in the clinical situation we are confronted with behavior (the dependent variable) and must guess as to what could have produced it–guess as to what independent variables it is a function of. Then, in the light of that guess, and with the aid of hypotheses as to how changes in contemporary or intercurrent factors in patients' lives may alter their behavior, we start to work and hope for the best. Because no two patients and no two situations are alike, we have fewer opportunities for reruns, for corrections of mistaken guesses. Further, with human subjects in clinical settings some socialization experiences and learning may be irreversible: impressions are formed, patients develop theories, skills are acquired in changes that are so enduring that the patient cannot be returned to a condition of naiveté that permits a replicated experiment with that patient in an own-control design. Thus, practical application remains largely in the context of discovery, with firm scientific data in the context of justification most difficult to come by.

In addition, we can control only a fraction of the important independent variables in patients' lives. Patients already have histories. Patients work and live in social contexts that have their own dynamics and necessities. Further, the human capacity and propensity for symbolic behavior and internalized representations permits a decoupling of important behavior from controllable external conditions, and over the broadest range. Such behavior includes expectations, interpretations, theories about the self and the world and the future, and fantasies. These are related in complex and poorly understood ways to external conditions and to other, more overt behavior. Indeed, the current excitement over problems of cognition, thinking, and language in psychology testifies to the importance of these covert aspects of human behavior and to our uncertainties about them.

None of this is to imply that a scientific understanding of therapeutic change and how to produce it is, in principle, impossible, but rather to indicate that it has not been achieved yet. To my mind, behavioral therapy, as well as other kinds of psychotherapy, is still an art form. Considerations from theory, seen within a framework that is at least quasiscientific, can help, however.

Neither behavior (learning) theory nor dynamic theory is a monolithic unity; each has many variants. Because of the contradictions among the theories, they cannot all be true. Most likely, none is wholly true, but that need not stop us. Actually, the main value of a theory is to help one think about a domain productively. Using several contrasting theories, one can think about a domain in several different ways. I would like to suggest that, rather than being competitors, behavioral and dynamic theories can be used as complementaries, giving a therapist more of a three-dimensional view of a patient's problem and revealing facets of the case that might be missed if one metaphor were followed exclusively. But just as a holograph requires coherent light from two laser sources to generate a three-dimensional image, complementary metaphors will be productive only if each is taken seriously and used consistently. Sloppy eclecticism, easy compromise, or premature synthesis will only blur the images.

The behavioral approach helps in other ways, as well. It is optimistic and specific: Something tangible can be accomplished, even in difficult cases, by application of procedures that can be described explicitly. In principle, failures only reflect errors in the behavioral analysis (equivalent to diagnosis) or in execution of treatment—all rectifiable. And the appearance on the scene of a consultant who looks as if he thinks that something can be done, and as if he knows how to do it, has powerful nonspecific effects on staff morale and on the patient.

WHAT ARE THE GOALS OF TREATMENT? TOWARD WHAT SHOULD IT BE DIRECTED?

Too often the target of treatment is "problem" behavior that traditionally is called "maladaptive." Interesting questions arise immediately: "Who is having the problem? "If so maladaptive and self-defeating, why does the behavior persist?

Without at least some reinforcing support from the environment, behavior will disappear by extinction. Thus, as a working hypothesis, it is useful to assume that the "maladaptive" behavior is getting something for the patient, even though it appears unattractive and self-defeating. Indeed, it should be seen as adaptive in some sense, but at a low level of efficiency and at a high cost in effort and self-respect.

Behavior control is a two-way street, so when misbehavior, noncooperation, and other behavioral problems occur at any level one looks first at what those in control of the major reinforcing contingencies in the situation are doing. Usually, the behavior of the patient reflects contingencies (or the lack of them, in frustration) imposed by the controllers. Often the contingencies (or frustrations) are so embedded in social roles and the ecology of the situation that they are essentially invisible except to a perceptive outside observer. Ironically, it is often the persons who have the "problem"—who are complaining about the patient's behavior—who also are unwittingly maintaining it.

The patient's problem behavior usually represents an extension and reiteration of symptomatic behavior that has had a long history of reinforcement in terms of impulse gratification, anxiety reduction, and/or secondary gain. Through the patient's experience, it has been shaped to become a powerful and coercive social stimulus to other people; it produces results in the hospital as it did elsewhere. A patient who cannot reinstate more acceptable prosocial behavior under the usual permissive and accepting hospital routine becomes the problem patient—the kind usually referred to us because his attempts at social control of the situation are so troublesome. The thrust of the request for help in management is usually directed toward control of symptomatic behavior—in effect, a thrust toward *containment* of the patient. Indeed, much that has been written about behavioral therapy seems pointed toward containment.

Simple elimination of troublesome behavior by extinction or punishment without providing alternative behavioral avenues to legitimate gratification for the patient only invites frustration and power struggles. Symptom substitution, once expected by dynamically oriented therapists, does not appear to be a serious problem. Power struggles crop up frequently, however. In the literature on behavioral therapy, such struggles usually surface under rubrics such as "the unresponsive patient," how to obtain patient cooperation, how to find effective reinforcers and why so many that should work do not, difficulties in preventing circumvention of the program by the patients, and so on (see Kazdin, 1973). These power struggles can be most difficult and unpleasant. They take place on the patient's "home turf" where he is the practiced expert who knows just how to vary or modulate his "presentation" for maximum social effect. The patient usually is playing for higher stakes than the therapist, too. What can a therapist throw into the balance that will compare, for example, with an anorexic's willingness to starve to death in order to win?

As a first approximation, at least, treatment efforts should be directed toward establishing (or, more commonly reinstating) effective, reward-producing prosocial behavior. This prosocial repertoire, not the symptoms, represents the real target. As it develops, the patient can relinquish marginally effective, high-cost symptomatic behavior without relinquishing access to legitimate gratifications or his self-respect. In effect, I would emphasize *enhancement* of the patient as an initial thrust to be followed as far as possible. Containment tactics and other direct behavioral procedures such as desensitization would than be reserved to deal with residual symptoms that do not "take care of themselves," that do not disappear as the patient's capacity to qualify for and use normal social reinforcement grows.

Unfortunately, limits in this capacity cannot be overcome for some patients, and only modest improvement appears. Even so, modest progress increases the flux of positive hedonic experiences for all concerned, and is preferable to deepening chronicity and further regression.

BEHAVIORAL ANALYSIS, PSYCHODYNAMICS, AND THE CREATION OF PROSOCIAL BEHAVIORAL REPERTOIRES

Behavioral analysis (as a diagnosis) customarily seeks to identify the behavior to be eliminated, to determine how it is controlled, and to devise a program that will weaken the unwanted behavior and provide for replacing it with a prosocial repertoire. Despite the specificity and clarity of general behavioral principles, this analytic enterprise is also still an art form. What you observe—the patient's behavior—represents the end-product of prior and intercurrent operation of important independent variables. What these may be must be guessed at. And if the behavior of interest is under adjunctive or some other anomalous schedule control, the relation between the behavior and its controlling variables may be quite indirect and obscure (see Hinde and Stevenson-Hinde, 1973; Hunt, 1975). One's power to manipulate even a fraction of the milieu in which a patient is behaving is decidedly limited, and his history is unreachable, so empirical determination of the effects of many variables may be next to impossible. In clinical guessing, one suddenly discovers how empty and paradigmatic even his nearest and dearest behavioral theory actually is. Stimuli, responses, reinforcers, and the like are abstractions, yet in clinical work when one has to act these abstractions have to be fleshed out with particulars. Specifics as to the content of what actually is done can make all the difference in the world. In the usual behavioral reports, common sense (or, often, cryptically employed clinical and dynamic sophistication) determines the selection of these specifics.

Some time ago, Dyrud and I argued that dynamic knowledge can help behavioral therapists make shrewd guesses as to what is controlling patient behavior, and how (i.e., what the patient is working to get, what circumstances he sees as favorable to getting it, and what constructive substitute behaviors and rewards he might be able to settle for) (Hunt and Dyrud, 1968). More recently, Feather and Rhoads (1972 *a,b*) suggested that explicit use of dynamic theory in conjunction with selection of specific behavioral procedures. Further, dynamic theory highlights the importance of the relationship between patient and therapist, touching on its vicissitudes and how it may be used therapeutically, yet this matter is rarely considered explicitly and in any figured manner in most behavioral literature. Yet without an effective relationship—an alliance, the therapist cannot be an effective social reinforcer for his patient. In addition, negative- and counter-transferential reactions may become critical, particularly in dealing with the ubiquitous conflicts over control and autonomy in today's adolescent and young adult patients. Unfortunately, when flirting with psychodynamics it is all too easy to slip into casual and imprecise use of concepts to arrive at easy "explanations" (e.g., motivational). Because of the imprecision, the lapse may not only contribute nothing new but may even undercut whatever rigors the behavioral analysis has brought to the clinical enterprise. Constant, critical vigilance is the only antidote I know.

The emphasis on *enhancement* rather than *containment* of the patient affects the focus of behavioral analysis and, I hope, the patient's long-term career. In addition to being concerned with the "problem" behavior that has led the staff to seek me out, I have found it *even more helpful* to start looking at those aspects of the patient's functioning that still operate effectively—to look at what is *going right* rather than at what is *going wrong.* This reveals areas of presumably intact, reasonably normal behavioral control and, more important, what the patient can and will work for, including activities that can be used as rewards. Once one begins to look, surprisingly large areas of reasonably intact functioning may become apparent. Indeed, except for two or three deeply anhedonic, categorically oppositional patients, almost all patients have turned out to have some things they do well and like to do that are neither illegal, immoral, fattening, nor unavailable. In fact, Blumenthal and Carpenter (1974), in a careful time-sampling study of patient behavior in the Psychiatric Institute's Intensive Care Unit, found that about two-thirds of all instances of patient behavior observed over a period of 4 months were socially appropriate and essentially normal rather than pathological. The other one-third, of course, was the basis for the patients' residence in the Unit, which houses the most disturbed psychiatric patients at the medical center. Even here, patients revealed a substantial base of prosocial behavior, however precariously poised, from which to work out.

For deeply regressed patients, desired performances and their rewarding consequences need to be rather concrete in the beginning, with fixed and specific tasks and tangible reinforcements. Expanded success experience not only extends the patient's functioning behavioral repertoire but also enhances self-esteem. As things start to work better and better for the patient, and the quality of his subjective life begins to improve with the increase in the positive hedonic flux, usually one sees improvement in behavior outside the direct scope of the program (ripple effect). Then the program's sights must be raised progressively through graded structure to aim toward expansion into broader realms of constructive, prosocial behavior. The program must be designed so that such behavior eventually comes under the operant control of the intermittent rewards that sustain all of us (e.g., real appreciation for a hard job well done, the development of interests and standards that make at least some achievements intrinsically reinforcing). The goal of restoring capacity for constructive choice and autonomous self-control cannot be achieved unless concrete rewards for simple performance are eventually faded out (progressively withdrawn) to be replaced by bigger, better, more adult and normal rewards for more complex self-management and effective planning. Alternatively, the program may shift the successful patient categorically to a new status that confers more privileges, access to a larger number of preferred activities, privacy, and so on, all contingent upon sustaining some minimum specified level of continued prosocial performance.

Patients should have as much responsibility as they can manage for choosing behaviors to change and for monitoring their own progress. The metaphor of choice and self-management is probably as important here as the fact. (Back-up monitoring by staff can provide for checks on corner-cutting. Surprisingly often, this is unnecessary; patients usually are quite honest once an alliance has been established.) It is often useful to set out the details of such a mutually agreed upon program in a contract, with all parties having a copy. (Regressed patients often use their copies, and the rating sheets they are to have the staff fill out, as transitional objects, carrying them around and wearing them out by frequent scrutiny.)

The patient's records of his performance can serve as a basis for direct social reinforcement by the therapist as well as a basis for points or other token rewards that can be exchanged for back-up amenities. Less regressed patients can keep diaries and be differentially reinforced for accuracy, perceptiveness, maturity, autonomous choice, and other qualitative aspects of performance. In fact, a sensitive and perceptive therapist can develop a flexible format that eventually leads to differential reinforcement of increasingly subtle self-regulation and social functioning and, thus, to substantial changes in intrapsychic processes through behavioral intervention. Here, the therapist not only differentially reinforces overt behavior, but also what the patient says (writes) to himself about his own behavior and feelings. Interestingly enough, patients reaching and going beyond these advanced stages often continue to keep diaries and use the language of points and rewards long after transactions with the therapist have become largely cognitive and verbal. The concrete rhetoric, based on the shared experience of the two participants, makes it possible for the patient to refer to things that may be difficult to verbalize abstractly. And usually, by this time, the patient has developed a theory about his own behavior that has considerable guiding significance, based largely on instruction by the therapist and first-hand experience with himself. This is a most important step in developing autonomy and self-control.

CREATION OF A RESPONSIVE ENVIRONMENT

The patient spends but a fraction of his total week in direct contact with his therapist. Because so much of behavioral therapy takes place on the ward and out of session, nurses, occupational and recreational therapists, aides, and other patients are the real "cutting edge" of therapy insofar as realistically influencing the patient's behavior is concerned. If the patient's prosocial behavior is to be supported and strengthened, the milieu must be differentially responsive to it. This requires more than just environmental enrichment; both rich and deprived environments can be functionally unresponsive.

Indeed, with the pressures of staff, many supposedly therapeutic environments actually turn out, on closer scrutiny, to be biased in the direction of reinforcing pathological behavior (Hunt, 1971). The necessities of life, and even such amenities as television, reading matter, and sympathetic emotional support—whatever is there—are available to all equally on a noncontingent basis, as a fundamental patient right. Symptoms and feelings, usually unpleasant ones, and symptomatic behavior are major objects of attention in patient and staff group meetings. Normal and constructive behavior, though fondly hoped for, often comes to be taken for granted when it occurs, probably because busy staffs must turn their attention more to things that are going wrong than to things that are going right. (After all, the staff itself lives in a reinforcing community that prizes smooth operation, and in which trouble causes trouble.) Diversions and activities with therapeutic overtones are planned by specialists, but planned fun loses part of its savor, particularly if it is prescribed "for the patient's own good." Often, instead of being a reward for prosocial behavior, such activities become the prosocial behavior that must be engaged in to earn a reward.

These milieus are neither malign nor callous. Rather, they may be too indulgent and unconditional and, thus, insensitive and unresponsive to small increments in constructive, normal behavior. The best of such milieus really may be quite boring and bland at times. Then, symptomatic behaviors that produce some action (e.g., attention, even punishment), diversion (e.g., watching everybody get into a flap at the community meeting), or even individual personalized human contact (e.g., a "sympathetic" conversation with a nurse or aide centered around pathology) gets differentially reinforced, however inadvertently.

To tilt a milieu toward differential responsivity to a patient's prosocial behavior usually requires skillful, practical social engineering. Enthusiasm, dedication, instructions, even orders produce only temporary effects on staff, at best. The changes required of staff cannot deviate too far from standard procedure or require much extra effort and initiative; the staff is busy and, besides, already has its own way of doing things. The new "minisystem" must change the patient for the better quickly, at least insofar as he constitutes a burden on the staff. Otherwise, reinforcement for the staff may be too sparse. Most important, the therapist must attend closely to staff efforts, responding positively, with discrimination, and frequently to the products of staff efforts (e.g., ratings, logbooks, nursing notes). If the therapist does not follow through in this way every two or three days at the least, and if the program does not work, the whole enterprise will quickly drift back to the *status quo ante.*

Actually, it usually turns out to be easier to plan for the patient than to plan for the staff. They are not oppositional, antiscience, or malign, but rather are controlled by the reinforcement *they* get, not by the reinforcement the therapist gets. To make the situation more responsive to good staff work is one more balance the therapist must tilt.

POSTTHERAPY CARRYOVER AND SELF-CONTROL

All therapies must solve the problem of producing favorable changes that endure and do not disappear when therapy stops. In some ways, this problem can be most acute in behavioral therapy using operant reinforcement in which changes are produced and maintained by reward programmed by the therapist.

(The standard procedure for showing that the reinforcement produced the beneficial change is to omit the reinforcement in extinction or reversal and note how the beneficial change disappears!) Fortunately, as indicated earlier, it is possible in a good program to fade control from one discriminative cue stimulus to another and from primary rewards to conditioned social reinforcers that sustain most of us in our day-to-day endeavors. In fading, one stimulus (cue or reinforcing, as the case may be) is gradually and progressively replaced with another, here a naturally occurring one that normally signals or rewards behavior in conventional human settings. In addition, behavior change produced by therapeutic intervention may be based on very lean, intermittent schedules of reinforcement. This makes it resistant to experimental extinction inasmuch as only occasional reinforcement may be required to sustain it. As a general, practical rule, behavior that is to be enduring should be based upon intermittent reinforcement and so shaped that it evokes conventional social reinforcers from other people. These others then become "allies in reinforcement" (Baer, 1968) and may be counted upon to keep up the good work. Often, relatives trained to continue the reinforcement program after the patient has returned home provide valuable supplements to treatment.

The problem of self-control is far more difficult to deal with within the framework of behaviorism which, in its extreme or radical form, argues that behavior is under the control of the environment. Skinner (1971), Goldiamond (1965), and Goldiamond and Dyrud (1968) have described most articulately and persuasively how one manipulates one's own environmental circumstances to change one's own behavior toward desired ends, and thus to control it. This kind of control involves arranging cues or discriminative stimuli in such a way as to make desired behavior more likely (or undesired behavior less so) and scheduling differential rewards for oneself on completion of performance requirements one has set for oneself.

This approach has been elaborated and refined with recent developments in cognitive approaches to the theory of complex human behavior and social learning (see Hunt, 1975). Given sufficient commitment to standards of performance and to goals, based on internal evaluative and cognitive templates, one can indeed define one's own success and failure, deliver rewards to oneself if and when deserved, and so on. Actually, as far back as Skinner's *Science and Human Behavior* (1953) we have had a pretty good psychology of the "controlled self" which has now become reasonably well developed and effective. What we lack is a psychology of the "controlling self" that determines what to control, the standards to set, and gets somehow "committed" sufficiently to its plan to hold the controlled self to its schedule, come what may. Current approaches to the psychology of cognition and decision-making probably are too rationalistic to take adequate account of the nonrational aspects of such commitment. Symbolic behaviors and internalized transformations that lie well outside of awareness and the usual scope of logical processes contribute decisively; the psychology of commitment probably cannot be understood until we learn how to study such behaviors more effectively.

REFERENCES

Baer, D. M. (1968). Some remedial uses of the reinforcement contingency. In *Research in Psychotherapy, Vol. III* (J. M. Shlien *et al.*, Eds.). Washington, D.C.: American Psychological Assn.

Blumenthal, R. L., and Carpenter, M. D. (1974). The effects of population density on the overt behavior of mental patients. *J. Psychiat. Res. 10*, 89–100.

Feather, B. W., and Rhoads, J. M. (1972*a*). Psychodynamic behavior therapy: I. Theory and rationale. *Arch. Gen. Psychiat. 26*, 496–502.

Feather, B. W., and Rhoads, J. M. (1972*b*). Psychodynamic behavior therapy: II. Clinical aspects. *Arch. Gen. Psychiat. 26*, 503–511.

Goldiamond, I. (1965). Self-control procedures in personal behavior problems. *Psychol. Rept. 17*, 851–868.

Goldiamond, I., and Dyrud, J. E. (1968). Some applications and implications of behavior analysis for psychotherapy. In *Research in Psychotherapy, Vol. III* (J. M. Shlien *et al.*, Eds.). Washington, D.C.: American Psychological Assn.

Hinde, R. A., and Stevenson-Hinde, J. (1973). *Constraints on Learning*. New York: Academic Press.

Hunt, H. F. (1971). Behavioral considerations in psychiatric treatment. In *Science and Psychoanalysiss, Vol. XVIII* (J. Masserman, Ed.). New York: Grune and Stratton.

Hunt, H. F. (1975). Behavioral therapy for adults. In Freedman, D. X. & Dyrud, J. E. (Eds.). *American Handbook of Psychiatry, Vol. V*, second edition (D. X. Freedman and J. E. Dyrud, Eds.). New York: Basic Books.

Hunt, H. F., and Dyrud, J. E. (1968). Commentary: Perspectives in behavior therapy. In *Research in Psychotherapy, Vol. III* (J. M. Shlien *et al.*, Eds.). Washington, D.C.: American Psychological Assn.

Kazdin, A. E. (1973). The failure of some patients to respond to token programs. *J. Behav. Ther. Exptl. Psychiat. 4*, 7–14.

Reichenbach, H. (1938). *Experience and Prediction.* Chicago, Ill: University of Chicago Press.

Skinner, B. F. (1953). *Science and Human Behavior.* New York: Macmillan.

Skinner, B. F. (1971). *Beyond Freedom and Dignity.* New York: Alfred A. Knopf.

The Affective Significance of Uncertainty

D. E. BERLYNE

During the last fifty years or so, the study of motivational processes has been dominated by three concepts: "drive," "arousal," and "stress." These concepts have been developed by three distinct currents of research. "Drive" comes from experimental psychology, particularly the experimental study of animal behavior and learning. "Arousal" has its origins in neurophysiology and in psychophysiology, and "stress" has grown out of developments in medicine. The term "anxiety," that Proteus of psychological literature, has been used at one time or another as a synonym for all of these concepts.

Each of the three concepts refers to a factor or variable descriptive of an organism's state, detectable or measurable through a variety of indices. On the whole, fluctuations in "drive" are measured through observable characteristics of behavior. "Arousal" is measured through psychophysiological changes, i.e., electrical or mechanical signals picked up from the surface of the body or changes in the electrical activity of the brain recorded through direct probes. "Stress" is measured principally through biochemical and pathological phenomena.

Drive, arousal, and stress are all induced or intensified by departures from the normal equilibrium of the organism, from the prerequisites of healthy, safe, adaptive functioning. Those who think in terms of "drive" tend to conceive of these disturbances as occasions for remedial action, including the acquisition of new behavior patterns when the existing behavior repertoire leaves the organism

D. E. BERLYNE · University of Toronto, Toronto, Canada.

in the lurch. Specialists in "arousal" focus largely on fluctuations in the efficiency with which information is gathered and processed and with which psychomotor processes are regulated. Specialists in "stress" look for transient or long-lasting breakdowns in adaptive functioning.

Despite these differences in emphasis, there are unmistakable overlaps among the phenonema that are recognized as indices of drive, arousal, and stress. For example, electrodermal changes are widely used to monitor both arousal and stress, and variations in drive and in arousal are traced through changes in the vigor or amplitude of motor responses, among other measures. There are likewise overlaps among the precipitating conditions of increases in drive, arousal, and stress, particularly since, as already mentioned, all three are commonly attributed to departures from an optimal psychosomatic state. Consequently, the entities denoted by the three terms have much in common. There is a great need for more investigation of the relations among them, and the relative lack of such research so far can be attributed to their issuance from distinct traditions. To refer to their common ground, we can appropriate adopt the old term "affect," a term which is broader in connotation than "motivation" and "emotion."

Until about twenty years ago, a rather limited view of the determinants of affect was prevalent. It was believed that affect could result from external irritants or from internal upheavals, particularly those due to deficits of vital substances such as nutriments and water. In addition, it was believed that previously neutral stimuli can become productive of affect if they have regularly coincided with external irritants or internal upheavals. This latter notion is found in the Freudian conceptions of "anxiety," the "wish," and "cathexis" and in the neobehaviorists' conceptions of "secondary rewards" (Skinner's "conditions of reinforcers") and "secondary drive stimuli" (Skinner's "conditioned aversive stimuli"). In other words, affect was linked to the needs of bodily tissues other than the nervous system. It came later to be realized that the brain and other neural structures also have prerequisites of effective operation whose nonfulfillment can be disturbing. Experimental research on curiosity, exploratory behavior and related phenomena gradually made it clear that affect can result from relations, harmonious or inharmonious, among processes simultaneously initiated in the brain, which can in their turn depend on relations among items of information coming in through sense organs. In other words, affect can depend on stimulus structure and, in particular, on variations along familiar-novel, expected-surprising, simple-complex, clear-ambiguous, and stable-variable dimensions, for which I myself (Berlyne, 1960) proposed the term "collative variables." Over the years, more and more areas of research in the behavioral sciences have produced evidence for the motivational importance of these collative stimulus properties. These properties have been discussed in terms of interresponse conflict and similar notions. Some of them at least can usefully be measured and given quantitative treatment with the help of the concepts introduced by information theory, such as uncertainty and amount of information.

The measure of uncertainty introduced by information theory is a strictly mathematical one. It is applicable to any situation in which any one of a number of mutally exclusive kinds of events can occur (or will occur or has occurred), to each of which a probability value can be attached. Uncertainty increases with the number of alternative possibilities and, when this number is held constant, with how near the probabilities come to being equal. The probabilities in question are usually interpreted as relative frequencies. Viewed in this way, uncertainty has nothing necessarily to do with how uncertain anybody feels. Motivational phenomena must, however, depend more directly on what we might call subjective uncertainty (Berlyne, 1960, 1974*a*). This is analogous to the objective uncertainty of the information theorists, except that subjective probabilities (or degrees of confidence) must replace objective probabilities (or relative frequencies) and the kinds of events that are actually possible are less important than the kinds that the subject conceives as possible or can represent to himself. When a subject has had a fair amount of experience with a particular situation, learning will tend to bring his expectations and subjective probabilities into line with actual possibilities and objective probabilities, so that subjective uncertainty will approximate objective uncertainty. But this is not always the case.

What I should like to do is to review very briefly some experimental projects carried out in our laboratory at the University of Toronto over the last few years. They provide support for the hypothesis that subjective uncertainty, induced by external stimulation, can be a source of arousal and of drive. If this hypothesis is valid, and if the indications of close affinities between drive and arousal on the one hand and stress on the other hand can be trusted, we can conclude that uncertainty deserves more attention than it has hitherto received as a source of stress, including pathogenic stress.

HEDONIC VALUE

Before considering the experiments, however, we must consider the relations between drive, arousal, or affect on the one hand and positive and negative hedonic value on the other. "Positive hedonic value" is a term used to cover pleasantness, satisfaction, reward value, and positive utility or incentive value (Berlyne, 1967, 1973). Negative hedonic value similarly covers unpleasantness, annoyance, punishment value, and negative utility or incentive value. The terms making up these groups do not by any means denote the same phenomena. What they stand for is gauged variously from verbal reports, facial and postural expressions, performance of motor responses, and learning. Nevertheless, there seems to be some tendency for the various phenomena subsumed under "hedonic value" to occur together, suggesting that they have a common substrate.

At one time, the view identifying increased drive with negative hedonic value and positive value with drive reduction was popular. It was vigorously championed by Hull (1952) and other members of the Yale group (e.g., Mowrer, 1938; Dollard and Miller, 1950) in the 1930s and 1940s. It was, however, challenged by advocates, notably Sheffield (1966), of the opposite view that reward (reinforcement) comes from drive induction. This early controversy was generally conducted within the framework of the learning theory of the time. In the meantime, evidence bearing on the issue is to be found in such varied research areas as neurophysiology, psychopharmacology, the experimental study of exploratory behavior, experimental aesthetics, verbal learning, and child development. Furthermore, the problem came later to be examined in terms of the newer concept of "arousal" rather than of the older concept of "drive," with which it has much in common. These various lines of evidence (Berlyne, 1967, 1971) suggest—although they are far from sufficient to settle the matter definitively—that the drive-reductionists and drive-inductionists were both right. The evidence seems to favor the tentative conclusion that positive hedonic value can arise in either of two ways. On the one hand, a moderate increase in arousal can be pleasurable, rewarding, etc. (the "arousal-boost" mechanism). A more severe increase in arousal is likely, on the other hand, to be aversive and punishing; in this case, conditions that reduce arousal can give rise to pleasure and reward (the "arousal-reduction" mechanism).

We can conveniently use the term "arousal potential" to refer collectively to the numerous variables that can influence arousal, including the collative variables that have received recognition comparatively recently. When hedonic value is examined in relation to any one of these variables or to arousal potential globally, nonmonotonic curves regularly emerge. These include inverted U-shaped curves, or occasionally curves with several peaks, and curves of the shape shown in Fig. 1. This is the shape of the curve that Wundt (1874) used to represent the way in which pleasantness-unpleasantness varies with stimulus intensity. The Wundt curve can actually be derived from the assumption that there are two antagonistic systems in the brain at work, one generating positive hedonic value when activated and the other negative hedonic value, provided that the latter curve has a higher threshold and a higher asymptote (Fig. 2). These two brain systems might very well be identifiable with two of the three systems governing hedonic processes that Olds (Olds and Olds, 1965) has described, increasing the number later to four (Olds, 1973). They are what he calls the "positive-reinforcement focus" and "periventricular system," respectively. When arousal is raised to an uncomfortable level, we may infer that a third system is implicated in any subsequent pleasurable relief. This is pre-

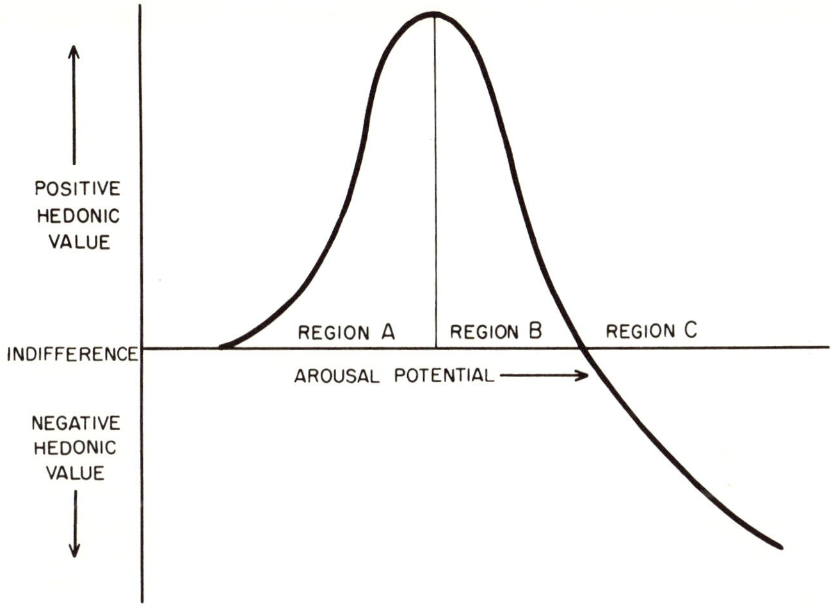

Fig. 1. Hypothetical curve relating hedonic value to arousal potential, derived from Wundt's curve (from Berlyne, 1967, 1973).

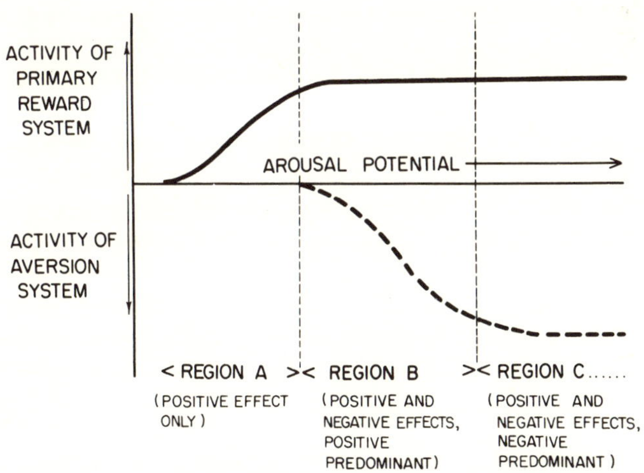

Fig. 2. Hypothetical curves representing activity of two antagonistic brain systems (from Berlyne, 1967, 1973).

sumably identifiable with Old's "positive-reinforcement field," which inhibits the periventricular system, just as this in its turn inhibits the positive-reinforcement focus.

Attribution of the hedonic-value curve to the interaction of brain systems (Figs. 1 and 2) leads to the conclusion that the dimension of arousal potential (or any of its component variables) can be divided into three regions (apart from the values that are below the absolute threshold and thus can have no affective impact). First, there is region A, where only rewarding and pleasurable effects of arousal increase are in evidence. Then, in region B, an increase and a subsequent decrease in arousal will both have positive hedonic value. Lastly, in region C, the rise in arousal is severe enough to generate predominant discomfort, punishment, and aversion, so that only a subsequent decrease can lead to positive hedonic value.

This formulation implies that arousal-heightening conditions can in some circumstances be welcomed and sought out, even though they are shunned and promptly corrected in other circumstances. This fits in with the recent accumulation of evidence that human beings and higher animals are not always aiming at quiescence and inactivity but spend much of their time actively providing themselves with stimulation, challenges to their capacities, and opportunities to learn. Since positive hedonic value is generally, but of course not invariably, associated with conditions that conduce to adaptation and well-being, we may conclude that moderate rises in drive or arousal can be beneficial, and possibly essential, with respect to psychological and physical health. This view is, of course, consonant with Selye's (1974) recognition of eu-stress as well as dis-tress.

Let us now return to uncertainty, the component of arousal potential on which we are concentrating, in relation to hedonic value. Uncertainty, objective or subjective, is associated with the amount of unpredictable variety furnished by the environment. We may assume that, like other components of arousal potential, its relation to hedonic value follows the Wundt curve. This means that slight influxes of uncertainty will be gratifying and therefore sought. This is most clearly the case at times of boredom or sensory deprivation (cf. Jones, 1966), but we seem also to like occasional mildly puzzling or disconcerting experiences in the course of normal everyday life. Then, there will be somewhat higher degrees of uncertainty, in which the puzzlement and the following enlightenment are both pleasurable. Finally, severe degrees of uncertainty are likely to be uncomfortable, distressing, and even intolerable, so that information removing the uncertainty will be gladly received.

UNCERTAINTY AND THE ORIENTATION REACTION

I should like first to mention some experimental results pertaining to the question of whether uncertainty induces arousal. It is by now well known that virtually all stimuli evoke the orientation reaction (Sokolov, 1958; Berlyne, 1960), a phenomenon which includes indices of a brief rise in arousal followed by a drop (sometimes accompanied by a long-term upward drift in arousal, corresponding to what is called the "tonic orientation reaction"). The only exceptions appear to be stimuli that are below the absolute threshold or stimuli that have been repeated several times in the recent past without biologically important accompaniments. So, the question that faces us is whether stimuli productive of more subjective uncertainty evoke more intense orientation reactions.

We have been following the assumption that subjective uncertainty is a special form of interresponse conflict, involving competition among response tendencies corresponding to the mutually exclusive events that are anticipated. So, it is appropriate to begin by mentioning an experiment (Berlyne, 1961) that was designed to find out whether degree of conflict affects one component of the orientation reaction, namely the conductance galvanic skin response (GSR). The subject sat facing a panel on which eight lights were arranged in the form of a diamond, two at each corner. His hand controlled a key that could be moved in any of four directions, corresponding to the corners of the diamond. High-conflict trials, on which two lights at different corners (dispersed stimuli) were illuminated and the response corresponding to either one of them had to be made quickly, were interspersed with low-conflict trials, on which two lights appeared at one corner (adjacent stimuli) and the key had to be pressed in the single corresponding direction. The lights were on for 10 sec and the key had to be moved as soon as they went off. This was to separate the GSR occasioned by the initial impact of the stimulus pattern (which was the one of interest, since it coincided with the onset of high or low conflict) from the GSR due to stimulus offset and the motor process. Before the phase of major interest just described (Phase II), there was an initial phase (Phase I), during which the subject was exposed to similar light patterns without any mention of a motor response. Apart from the measures just reviewed, efforts were also made to control for other factors that might be confounded with degree of conflict.

As Fig. 3 shows there was no significant difference between the GSRs to adjacent and dispersed stimuli in Phase I, when conflict played no part. Similarly, there was no significant difference in the GSRs occurring just after the extinction of the lights and the performance of the response in Phase II. But the result of main interest, namely the difference between GSRs to the onset of dispersed (high-conflict) and adjacent (low-conflict) stimuli in Phase II, was significant in the direction indicative of a tendency for the intensity of the orientation reaction to increase with conflict.

A later experiment (Berlyne and Borsa, 1968) was directly concerned with effects of subjective uncertainty, induced through exposure to blurred colored slides showing familiar objects on plain backgrounds. The duration of EEG desynchronization was used this time to measure the intensity of the orientation

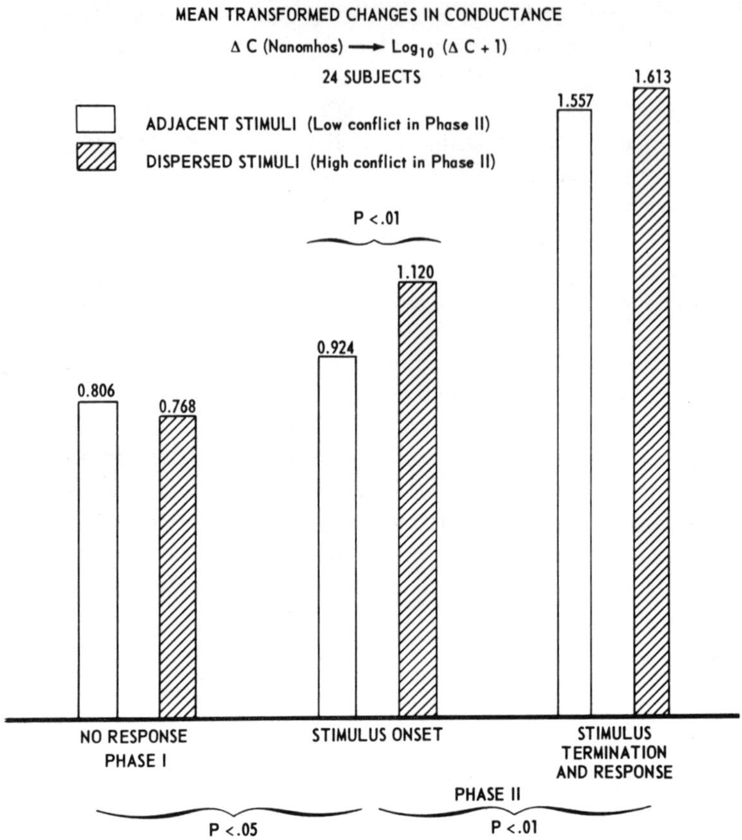

Fig. 3. Mean GSR scores in experiment on effects of degree of conflict (data from Berlyne, 1961).

reaction. In the first experiment (see Fig. 4), subjects were exposed to blurred versions of 15 pictures and clear versions of 15 others for 4 sec each. The mean duration of desynchronization was about 10% greater for the blurred pictures, a difference that was statistically significant. A second experiment was carried out to verify that the difference was due to subjective uncertainty rather than to blur as such. It was necessary to find a way to rob the blurred pictures of power to induce uncertainty. This could be done by presenting the corresponding clear picture before each blurred picture, so that, on seeing the blurred picture, the subject knew what object it depicted. As well as a sequence with this condition (the C-B condition), every subject had another sequence in which the blurred slide appeared before its corresponding clear slide (the B-C condition), so that the association between blur and uncertainty was here retained. Half of the

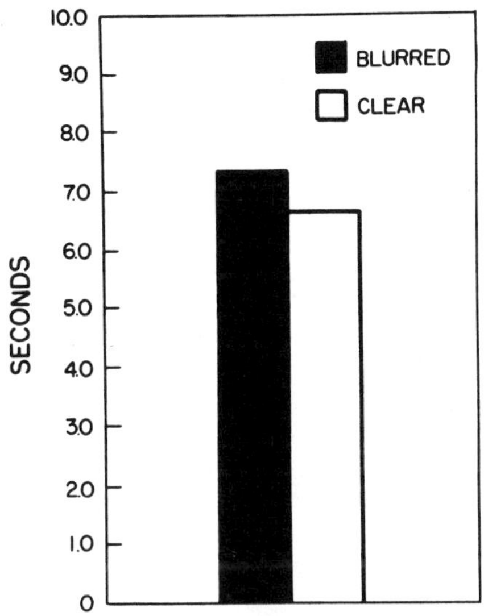

Fig. 4. Mean EEG desynchronization durations for blurred and clear pictures in experiment 1 (data from Berlyne & Borsa, 1968).

subjects had the B-C sequence first and the C-B sequence second, and the other half had the reverse arrangement. As Fig. 5 shows, the blurred slide once again produced longer desynchronization than the clear slide in the B-C sequence, whether this response came first or second, and the difference was again significant. In the C-B sequence, however, the mean duration was actually slightly greater for the clear slides, but not significantly so. So, we have confirmation that longer desynchronization results from blurred slides only when they induce subjective uncertainty.

EXPERIMENTS ON LEARNING MOTIVATED BY UNCERTAINTY AND REINFORCED BY UNCERTAINTY REDUCTION

Blurred Pictures and Manual Responses

The term "drive" has a number of different connotations (see Berlyne, 1960). Some of them (relevant to the so-called "energizing" or "activating" effects of drive) are virtually identical with those of "arousal." So, the experiments just mentioned confirm that subjective uncertainty has one of the effects

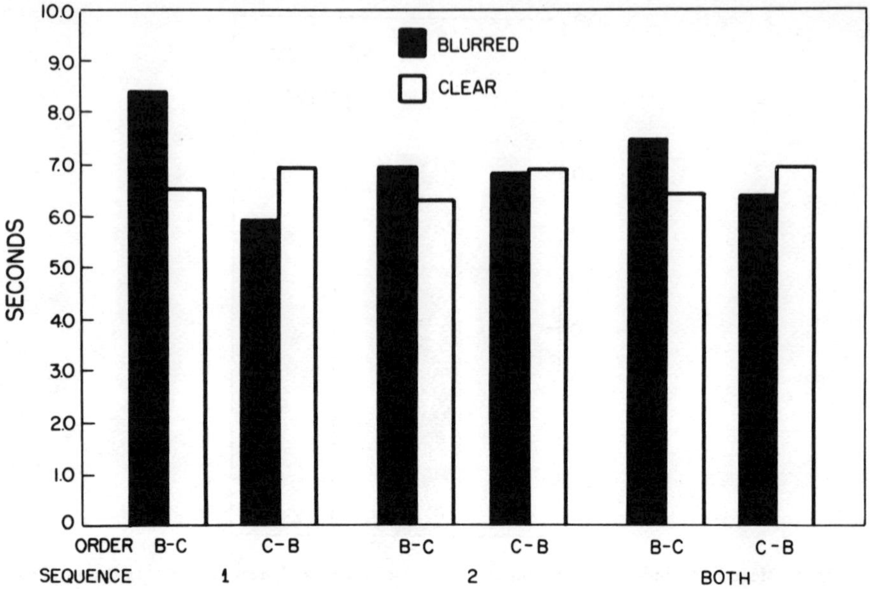

Fig. 5. Mean EEG desynchronization duration for blurred pictures in experiment 2, with and without uncertainty, and clear pictures (data from Berlyne & Borsa, 1968).

connoted by an increase in drive. According to another connotation of the term, heightened drive is an aversive state, which means a state whose relief is rewarding and can therefore promote the learning of a response that precedes it. The next experiments to be reviewed were devoted to the question of whether subjective uncertainty enhances the reward value or reinforcing effect of subsequent information by which it is relieved.

In the first project, carried out by Nicki, (1968, 1970), black and white blurred and clear slides were used. In his initial experiment, Nicki put the subjects through a sequence of 20 trials. Each began with a 10-sec presentation of a blurred slide on a screen, after which a buzzer sounded. The subject had then to press either of two telegraph keys, one of which caused a clear version of the same slide to appear and the other an unrelated clear picture.

As Fig. 6 shows, the probability of pressing the key exposing the related picture (the one relieving the uncertainty generated by the previously exposed blurred picture) began at 0.5 but gradually increased to about 0.8. This confirms that exposure to uncertainty-reducing information is more rewarding (reinforcing) than exposure to other information.

Again, it was desirable to verify that uncertainty is really the responsible factor by seeing what would happen if the blurred picture were preceded by its clear counterpart and thus dissociated from uncertainty. Figure 7 shows the two conditions that were used in a later experiment of Nicki's and their results. The

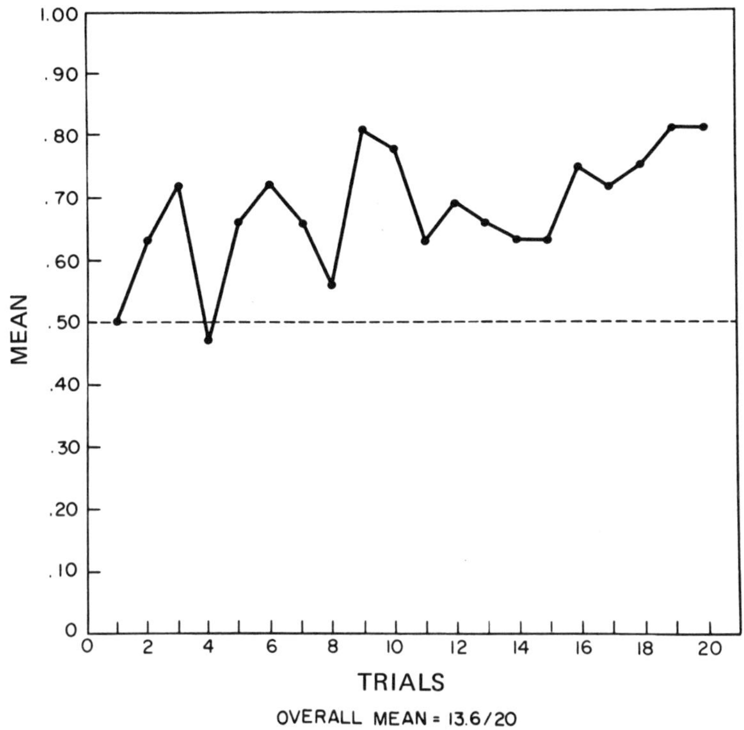

Fig. 6. Mean proportions of responses on key exposing clear version of blurred picture just seen (from Nicki, 1968, 1970).

left-hand part of the figure refers to a condition in which a clear slide (C_1) is followed by its blurred counterpart (B_1) and then by a choice between a response producing C_1 and a response producing a different clear picture, C_2. In this case, the effect found in the first experiment does not appear: The subject favors the response producing C_2, which is a novel picture for him. In the control condition, to which the right-hand part of Fig. 7 refers, a third clear picture, (C_3) appears before B_2, which thus induced uncertainty, and, as before, the uncertainty-relieving response occurs more often.

Blurred Pictures and Incidental Free Recall

For the next project (Berlyne and Normore, 1972), we reverted to colored blurred and clear slides as means of inducing and then reducing certainty. But this time, the phenomenon under study was verbal learning and in particular, incidental free recall.

In the first experiment that will be mentioned (Experiment 1), subjects were

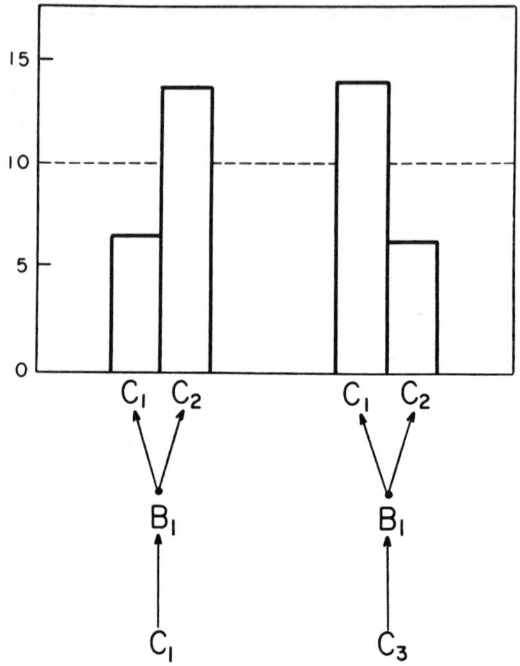

Fig. 7. Mean number of responses (out of 20) to keys exposing different clear pictures in experiment 6 (data from Nicki, 1968, 1970).

exposed to a sequence of 24 items, eight belonging to each of three classes. The three classes, which were randomly interspersed, consisted of (a) Single-Clear (exposure of a clear picture for 5 sec), (b) Double-Clear (presentation of a clear picture for 10 sec), and (c) Blurred-Clear (presentation of a blurred picture for 5 sec followed by 5 sec of the corresponding clear picture). Immediately after the conclusion of these items, subjects were asked to write down, in any order, the names of as many depicted objects as they could recall. But they had been given no prior warning to this recall test and had not been instructed to commit the objects to memory, so that any learning that occurred was incidental learning. On the contrary, GSR electrodes were attached to their hands to support a false explanation that the experiment was concerned with psychophysiological recording. As the first column of Table 1 shows, the Blurred-Clear condition produced significantly more recall than either of the conditions in which the blurred slide was absent.

Experiment II (see the second column of Table 1) had the same procedure except that the recall test took place 24 hr after exposure to the pictures. The same effect appeared, although, as one would expect, all three means were lower than in the immedate test of Experiment I. Then, Experiment III (third column of Table 1) repeated the procedure of Experiment I, except that subjects were

Table 1. Mean Number of Items Recalled, Experiments I–V[a]

	Experiment				
Condition	I (imm., inc.)	II (del., inc.)	III (imm., int.)	IV (imm., inc.)	V (imm., inc.)
Single-Clear	2.5	1.6	4.6		
Double-Clear	3.4	2.6	5.2	2.8	4.0
Blurred-Clear	4.0	3.2	5.2	3.8	4.2
Clear-Blurred				2.7	
Unrelated Blurred-Clear					3.0

[a]Imm. = immediate recall, del. = delayed recall, inc. = incidental learning, and int. = intentional learning. Total no. of items recalled = 8 (From Berlyne and Normore, 1972).

told in advance that they should try to remember as many as possible of the objects depicted and that a recall test would be given immediately afterwards. The superiority of the Blurred-Clear condition does not appear in this case, which involves intentional rather than incidental learning.

So, these results are compatible with the hypothesis that learning is facilitated when the information to be retained relieves previously induced uncertainty, at least as far as incidental learning is concerned. It was necessary, as in the projects mentioned earlier, to see what would happen if either the uncertainty-inducing phase or the uncertainty-reducing phase were absent. So, in Experiment IV, there was incidental learning and an immediate recall test, but, in contrast with Experiment I, the Single-Clear condition was replaced by a Clear-Blurred condition, i.e. a condition in which each clear picture appeared for 5 sec immediately before its blurred counterpart, which consequently produced no uncertainty regarding the depicted object. As Table 1 shows, the Blurred-Clear condition once again surpassed the Double-Clear condition, but the Clear-Blurred condition did not. Then, Experiment V replaced the Single-Clear condition with an Unrelated-Blurred-Clear condition. Here, the clear picture was preceded by the blurred version of a quite different picture, which therefore induced uncertainty that the clear picture did not remove. This condition likewise made for less recall than the Double-Clear condition (see Table 1). So, our results confirm that the Blurred-Clear condition facilitates incidental recall only when uncertainty is first generated and then eliminated.

Prequestioning and Paired-Associate Learning

Another way to induce uncertainty is to present a question to which the subject can think of several possible answers without knowing which is the correct one. This method was used in a project on paired-associate learning

(Berlyne *et al.*, 1968). In the one experiment that there is space to mention, an experimental group went through two sequences of ten items. One sequence consisted of Partial-Guess items, in each of which a Turkish word (or what passed for a Turkish word) appeared on a screen for 4 sec next to the name of a category in parentheses (e.g., "tree," "food"). It was explained to the subject that the Turkish word meant something belonging to the category with which it was paired and that he was to voice a guess with respect to its meaning. Immediately afterwards, the Turkish word reappeared for 4 sec paired with an English word belonging to the category. For the other sequence, the Double-Presentation condition was used: The Turkish word appeared next to its purported English equivalent for 8 sec. It was hypothesized that the guessing task used for the Partial-Guess items would induce uncertainty, which would be relieved by the subsequent presentation of the English word that the subject had just been seeking. Immediately after exposure to the two sequences, the Turkish words were presented, and the subject was asked, having had no previous warning, to supply the corresponding English words when he could.

As Fig. 8 shows, incidental recall was much better for the Partial-Guess items than for the Double-Presentation items, a finding compatible with the hypothesis that the succession of uncertainty induction and uncertainty reduction

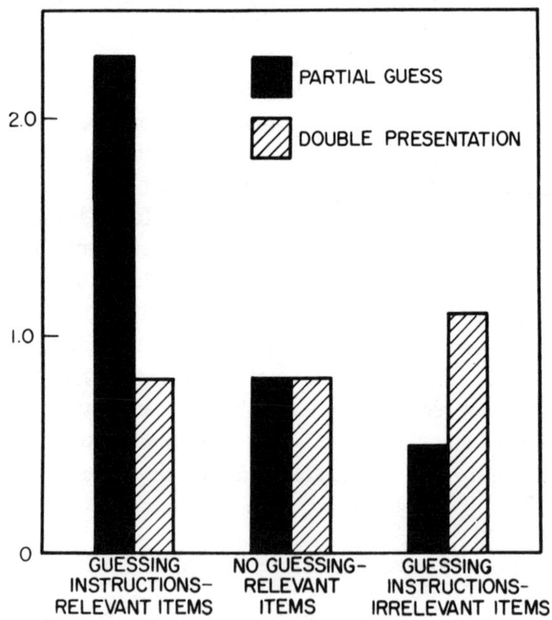

Fig. 8. Mean incidental-recall scores for paired-associate verbal learning under different conditions in experiment 6 (data from Berlyne, *et al.*, 1968).

facilitates incidental learning. The effects had been absent in an earlier experiment using intentional learning.

Two control groups also took part in the experiment. One group saw the same slides as the experimental group, but they were not told to guess the meaning of the Turkish word when it appeared next to its category word. One would expect the equivalent of the Partial-Guess condition to produce no uncertainty in their case, and, as can be seen from Fig. 8, the scores for the two conditions were equal. The second control group had the Partial-Guess items with instructions to guess, but each of these was followed by a quite different Turkish word paired with its English equivalent, so that any uncertainty due to the guessing was not relieved. In their case, the Double-Presentation condition produced better recall, reversing what was found with the experimental group. To sum up, therefore, we once more have evidence that a condition in which uncertainty is presumably aroused and then removed facilitates incidental recall but conditions in which one of these occurs without the other do not.

EXPERIMENTS WITH PATTERNS COMPOSED OF RANDOMLY SELECTED ELEMENTS

A rather different way of manipulating subjective uncertainty is to expose a subject to a sequence of stimuli, each of which is randomly and independently selected from a set (sample space, ensemble) of alternatives (Berlyne, 1974*b*). If all the alternate possibilities are equally likely to occur, the objective uncertainty, measured in bits, will be equal to the logarithm to base 2 of their number. We may presume that, once the subject has experienced a sample of such elements, his subjective uncertainty regarding what form the next element will take will approximate the objective uncertainty. In everyday language, this uncertainty will reflect the amount of variety of unpredictability of the pattern.

A set of 20-sec 40-tone sound sequences, representing six different levels of uncertainty in accordance with the principles just discussed, has been prepared by Vitz (1966). The number of possibilities in Vitz's sequences, defined in terms of frequency, duration, and loudness of tones, ranges from two at uncertainty level one (1 bit per tone) to 576 at uncertainty level 6 (9.17 bits per tone). Crozier, in a doctoral dissertation project carried out in our laboratory, used Vitz's sequences and others that he constructed in a similar manner to probe verbal and nonverbal indices of hedonic value (Crozier, 1973, 1974).

Crozier asked psychology and music students to rate the sequences on several 7-point scales. The curve relating mean judged complexity to uncertainty level came close to a perfect rising straight line (Fig. 9). Judged interestingness also rose monotonically with uncertainty, the steepness being greater among

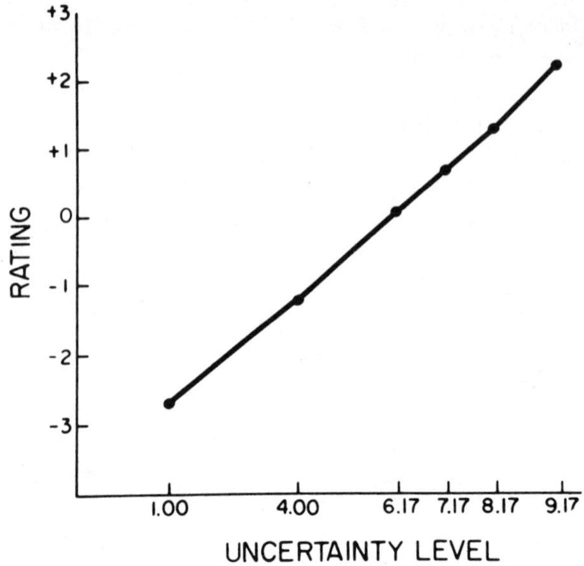

Fig. 9. Mean ratings of sound sequences on simple-complex scale (from Crozier, 1973, 1974).

music students whan among psychology students (Fig. 10). Nonmonotonic curves emerged, however, when the sequences were rated on displeasing-pleasing and ugly-beautiful scales (Fig. 11 and 12). There, we find inverted U-shaped functions, and, as can be seen from the right-hand panels in the figures, music students judged the higher uncertainty levels more favorably, and the lower uncertainty levels less favorably, than the psychology students on these scales.

Two nonverbal measures of exploratory behavior were also used. In one procedure, Listening Time is recorded: The subject listens to each of the sequences, presented in a random order, for as long as he wishes, pressing a button to replace each by the next when he chooses. Figure 13 shows the mean Listening Times for a group of psychology students. In the Exploratory-Choice procedure, the subject hears the first 10 sec of each of two sequences and then has to decide which of the two he wishes to hear continued, pressing an appropriate button to cause one to resume. All possible pairs of uncertainty levels appear equally often, and Fig. 14 shows their mean probabilities of being chosen in an experiment by Bragg and Crozier (1974). Listening-Time curve tends to rise monotonically with uncertainty and thus to follow judged interestingness. Exploratory Choice, on the other hand, shows a tendency to reach a peak at intermediate uncertainty levels and then to decline, thus showing a closer relation to judged pleasingness and beauty. Crozier found, however, using recently developed statistical techniques, that subjects differ in the degree to

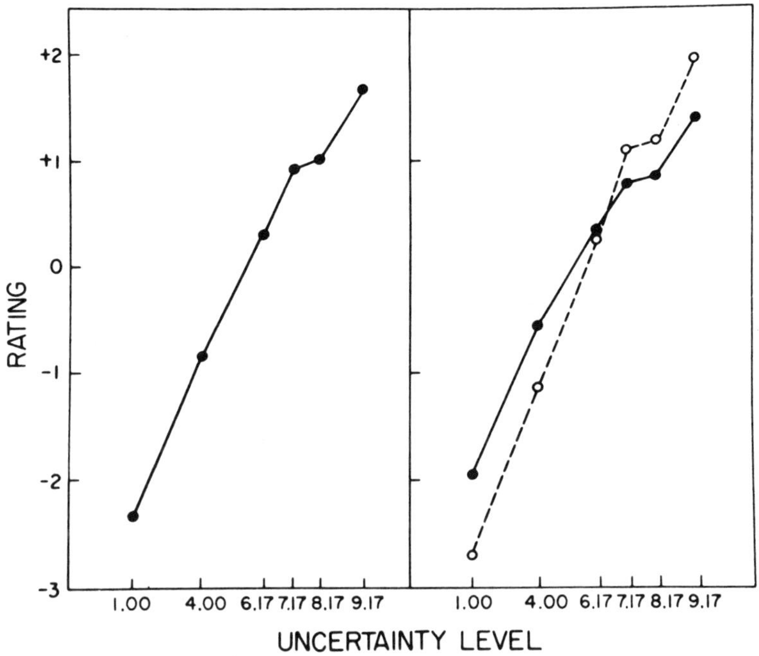

Fig. 10. Mean ratings of sound sequences on uninteresting-interesting scale. Left-hand panel: all 24 subjects. Right-hand panel: curves for psychology (continuous) and music (broken) students (from Crozier, 1973, 1974).

which judged interestingness and judged pleasingness beauty determine their Exploratory Choices.

In general, these results resemble what has been found repeatedly in experiments with a variety of visual material differing in complexity, (See Berlyne, 1971, Chapter 13), although multipeak curves sometimes emerge when visual complexity is related to judged pleasingness.

Since Crozier completed his initial project, we have extended this line of work in several directions. Visual counterparts of Vitz's sound sequences have been constructed, both static and dynamic. The static patterns consist of randomly chosen elements differing in shape, color, and size (Berlyne, 1974*c*). The dynamic equivalents consists of films showing a light spot jumping horizontally from side to side on a dark background (Normore, 1974). The positions occupied by the spot and its brightness correspond to the pitch and intensity of the tones in Vitz's sequences; duration is retained as the third parameter. With both of these kinds of visual pattern, judged complexity and interestingness show a tendency to increase with uncertainty. Ratings for pleasingness and beauty give rise to a multipeaked curve with static patterns but a

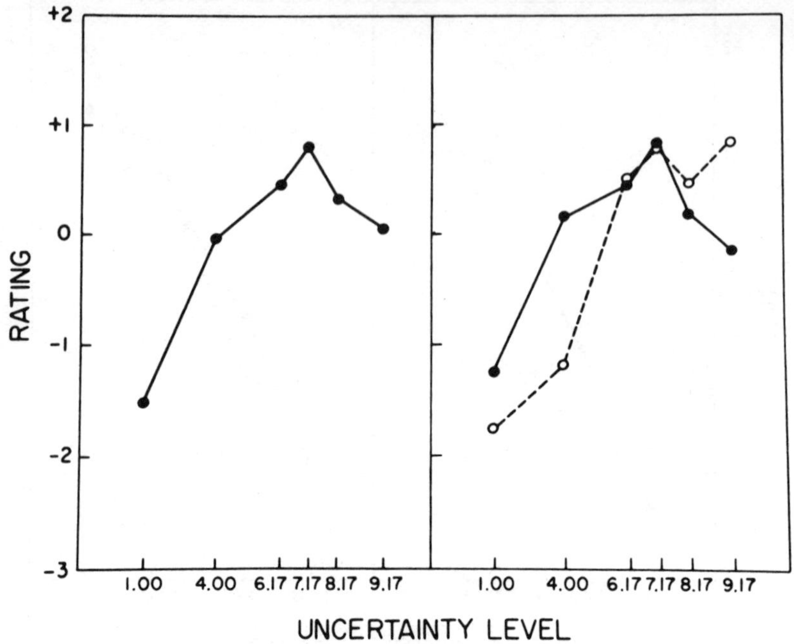

Fig. 11. Mean ratings of sound sequences on displeasing-pleasing scale. Right-hand panel: all 24 subjects. Left-hand panel: curves for psychology (continuous) and music (broken) subjects (from Crozier, 1973, 1974).

monotonically rising curve with dynamic patterns. This latter result suggests that the optimal or preferred uncertainty level may be greater in the visual than in the auditory modality. This would not be surprising, in view of the fact that we take in much more information through the visual sense, so that, whereas Vitz's high-uncertainty sound sequences resemble some of the most complex melodies heard in twentieth-century avant-garde music, the high-uncertainty films are, of course, much simpler than most of the visual displays that life offers.

In the Vitz sequences, uncertainty was manipulated by varying the number of possibilities from which each tone was chosen. There are, however, other ways of manipulating this variable. In the experiment with static visual patterns, the number of elements, as well as the number of possible forms each element could take, was varied. And in subsequent experiments with sound sequences, not only the number of possibilities per tone but also distributional redundancy (i.e., inequalities among the relative frequencies of different pitches, loudnesses, and durations) was varied. The results of both kinds of experiment vindicated still further the importance of uncertainty, since mean ratings and measure of exploratory behavior appear to follow simple (monotonically rising or inverted U-shaped) functions of uncertainty, however produced.

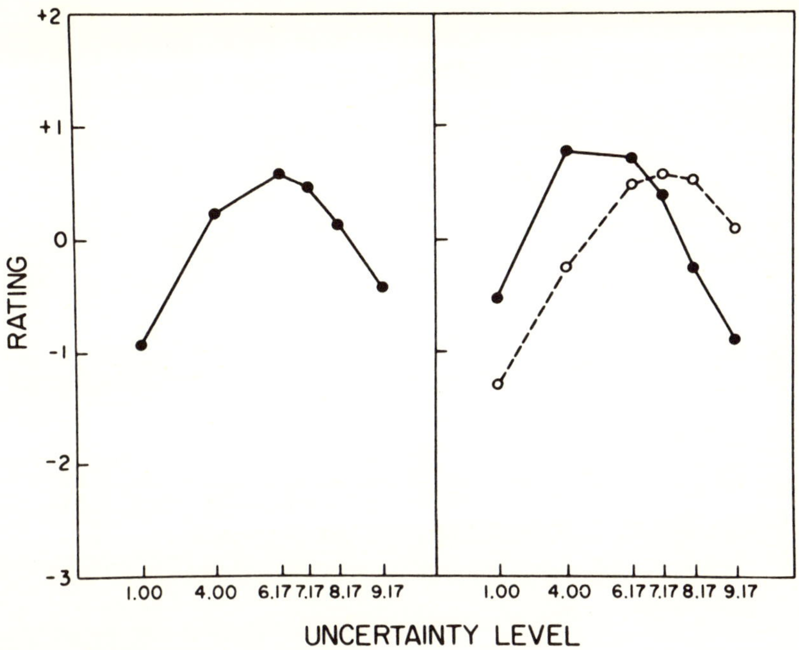

Fig. 12. Mean ratings of sound sequences on ugly-beautiful scale. Right-hand panel: all 24 subjects. Left-hand panel: curves for psychology (continuous) and music (broken) subjects (from Crozier, 1973, 1974).

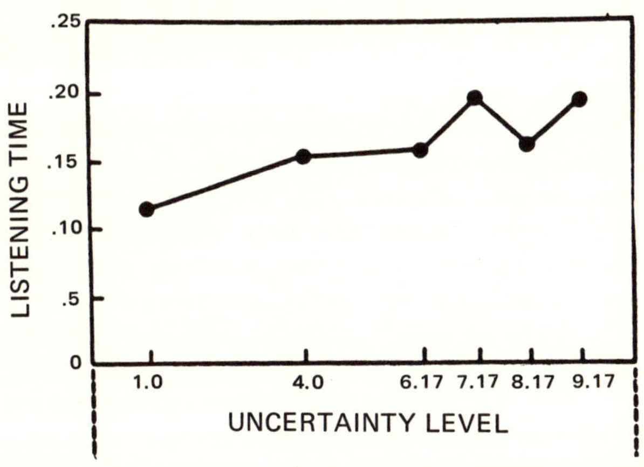

Fig. 13. Mean proportional Listening Times (from Crozier, 1973, 1974).

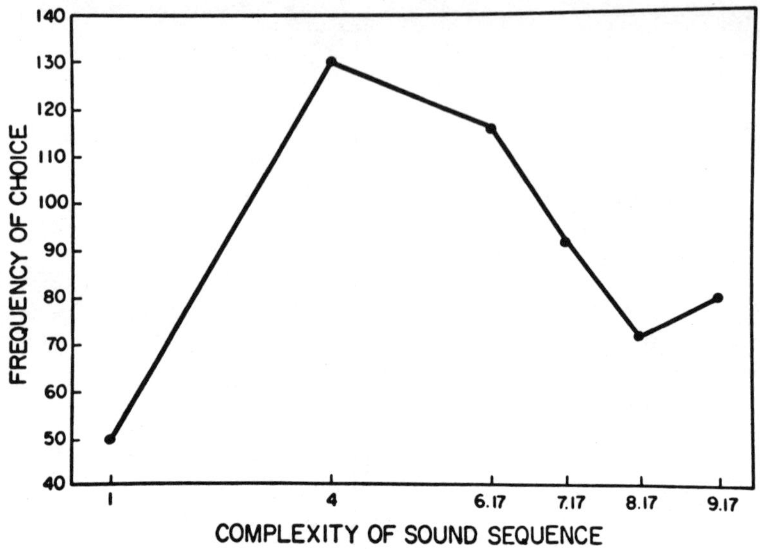

Fig. 14. Mean probabilities of Exploratory Choice (from Bragg & Crozier, 1974).

Bragg and Crozier (1974) have compared adults with 8-year-old and 14-year-old children, using sound sequences constructed according to the principles introduced by Vitz. There were no significant differences among age groups with respect to Listening Time and Exploratory Choice. But the 8-year-olds' ratings for interestingness, pleasingness, and beauty tended to vary inversely with uncertainty level. This meant that the youngest children's judgments gave rise to curves of a different shape from those found in adolescents and adults. It also meant that they failed to distinguish pleasingness and beauty, on the one hand, from interestingness, on the other hand, as older subjects are invariably found to do.

Finally, research with sound sequences has been extended cross-culturally (Berlyne, in press). Data have been gathered from subjects in the Ivory Coast, India, and Japan, as well as Canada. Some uniformities, as well as differences, appear when the different populations are compared. As far as complexity ratings are concerned, all groups show a rising trend with uncertainty level. Some groups show the same kind of trend in their interestingness judgments, whereas others appear to find intermediate uncertainty levels most interesting. Translation problems might very well be more serious with the word "interesting" than with the other words used to label scales. All groups show inverted U-shaped curves for ugly-beautiful judgments, although the locations of the peaks differ. The same applies to displeasing-pleasing judgments, except that Indian villagers, like Canadian 8-year-olds, contrast with the other groups in producing a monotonic downward curve. When Exploratory-Choice scores were obtained from Canadian students, Indian students, and Indian villagers, all three groups produced inverted U-shaped curves appeared with peaks at uncertainty level 2.

CONCLUSIONS

We have, therefore a fair amount of support for the view that subjective uncertainty can be an important source of affect. More particularly, degree of uncertainty can influence arousal (the intensity of the orientation reaction), drive (conceived as an aversive condition whose alleviation can reinforce learning), verbal judgments indicative of hedonic value, and measures of stimulus-seeking (exploratory) behavior. The close relations, direct or inverse, linking these phenomena with the concept of "stress" invite the conclusion that uncertainty can also be a source of stress. This is attested by much everyday experience. There is also experimental work, using acknowledged indices of stress, to bear it out. Examples are to be found in the findings reported in this Symposium by Frankenhaeuser and by Mason. What is often referred to in experimental reports as "novelty" or degree of "stimulation" usually turns out to be uncertainty. If uncertainty can cause stress, this, it must be reiterated, does not imply that uncertainty always is or should be shunned. There are times when mild doses of uncertainty are gratifying and presumably beneficial, and lack of uncertainty, which means lack of variety or stimulation, can, as so many experiments on sensory or perceptual deprivation have shown, become intolerable and detrimental to a variety of psychological functions.

Nevertheless, there is reason to believe that immoderate subjective uncertainty can be severely disturbing. The fact is that human adults, unlike children or lower animals, do not respond to momentary stimuli that come and go. We normally live in a world of stimulus events that have grown out of specifiable previous stimulus events, that will lead to specifiable subsequent stimulus events, and that have specifiable hidden accompaniments, all of which we can represent to ourselves. And our responses depend on these represented antecedents, consequences, and accompaniments, as well as on the properties of the stimuli that are momentarily exciting our receptors. Unexpected happenings that come from nowhere and disappear into nowhere can be terrifying. We are discomforted and distressed whenever we are at a loss to identify what preceded a situation in which we find ourselves and what will follow from it.

There are strong reasons to suspect that, in extreme forms, such situations can provoke breakdown and illness. But it is worth noting that any illness is itself apt to be a case in point. The discomfort and pain that are inseparable from a symptom can surely be aggravated by the patient's bewilderment regarding whence it arose, what its outcome will be, and what is going on inside him while it is in evidence.

ACKNOWLEDGMENT

The preparation of this paper and research mentioned in it were supported by Research Grants A–73 from the National Research Council of Canada and S72-1405-X2 from The Canada Council.

REFERENCES

Berlyne, D. E. (1960). *Conflict, Arousal and Curiosity.* New York: McGraw-Hill.

Berlyne, D. E. (1961). Conflict and the orientation reaction. *J. Exp. Psychol. 62*, 476–483.

Berlyne, D. E. (1967). Arousal and Reinforcement. In *Nebraska Symposium on Motivation* (D. Levine, Ed.). Lincoln, Neb.: Univ. Nebraska Press, 1–110.

Berlyne, D. E. (1971). *Aesthetics and Psychobiology.* New York: Appleton-Century-Crofts.

Berlyne, D. E. (1973). The vicissitudes of aplopathematic and thelematoscopic pneumatology (*or* The hydrography of hedonism). In *Pleasure, Reward, Preference* (D. E. Berlyne and K. B. Madsen, Eds.). New York: Academic Press.

Berlyne, D. E. (1974*a*). Information and motivation. In *Human Communication: Theoretical Explorations* (A. Silverstein, Ed.). Hillside, N.J.: Earlbaum.

Berlyne, D. E. (1974*b*). *Studies in the New Experimental Aesthetics: Steps Toward an Objective Psychology of Aesthetic Appreciation.* Washington, D.C.: Hemisphere.

Berlyne, D. E. Psychological aesthetics. In *Handbook of Cross-Cultural Psychologyy, Vol. II* (H. C. Triandis, Ed.). In press.

Berlyne, D. E., and Borsa, D. M. (1968). Uncertainty and the orientation reaction. *Percept. Psychophys. 3*, 77–79.

Berlyne, D. E., Carey, S. T., Lazare, S. A., Parlow, J., and Tiberius, R. (1968). Effects of prior guessing on intentional and incidental paired-associate learning *J. Verbal Learning and Verbal Behavior 7*, 750–759.

Berlyne, D. E., and Normore, L. F. (1972). Effects of prior uncertainty on incidental free recall. *J. Exp. Psychol. 96*, 43–48.

Bragg, B. W. H., and Crozier, J. B. (1974). The development with age of verbal and exploratory responses to sound sequences varying in uncertainty level. In *Studies in the New Experimental Aesthetics: Steps Toward an Objective Psychology of Aesthetic Appreciation* (D. E. Berlyne, Ed.). Washington, D.C.: Hemisphere.

Crozier, J. B. (1973). Verbal and exploratory responses to sound sequences varying in uncertainty level. Unpublished Ph.D. thesis. University of Toronto.

Crozier, J. B. (1974). Verbal and exploratory responses to sound sequences varying in uncertainty level. In *Studies in the New Experimental Aesthetics: Steps Toward an Objective Psychology of Aesthetic Appreciation* (D. E. Berlyne, Ed.). Washington, D.C.: Hemisphere.

Dollard, J., and Miller, N. E. (1950). *Personality and Psychotherapy.* New York: McGraw-Hill.

Hull, C. L. (1952). *A Behavior System.* New Haven: Yale University Press.

Jones, A. (1966). Information deprivation in humans. In *Progress in Experimental Research, Vol. 4* (B. A. Maher, Ed.). New York: Academic Press.

Mowrer, C. H. (1938). Preparatory set (expectancy): a determinant in motivation and learning. *Psychol. Rev. 45*, 62–91.

Nicki, R. M. 1968). The reinforcing effect of uncertainty reduction on a human operant. Unpublished doctoral dissertation, University of Toronto.

Nicki, R. M. (1970). The reinforcing effect of uncertainty reduction on a human operant. *Can. J. Psychol. 24*, 389–400.

Normore, L. F. (1974). Verbal responses to visual sequences varying in uncertainty level. In *Studies in the New Experimental Aesthetics: Steps Toward an Objective Psychology of Aesthetic Appreciation* (D. E. Berlyne, Ed.). Washington, D.C.: Hemisphere.

Olds, J. (1973). Brain mechanisms of reinforcement learning. In *Pleasure, Reward, Preference* (D. E. Berlyne and K. B. Madsen, Eds.). New York: Academic Press.

Olds, J., and Olds, M. E. (1965). Drives, rewards and the brain. In *New Directions in Psychology II* (F. Barron, W. C. Dement, W. Edwards, H. Lindman, L. D. Phillips, J. Olds, and M. Olds). New York: Holt, Rinehart & Winston.

Selye, H. (1974). *Stress without Distress.* New York: J. B. Lippincott Co.

Sheffield, F. D. (1966). A drive-reduction theory of reinforcement. In *Current Research in Motivation* (R. N. Haber, Ed.). New York: Holt, Rinehart & Winston.

Sokolov, E. N. (1958). *Vospriiate i uslovny refleks.* Moscow: University of Moscow Press. (*Perception and the Conditioned Reflex.* New York and London: Pergamon, 1964.)

Vitz, P. C. (1966). Affect as a function of stimulus variation. *J. Exp. Psychol. 71*, 74–79.

Wundt, W. (1874). *Grundzüge der physiologischen Psychologie.* Leipzig: Engelmann.

WORKSHOP III. Clinical Modification of Behavior

Edited by WAGNER H. BRIDGER

Some very important issues at various levels have been discussed. If what has been reported is valid—and this validation must await replication—we have made some fundamental breakthroughs in terms of health. McClelland claimed that he was able to identify certain personality characteristics which are present in Western society and which make many people prone to various major medical illnesses. Among these illnesses there are primarily cardiovascular disturbances, but also peptic ulcer and according to some investigators an increased susceptibility to cancer. McClelland states that this specific personality characteristic which makes an individual susceptible to illness is what he calls, "the need for power." Though McClelland didn't mention the studies of Friedman and Rosenman (1959), these investigators described a type A personality which is very similar to the characteristics described by McClelland as "the need for power." Friedman and Rosenman state that type A personality is a coronary-prone individual who also has high levels of catecholamines. Furthermore, these findings are similar to what has been reported from Stockholm by Dr. Levi (1971) who has described the factors that lead to increased catecholamine production in terms of response to stress. However, McClelland is going further and has specified the specific personality that predisposes individuals to a hyper-catecholamine response and in addition he has described the kind of environment that leads these susceptible individuals to give these responses. He states that if you have a competitive society which emphasizes drive for power and if

WAGNER H. BRIDGER · Albert Einstein College of Medicine, New York. Workshop moderated by Chester M. Pierce.

you rear individuals in that society with an emphasis on drive for power, you are very likely to have a high incidence of chronic medical illness. McClelland, however, did not discuss why some individuals in that society are more likely to develop such a personality and medical illness than others. He did not deal with temperament, social class, or ethnic background. McClelland after describing this situation offered two approaches to dealing with it: the development of specific treatments and, perhaps, that we should restructure our society so that our environment will no longer elicit this drive for power. Among the specific treatments which have been mentioned by McClelland and Levi are nicotinic acid which diminishes the triglyceride response to stress, prolonged treatment with antianxiety drugs, or training in meditation. Levi apparently has the same approach as McClelland because he suggests that specific treatments will not be as effective as changing the basic rationale of our society with its emphasis on competition which further leads to distress and disease. He finally quotes a Chinese author, "efficiency is a means not an end. If we are asked to choose between producing more and better goods at the expense of our social integrity and producing fewer goods of lower quality we will unhesitatingly choose the latter if by such a choice we would avoid pain and disaster to our people" (Levi and Kagan, 1971). Thus what is suggested is that the only way we can do away with chronic diseases is to change our social system. Of course this approach is based on a series of premises. The most important of these premises is that the main determinant of this hyperresponse to stress is the power need of certain individuals. McClelland has developed this main thesis on the basis of experiments which involved subjects making up stories in response to various stimuli. The content of these stories is then rated in terms of power need and is correlated with catecholamine secretion. McClelland who is well known for his tests of achievement motivation states that the power need is much more important in respect to catecholamine activation than achievement motivation. I think this conclusion is not very compelling. It is very hard to equate the stimuli used in his measurement of power need with the stimuli used in his achievement motivation. He used a recording of Churchill's speeches as a stimulus for measuring power need and he claims that this produces more of an epinephrine response than the cognitive test that he used in measuring achievement motivation. Unless a complete parametric analysis is done with varying stimuli, I think it is somewhat premature to assume that power drive produces greater epinephrine secretion than achievement motivation. Frankenhaeuser reported in a study performed in Sweden that mental arithmetic provides an increase in epinephrine. There are two approaches in scientific research. One is to formulate a hypothesis and accumulate data to see if that hypothesis can be disproved. The second is to formulate a hypothesis and collect data in a sense to illustrate it or to confirm it. I think that McClelland utilizes the latter approach. He did not postulate alternative possibilities to explain his data. Among the variables which he does not mention are genetic, diet, ethnic, racial, etc., which may be confounded with his so-called personality factor. The mechanisms involved in producing a correlation between personality and illness may be complicated. For example, individuals with a great need for power may have a style of living in relation to diet, smoking, and alcohol which may lead to the increased incidence of illness without involving specific kinds of response to power stimuli which increases catecholamines. While it is true that some studies have shown that patients who have coronary heart disease have a higher excretion of epinephrine than subjects who do not, this is still a correlation rather than a causal relation. We do not know the exact relation of epinephrine to the incidence of atherosclerosis or even hypertension.

There is a basic contradiction between what McClelland was talking about and the data presented somewhat later by Marks, who mentioned that one of the techniques he uses in treating patients with compulsions and phobias is to expose these patients to noxious stimuli until they get used to it. They habituate or extinguish their anxiety reaction. In McClelland's point of view exposing individuals to power stimuli creates what he called a "power need." This disposition makes the individual prone to react to power-need stimuli; thus we get an individual who is chronically a power-need-disposed individual who therefore becomes involved with environmental situations which evoke his need for power and which therefore leads to high levels of stress. In a sense the individual becomes sensitized to these kinds of stimuli. Now in Marks' presentation dealing with phobic and compulsive patients he said that only 3% become sensitized. Most of the patients become habituated, getting used to the stimuli. So in a sense what he is describing is contradictory to the concepts mentioned by McClelland. Of course, on the one hand you are dealing with phobic or compulsive patients; on the other hand, with personality characteristics in general. In McClelland's description you get increased sensitization and the production of predispositions. In Marks' approach you get habituation and extinction. This could be related in some way to what Miller described when he mentioned Weiss' experiments where he exposed rats to chronic unavoidable shock and stress. After many such exposures they showed tolerance rather than sensitization. This is more in keeping with Marks' description of individuals becoming tolerant to noxious stimuli. McClelland emphasizes one specific personality trait, the need for power. Atkinson on the other hand talks about the fear of failure. I am not sure why the need for power should be more related to stress than the fear of failure. As I mentioned before there is no quantitative way of making these psychological states parametrically equivalent, and therefore it is premature to say that one personality trait or the stimulus that elicits such reactions is more important than another stimulus in respect to eliciting stress or cardiovascular disease. However, I think with a complex parametric analysis of all these variables it may be possible to put these various hypotheses to empirical testing. Included here would be the problem of a general adaptation syndrome and the patterns of the various hormones and whether there are specific features of the stimuli which inhibit or facilitate excretion of this or that hormone. These are all questions that can be answered with future research and they are important in terms of understanding the mechanism of the interrelationship between subjects and their environment. Perhaps I should mention a problem of the use of the concepts eu-stress and distress in McClelland's theory. People who have this need for power and thus the predisposition for seeking power will if they are promoted to a position of authority where they can exercise their power more efficiently have what is called eu-stress. However, the kind of hormonal changes associated with this increased stress should produce a higher incidence of cardiovascular disease. On the other hand, there may be an individual with a depressive type of personality who doesn't have this need for power and when he gets promoted, he has difficulty handling this new position and thus has distress which produces an increased excretion of catecholamines. However, he has the kind of personality that apparently is less prone to cardiovascular disease. Thus it is rather complicated to pull together the concepts of eu-stress or distress and the life changes leading to increased incidence of various illnesses. I should mention here some recent experiments which are perhaps related to what Selye mentioned about the role of conditioning in the general adaptation syndrome. Ader at the University of Rochester and Ackerman of Albert Einstein College of Medicine have shown how early experiences in rats can either facilitate or prevent the occurrence of gastric ulcer. Apparently early life experiences (interacting with stress) leads these subjects to either be prone or resistant to psychosomatic illness. The ulcers are produced when the rats are somewhat older and are placed in a stressful situation. Dr. Frankenhaeuser stated that women do not show an increased excretion of epinephrine in response to psychological stress and it is assumed that this response of increased epinephrine excretion is related to the increased incidence of various illnesses. By chance I noticed in the newspaper that a study at the University of Michigan revealed that when women and men who had just lost their jobs in the current recession are compared, the women showed significantly more psychosomatic complaints than the men. According to McClelland their need for power is less than men and according to Frankenhaeuser they respond to stress with a lesser amount of epinephrine excretion. However, their loss of jobs produced more psychosomatic illnesses so it seems that the relationship between life change, adrenalin excretion, and the incidence of illness is still rather complicated. A further confusing bit of information is that when scientists have studied the role of life change in the incidence of illness, among the items mentioned are not only things such as divorce but also marriage. Also the loss of loved ones is a very powerful factor related to the increased incidence of illness. However, as is well known, the loss of a loved one leads to depression, which has been hypothesized to be correlated with a decreased level of catecholamine excretion. So here again we have a confusing situation, a loss of a loved one leading to grief and bereavement with a decrease of catecholamines and an increased incidence of psychosomatic illnesses. However, we do have an increased cortisol production in depressed patients. So the whole complex relationship between stress and the various hormonal changes is still unclear. We have decreased norepinephrine with increased cortisol in depressed patients. Holmes and his coworkers reported that with high life-change scores there is an increased incidence of accidents; in one of his studies he found that professional football players had more accidents on the field when they had high life-change scores. It is rather difficult to see how this increased incidence of accidents is related to increased adrenalin or other hormones. Stress changes psychological functioning independent of its effect of biological functioning. Attention is affected, distractability is increased, and there may be preoccupation with various ideas, and this change in psychological aspects of behavior may lead to an increased incidence of accidents and also may effect the incidence of various medical illnesses. When a person is psychologically disturbed by a stress, he may not treat the initial symptoms of his medical illness and thus what may start off as a cold may develop into pneumonia independent of the affect of stress on biological functioning. Thus one has to pay attention to all levels of the stress reaction from the psychological to the biological.

The articles about the effects of behavioral therapy are very dramatic. Recently there have been important advancements in the treatment and diagnosis of schizophrenia and the primary affective disorders. However, severe disabling neuroses have been avoiding our therapeutic efforts. It appears at least from the reports we heard, that behavioral therapists have been very successful in treating these severely disabling illnesses; just as successful as psychiatrists have been with the use of psychotropic drugs and electroconvulsive therapy. I think that if this is replicated it's something that has to be publicized to the highest degree because the severe compulsive disorders and the severe agoraphobic disorders are extremely disabling and if behavioral therapists have such an effective form of treatment, psychiatrists should immediately adopt it. However, I still have my reservations. They report that they get 40% recovery without treatment and, of course, they present no control groups. The patients who are being followed without treatment are followed for the same number of years (2 to 4 years). As is well known, placebo therapy has always been very effective and new forms of therapy are always more effective when they are first applied; but with years of use placebo therapy becomes less effective. Perhaps these good results with behavioral therapy are primarily related to the power of suggestion present with a new form of treatment. If we assume that these dramatic results are replicated and control groups are used which demonstrate that this is a specific form of treatment rather than suggestion, what is the mechanism of this form of treatment? This is where Wolpe and Marks seemingly do not agree. Their disagreement is not trivial. Marks states that his technique is somewhat related to what is called extinction, while Wolpe claims that behavioral therapy must involve counterconditioning, i.e., in order to diminish anxiety we have to have an incompatible reaction such as relaxation, or something else which is incompatible with the anxiety. I think this dispute can be settled by empirical research but I do not think the answers are definitive as yet. Wolpe started his basic research with animals and developed his model for reciprocal inhibition on the basis of this animal work. Of course he deserves tremendous credit for popularizing the importance of behavioral therapy as a mode of treatment in psychiatry but it very well may be that his model of reciprocal inhibition is not the whole answer and there may be other modes and mechanisms involved in behavioral therapy. However, I don't think that Marks has proven conclusively that his technique uses only extinction and there is no reciprocal inhibition involved. As Wolpe mentioned, Marks' subjects did have a psychiatrist or therapist with them while they were exposed to noxious stimuli and thus therapists may be the mechanism of counterconditioning which diminishes anxiety and leads to change. This may be a valid point because counterphobic activity has been used by many patients either on their own volition or through the suggestions of therapists, or pressure from a family, and it has been rather unsuccessful. There are many people who have an airplane phobia but, however, continually fly in airplanes and have an unpleasant reaction while doing so. These people never seem to extinguish their responses—somewhat contrary to Marks' results. Perhaps this is due to the fact that they were not in treatment, just undergoing exposure. Which means that Wolpe's hypothesis that Marks' success is related to the therapist being involved may still be valid. Leaving Wolpe's and Marks' dispute aside we should mention that a whole new area of behavioral therapy is developing which has to do with the use of operant techniques. This is primarily used in institutions such as classrooms or hospitals. This was emphasized by Hunt. It is an especially useful technique for helping disabled individuals—the hyperactive children in classrooms, chronic schizophrenics, or mental retardates. Hunt mentioned how important it is to know the exact forms of reinforcement and the contingencies, and how careful analysis of what is important in terms of reinforcement and contingencies will lead to success while a sloppy approach may lead to failure. At Bronx State Hospital which is connected with Albert Einstein College of Medicine we have been using behavioral therapy operant conditioning with chronic schizophrenics. However, here we use a variant of the technique: Rather than giving reinforcements to the patients, we give reinforcements to the attendants who take care of the patients. Specifically, we give them green stamps every time the behavior of their patients improve. The behavior which we rate are making beds, going to the cafeteria, and such normal functions as those. When we ran out of green stamps, the patients' behavior deteriorated. I have a small bone to pick with Hunt. He mentioned that symptoms apparently always have some sort of gratification or reward; I think many times they do. However, I don't think it is necessarily always true. Certain symptoms have their primary function as defenses against anxiety or in a sense using the psychoanalytical concept of compromise between the wishes and the punishment for this wish; symptom formation is what the psychoanalysts call it. In many symptoms we have secondary gratification, however, at this point it is premature to say that all symptoms serve the purpose of diminishing anxiety, or serve some sort of gratification. A good example of the problems that are involved is the fact that with amphetamine-psychosis we get paranoid delusions and hallucinations. It seems unlikely that these symptoms are related to defense mechanisms for the diminishment of anxiety. Finally, I would like to mention one other aspect of behavioral therapy: the use of behavioral therapy for the treatment of affective disorders. specifically depression.

Beck of the University of Pennsylvania is using a technique which he calls cognitive therapy. This is based on the concept that depressed patients have a very low self-esteem and his task is to make them increase their self-esteem. He has various simple tasks and he asks the patients to estimate how well they will do on a specific task, whether it is cutting papers or using a paper and pencil. Since they have such a low estimate of themselves, they always do better than they think they will do and he gradually increases the difficulty of the task and they always do better than they initially think they will. He claims he gets a marked improvement. Apparently this is a form of behavioral therapy which is useful in treating affective disorders that are traditionally treated with biological modalities–antidepressant drugs and electroconvulsive treatment. If one wants to integrate biology and psychology one could hypothesize that when self-esteem is increased with active behavior the catecholaminergic reward system is activated.

Let us now return to the problem of behavioral therapy and the differences between Wolpe and Marks. Wolpe states that the evidence is very much against the idea that change occurs in treatment because of exposures per se. He states that in animal experimental neuroses exposure does not result in response decrement and he feels that if a basic mechanism is postulated it should be manifest not only in humans but also in animals. He emphasizes that exposure works only when it is accompanied by something else that can compete with the anxiety. He reported on experiments performed in Russia where sexually receptive females were in proximity to male neurotic animals in the cage in which they were made neurotic, and this was effective in reducing the anxiety response. Grinker and Spiegel in their examination of war neuroses claimed that if pentathol were given to soldiers they would abreact their battle experiences and quite a proportion of these soldiers became symptom-free. There were numerous reports of soldiers who abreacted by themselves in a pub and these abreactions outside the presence of a therapist were nontherapeutic. Thus, unless the role of the therapist can be controlled for in Marks' research, we cannot say that exposure by itself is a therapeutic instrument. There is also evidence given by Wolpe that tranquilizing drugs help behavior therapy. He quotes a study by Miller, Murphy, and Mirsky at the University of Pittsburgh in which they produce a severe anxiety reaction in rats and then expose them for fairly prolonged periods (for five successive days) to the situation in which they were made fearful. Those who were exposed with no drug tend to get worse. The other animals were exposed to this situation and simply given chlorpromazine significantly improved their behavior. The animals that were simply given chlorpromazine and not brought into the experimental situation showed no change. Thus Wolpe mentions the need for exposure to produce his change but exposure by itself is not enough; reciprocal inhibition must be evident. However, he does not exclude the possibility that there are other mechanisms. He just states that he is not convinced by Marks' statement that there are no competitive responses in his therapy. While Marks' data showed a very rapid therapeutic response, Wolpe feels that there are other approaches which are also very successful. He mentions a study by Rubin, from his own department, who counteracted anxiety with hypnotically induced suggestions and obtained a rapid deconditioning of anxiety in a very high percentage of cases. Wolpe is also concerned that the concept of mere exposure simplifies the work of behavioral therapy. It makes it appear that behavior therapy is simple minded. Wolpe feels it is important to do a very careful behavioral analysis, developing a hierarchy of stimulus response relations to guide the therapy.

Let us return now to McClelland's presentation. Mark feels the concept of stress as a mechanism involved with psychosomatic illness is more complicated than mentioned by McClelland. For example, the black population in Africa has an extraordinarily high incidence of hypertension but not of coronary artery disease. Thus he feels that stress may make a contribution to illness but only accounts for a small part of the total pathology. He feels that we must keep in mind that we are dealing with multifactorial models; clearly no treatment acts by one principle alone. Marks feels that in dealing with therapy and especially behavioral therapy, one must develop a framework in which we can construe the various differential therapeutic components. The first component is motivation. Unless the patient is motivated to seek help and to follow the directives of a therapist, the therapeutic situation will not even start. Marks feels that in this area social pressures play a role. This is the first stage. However, he states that this motivation by itself does not produce improvement. The next stage is the execution stage: the execution of the therapeutic actions. Marks feels that for most patients but clearly not for all, what is necessary in the phobic and obsessive domain is exposure to the phobic situation. Now there are some patients for whom exposure by itself does work and there are others who seem to get better with none. For example, a depressed phobic will improve on imipramine or some other tricyclic without any exposure to the phobic situation at all. There are also some patients who sometimes dramatically produce intense abreactions, not necessarily related to their problems, and they show drastic improvement but this is unpredictable and only in a small minority of cases. Another technique is role rehearsal in psychodrama or a social skills training group. Exactly what is going on here is difficult to define. Is it exposure? Is it skills learning? Is it the learning of competing responses? Probably a wide variety of mechanisms are involved in such a situation. Marks has an illustration of how his model can help us explain patients. A woman of 35 is preoccupied with various rituals. She spends her whole day in checking which direction the light switches are, up or down; on what side the baby is lying when asleep. She checks the locks all the way around the house all day long and at night she wakes up already so tired from all the checking that she asks her husband on which side the baby is lying, left or right. So her husband obediently trots off because he does everything she asks and he checks for her. This goes on throughout the night. Marks is called in for consultation. What should he opt for? Marital intervention or some sort of exposure paradigm? It's the same as tossing a coin so he just opts for exposure. However, since he is leaving on a trip, he will not be able to engage in this but he informs her that it is going to be difficult. For example, one of the things she needs to do perhaps is have her husband lock her in the bedroom at night in order to prevent her from checking. And who's going to look after baby? After all, she's going to have to be up all night worrying about what she hasn't checked. She would have to put herself in a situation where she would be tempted herself to check and not be able to. Marks states he would begin his treatment after three weeks. As Marks states, three weeks later she said she was 70% better. She exposed herself to all the things he said he would do with her and every day she knocked down a new ritual. Her husband did lock her up each night, and on a six-month follow-up she was still well but the marriage was as bad as ever.

Marks emphasizes that when a patient went through exposure, he himself showed no personal involvement and this seems contrary to Wolpe's model.

Hunt states that the example given by Marks while showing that a therapist does not have to be present, in no way rules out the very potent doctor–patient relationship as being the mechanism involved in the treatment.

Hunt feels that the initial approach with the patient, dealing with what Marks called the first motivation stage, is a universal one and that one has to use an oblique approach. He feels that many symptoms are involved in a power struggle and if one deals immediately with the patient's symptoms one will not be very successfuf in starting therapy. Hunt feels that most symptoms probably involve some gratifications. He agrees with Bridger that there is probably no gratification involved in some symptoms which may be of biological origin.

A good example of behavior therapy that could explain irrational behavior is this: In animal research if one makes a shock signal for a shock-free period, then the animals will work so they will get the shock and get into a shock-free period. Obviously, human behavior is much more complex but we have to keep trying. Dyrud from the University of Chicago feels that the results of behavioral therapy are in the same ball park as all other forms of therapy. Two-thirds improved, one-third unimproved, if you include the 25% of patients who refuse to engage in therapy. Dyrud also emphasizes that operant training for psychotic children seems to be the best approach present at this time.

Mowrer commented that the issues raised by McClelland are very similar to those raised by Sullivan when he gave his presidential address before the American Psychiatric Association in 1958. He stated that present methods of dealing with people suffering from mental illness in large impersonal institutions were a failure. He stated further that these individuals were already marginal, alienated people and the institution just completed the process and he suggested something we now call community psychiatry where many functions that were formerly carried out at the state level are brought back into the community.

McClelland mentioned that there is an interesting Italian slate-mining community in eastern Pennsylvania that has had a remarkable health record for many generations. The health record for this group is now breaking down among the members of it who have moved into Philadelphia. But the community itself is becoming Americanized; they're losing their ethnic ways and developing the same stress syndromes that we have in this country generally. Mowrer emphasized that in that community, decision-making processes were shared, that decisions were made by the community as a whole. This appears to go a long way toward relieving stress. Mowrer stated that this is also true for such other communities as Synanon, Daytop Village, and Alcoholics Anonymous.

Mowrer stated that we have to rediscover tribal psychology where people are interdependent, have low individualism, and are ready to lay down their lives for the welfare—the survival—of the tribe; that one cannot act individually because he might risk the welfare of the tribe as a whole. With the development of cities we can get increasing alienation, separation, anomie, and isolation. We have to move toward a kind of therapy or rehabilitation where the emphasis is upon the relationship between an individual and one or more groups or support systems. Mowrer mentioned that in order to be normal a person needs to work, and in our society we do not have work for everybody. And we have a society that doesn't need and doesn't want us all. He mentions a comment attributed to Levi, from a Chinese source—that efficiency is a means not an end. Mowrer mentioned that in our society it has become kind of an end that has boomeranged on us, in that we produce more and employ fewer people. It seems that as we approach the wider ramifications of some of these problems, we are going to have to look at some very unpleasant things, namely the economy and the political nature of our society.

Let us return now to McClelland's concept of power and its relationship to illness. Klinger had certain reservations both in terms of methodology and interpretation. As he mentioned in his presentation, Klinger interpreted the telling of stories in response to the TAT card as aspects of current concern rather than as continuing personality dispositions. He thus felt that the need for power scores that McClelland talked about are not necessarily interpretable in terms of long-term personality traits. He felt that they are just as easily interpretable along the dimension of "blocked strivings." McClelland's own research lends itself in support of this interpretation. One of McClelland's students found that the relationship between need for power and alcohol use did not hold up for woman subjects. For women the use of alcohol tended to be related to the need for womanliness as related to the standard American sex-role prescription—particularly, failure in establishing oneself in that role. Thus Klinger interpreted the relationship between power and alcohol as not a definite one but rather as a relationship between blocked strivings and alcohol. What it is that produces the strivings is different between men and women. Klinger applied this technique to the problem of comparing need for power with achievement motivation, both concepts explored by McClelland. Klinger pointed out in respect to methodology that in achievement motivation the individual is stimulated to perform a task and allowed to perform that task. In testing the need for power the individual hears tape recordings of stirring speeches and then fantasizes but cannot carry out any activity. In the latter situation, then, one would have blocked strivings; in achievement motivation, these strivings would not be blocked. Klinger used his concept of blocked strivings in dealing with the problem of cross-cultural differences. For example, he stated that in certain societies which have instrumental dependency, that is where adults rely on other adults to help in solving their problems, a cooperative kind of society, there is much less alcoholism. However, in societies which do not allow this cooperativeness between adults, this dependency between adults, there are higher levels of alcoholism. Where there is little instrumental dependency, people are more likely to find themselves blocked. They are more likely to find themselves in situations in which their goals cannot be attained or certainly in which they have more to worry about in that respect. Thus they have a situation that blocks strivings which may increase adrenaline secretion and thus cause more illness in that type of society. Klinger's final criticism of McClelland's need for power hypothesis had to do with the studies of Rahe, Holmes, and Masuda, demonstrating that life changes produce great increases in physical breakdowns of various kinds. Klinger felt that most of these life changes, though not all, involved major losses: being divorced, going to jail, being fired, etc. These losses are not situations where the person is put into a kind of striving stress power drive; they are, rather, simple grief situations. However, alternatively, it still may invoke blocked strivings and these loss situations may activate this need for power which is obviously blocked. The answers are surely not in yet. Finally in respect to behavioral therapy, Klinger felt that cognitive reorganization is just as good a model as aspects of learning theory. Placed in a situation with a phobic stimulus, the patients develop a cognitive interpretation and then realize that nothing terrible is going to happen to them; they are survivable and they develop a new cognitive structure which leads to better adaptation. This concept of cognitive reorganization can apply to both of Wolpe's and Marks' models of behavioral therapy.

Kolb, coming from the viewpoint of clinical and psychodynamic psychiatry, had a different interpretation of the results of behavioral therapy from both Marks' and Wolpe's behavioral therapy. This interpretation relates to the previous meeting of the Kittay Foundation directed toward examining the problem of attachment, even though attachment plays a very important role in early childhood. Psychodynamisists have stated that there will exist desire for attachment throughout life and there will be a reward for an attachment mechanism. Kolb stated that placebo studies show good evidence that the appearance of another human being offers surcease to 30 to 40% of people. Merely the appearance of someone else who has authority produces immediate relief. Kolb stated that one of the important elements in the behavioral therapies is the patient's expectation of a therapeutic attachment to another person. The main problem in behavioral therapy is that 25% of the patients are not motivated to develop a therapeutic attachment to another person; probably because of negative transference that may have come from early family conditioning or early cultural conditioning. A problem facing therapy is to arrange for these distressed people to make contact with a therapist who might be helpful to them. Kolb felt that clinical psychiatrists, although not calling it behavioral therapy, have been treating phobics in the same way using the attachment to the therapist who is an authoritative physician, or ancillary physician, who reassures the patient and thus diminishes his symptoms. Furthermore, as he quoted his colleague Hunt, Kolb stated that it's not only the physician but the institution that may carry out this transference and it's not necessary for the individual always to be around but the institution being available or in contact through letters, phone calls, etc., can be successful for the therapeutic process. A final comment can be made about behavioral therapy and the apparent contradiction between Marks' and Wolpe's position. Mowrer pointed out that extinction can be interpreted as counter-conditioning. When the subject expects punishment and fears this, there is exposure and nothing happens; there is relief which is a form of reward, a pleasant experience which would tend to counteract the fear. Thus extinction is probably one aspect of counter-conditioning which has been proposed by Wolpe and thus Marks and Wolpe are not contradictory to each other, according to Mowrer. Malitz felt that the important group of patients in both Wolpe's and Marks' studies are those who do not respond to treatment. He claimed that they may be similar to those patients who are treated with psychopharmacological approaches and who do not respond because their metabolism of the drugs is different from the average patient. That is they do not develop high enough blood levels to give a response. Perhaps the patients who do not respond to behavioral therapy need a modification of behavioral therapy; many more sessions in order to bring them to the same level of extinction—or whatever process is involved—as the other patients.

Pierce emphasized the role of chronic stress in certain societies that he has had experience with. For example, at the geographical South Pole, men are hypervigilant and obviously under stress. They also share something with people who live in a ghetto, mundane stressful environment. Both participants live in a forced socialization and spatial isolation and are hypervigilant in both these environments. These people have to stay alert in order to survive and their preception is mostly about how helpless they are and about their passive state. Thus there is quite a bit of sleep disturbance in the South Pole and the people get depressed both physiologically and psychologically. They lose their delta rhythm within the first three months and it doesn't return for at least a year upon return to temperate zones. They are depressed at all physiological levels—depression of the antibody mechanism, of blood pressure, of temperature, etc. Their individual depression interferes with social dynamics and group interaction. People in the ghetto have the same hypervigilance, the same subjective depression, and because of their living conditions there is a sleep disturbance which unfortunately hasn't been well studied. Pierce stated that the work of Marks and Wolpe is similar to the kind of training given in the Peace Corps where people imagine and conceptualize problems they might have to solve before facing them. Pierce stated that perhaps some of the findings in this conference should be related to social problems, that the techniques of treating phobia may be applied to social problems; for example, the major difficulty of oppressed people is how and what they have to fear from their oppressors. As Pierce stated, Phobus in Greek mythology runs before Aries the god of war to help create confusion and terror. In our society there are microaggressions which constitute the way people are assaulted and aggressed in their average environment in a ritualistic sense. A white man walking down the street toward a black man doesn't expect to have to move his face and move his shoulder. The black man will move his shoulder and dip down. These are ritualistic stresses in the sense that we are talking about here and now. Perhaps research money will go into behavior modification instead of into psychosocial research. In the next decade there will be many applications for such studies (how to control a tenement building without the use of policemen, for example) and as an outcome of this there is going to be explosiveness among social groups. I think the experimenters who were involved in this kind of research will have to take a social position and begin to anticipate what they can do about it. Hunt feels that a systems approach is necessary in treating severe disorders. It is probably impossible to make up the loss of early ego development. They will have to be given compensatory devices and corrective social learning but there will always be an instability and thus the therapists will have to make themselves available or the system will have to make itself available to help the patient in need. The therapists must be open to phone calls, being able to intervene in small ways that are very critical in keeping the patients out of the hospital. Marks feels that perhaps cognitive reorganization would be useful but many times the cognitive statements of patients do not apply to their situation. For example, he had an agoraphobic patient whom he treated by asking her to imagine herself being chased by a tiger. It turned out that this was her main form of treatment. If she put a picture of a tiger in her room whenever she was fearful, she would go through the imaginary scene of the tiger chasing her and eating her. There is no clear explanation as to why this kind of imagination was effective in treating her agoraphobia. Marks also felt that group support may be helpful in behavioral therapy. Thus treating patients on cohesive groups is sometimes more effective when added to individual therapy. What he has mentioned as being successful is "acting." That is, he tells the mother, even though she wants her 19-year-old child to stay with her and not move out, to try to say, "It's OK if you want to, it's fine." The mother is at first reluctant but on the prodding of the therapist she says "OK, it's fine." After a while the mother acting the part of the supportive mother can let her daughter become independent. Then the patient leaves home and the problem ceases. He stated that this is like Diaghilev in ballet, "Act and you will feel." In respect to applying these concepts to social problems, Marks felt that it is important to be very careful in extrapolating principles from one system of organization to another. Stress in tissues is different from stress in individuals which in turn is also different from stress in social groups. In respect to the concept of extinction being counter-conditioning, Wolpe emphasized that reciprocal inhibition is a key thing in behavioral therapy. Even in the situations where imipramine was given to phobic patients he stated that imipramine had a tranquilizing effect and thus diminished anxiety. That is why it was effective in helping patients. However, Klein said that antianxiety drugs are ineffective. Wolpe also stated that in terms of the numbers of patients who did not go into behavioral therapy, he did not have the same kind of problem as Marks who stated that 25% of his patients did not enter into therapy. Wolpe found that very few patients did not accept his offer of behavioral modification, probably because he did not use overt high-intensity exposure which probably frightened patients away. His gradual desensitization program, however, allowed them to diminish their anxiety and they did not refuse treatment. As to the 10% who did not respond, perhaps he had what the analysts call negative transference. Wolpe also felt that cognitive reorganization is not sufficient. There are many patients who know that there is nothing wrong with an elevator but who are still fearful. It's only when the fear is diminished that they show a cure.

Dr. Bridger felt that all presentations at this conference had something in common: They dealt with the concept of stress as a situation that taxes the organism. It makes demands that the organism can't cope with. Biologists are accumulating data as to the various mechanisms in the hypothalamus and the dopaminergic and nonadrenergic system related to motivation and the mechanisms mediating stress. He pointed out that there is obviously no closure on these mechanisms. For example, stress releases prolactin whereas dopamine by itself blocks prolactin while stress increases dopamine turnover. The resolution of this controversy is not present. However, while the scientists do not understand fully the mechanisms mediating stress, there seems to be agreement that when an animal can cope with a noxious stimulus, as Miller pointed out, that stimulus becomes less noxious in terms of all the indicators of neurohumoral translation. As Grossman pointed out, an intact dopamine system is needed for an animal to cope. Thus there is an integration between the ability to cope and having an intact neurohumoral, neurophysiological system and perhaps differences between individuals and their coping behavior. Aside from these individual differences, people live in a psychosocial environment and that environment, when it produces stress, has one thing in common, that it produces stress in the terms of blocking of strivings. Whenever a person has to do things which he feels he can't, he is under stress and this leads to all the pathology of stress. Finally, when an individual therapist helps the person to cope through behavioral therapy, he is teaching that person to thus diminish the effect of stress. This is one technique that has been emphasized at this symposium.

REFERENCES

Friedman, M., and Rosenman, R. H. (1959). Association of specific overt behavior pattern with blood and cardiovascular findings; blood cholesterol level, blood clotting time, incidence of arcus senilis, and clinical coronary artery disease. *JAMA 169,* 1286–1296.

Levi, L., Ed. (1971). *Society, Stress and Disease.* London: Oxford University Press.

Levi, L., and Kagan, A. (1971). A synopsis of ecology and psychiatry: Some theoretical psychosomatic considerations, review of some studies and discussion of preventive aspects. Psychiatry (Part 1). *Proceedings of V World Congress of Psychiatry.* Mexico City, 25 November–4 December 1971, p. 379.

Concluding Remarks

BORJE CRONHOLM

The field covered by the authors and discussants at this conference is very wide indeed and thus, any attempt to integrate the various issues meets with difficulties. The wide variations in background and type of training among the participants may have led to some communication gaps. Conferences like this one offer special challenges for the participants. The psychiatrist finds himself confronted with the highly specific and technical terminology of the neurophysiologist, and the neurophysiologist has to try to understand the complexities and subtleties of the psychiatric language. But it may be these very difficulties which make conferences with participants from many fields particularly rewarding and potentially fruitful, widening the worlds of the researchers, making them appreciate that there are different ways of solving related problems. For that reason, cross-scientific conferences like this one are necessary, and the need will only increase with increasing specialization within each field of inquiry.

In making a short review of the contributions I have to be restrictive and pick out only a few themes from the overwhelming material that has been presented.

A central theme during the conference has been the "stress" concept. The grand old man within the field of stress research, Selye, strongly advocated the view that "stress" means "the nonspecific response of the body to any demand." It may be combined with disagreeable ("distress") or pleasant ("eustress") feelings. To Selye, "stress" is thus a state or a process within the organism. For

---

BORJE CRONHOLM • Department of Psychiatry, Karolinska Sjukhuset, Stockholm, Sweden.

the conditions or stimuli eliciting "stress" he uses the term "stressor." However, it was quite evident that these definitions are not generally accepted and that the term "stress" is used also in other ways.

The discussion between Selye and Mason was both stimulating and enlightening. Mason argued that there is no general, nonspecific hormonal reaction to all stressors; on the contrary, different types of stressors elicit different patterns of hormonal responses. He also argued that the "nonspecific," pituitary-adrenal response described by Selye may be interpreted as a reaction to the anxiety-provoking characteristics, common to so many "stress" experiments and situations. Both discussants presented interesting data in line with their respective hypotheses. The discussion will probably continue to occupy the scene for some years. More empirical data about various patterns of reaction to different stressful situations will certainly contribute to a more fruitful conceptualization of "stress" and "stressors."

Basic research concerning the functions of the CNS, relevant for our understanding of behavior in various situations, was presented by Grossman and Olds. They reported findings concerning the neuroanatomy and neurophysiology of cerebral drive and reward systems by means of mechanical or chemical destruction of cell groups or of tracts, and by local stimulation, using microelectrodes. Their results are of great importance for the understanding also of human behavior–it has, e.g., been assumed that disturbances within the reward system may be of importance for the schizophrenic anhedonia.

Miller underlined that different physiological responses to "stress" (here in the sense of Selye's "stressors") may be learned and that this holds true both for adaptive and for maladaptive reactions. Thus, interindividual differences in behavior are dependent on environmental factors. However, the findings reported by Corson indicate that genetic factors are also responsible for such interindividual differences. He had studied Pavlovian learning in dogs and found that they displayed two types of reaction in stressful laboratory situations. One group adapted well; the other had difficulties in adaptation and also showed a characteristic, antidiuretic reaction. Studies of interindividual differences in autonomic reactions to stressors are of course highly relevant for better understanding of the etiology and pathogenesis of psychosomatic disorders. It seems evident that both genetic and environmental factors (that lead to "learning") are of importance for these differences.

Frankenhaeuser presented data from her studies on urinary catecholamine excretion as an estimate of "stress" in different groups of individuals and in different situations, both in the laboratory and in real life. Her studies thus are examples of fruitful integration between basic and goal-directed research. She has explored various situational factors which are of importance for the catecholamine excretion. For example, the excretion is lower when the individual is able to control a stressful situation than when he is not. Those individuals who excrete high amounts of catecholamines perform better than others, but possibly they have to pay for their efficiency with higher morbidity. A study of sawmill workers lends some support to the view that a monotonous coercive and physically strainful job results both in a high level of catecholamine exretion and a high frequency of psychosomatic symptoms. Women excrete far less catecholamines than men in stressful situations. Frankenhaeuser speculated that this difference may be due to cultural factors and that women will in the future perhaps react more like men (and run the same risk of psychosomatic disorders!) if they adopt the same attitudes to work and competition.

Interindividual differences in reaction do not depend on sex only, but on many factors. It seems that every individual has an optimal level of arousal and also tries to reach that by seeking an optimal stimulation level.

Atkinson also reported on sex differences; contrary to men, women have a negative attitude to academic success, evident in fantasies induced by a short sentence describing a successful person of the same sex. This was interpreted to imply a much more stressful situation for female than for male students.

Klinger described an experimental study of the influence of motivational factors on cognitive processes. He found that when listening dichotically to two different texts, subjects more often switched their attention to concern-related than to other passages and also remembered them better. Berlyne gave a report of experimental studies on uncertainty as a stressor. He displayed blurred slides to his subjects who were asked to interpret them. This led to increased arousal as demonstrated by the EEG.

The last part of the conference was devoted more directly to problems of psychopathology and to its treatment.

As a psychiatrist involved in research on depressive diseases, I found it very difficult to follow Mowrer's arguments. If I didn't misunderstand him—and I am afraid I didn't—he meant that guilt feelings (always?) represent "real" guilt and that conscience (always?) represents "the voice of God"! Of course, the guilt feelings of a depressed patient, suffering from unbearable anxiety and self-reproaches, are real, as real as any mental events. But his aggressive conscience certainly does not represent "the voice of God"—if anything, it stems from the other one! And that voice must not be supported by the therapist—that would be malpractice—but should be expurgated by adequate means, by antidepressant drugs or by electroconvulsive treatment.

Both Marks and Wolpe presented convincing data showing that behavioral psychotherapy may be efficient in treatment of such types of neurosis as phobias and obsessive-compulsive states, more so than some other psychotherapeutic methods. Hunt gave a review of the difficulties and unsolved problems related to behavior therapy.

At the beginning of this conference Kittay underlined that the situation in the present world is apt to produce harmful stress in many of us. It is the duty for scientists in relevant fields to find methods to prevent noxious conditions and to help those who have become the victims of stress, whether the result is physical or mental. But we need more knowledge about the impact on the organism of different conditions and about the varying reactions of different individuals. This means that further research in the field in necessary. When doing that research it is important to remember Sabshin's warning against the anti-intellectual trends of today. We need facts, not prejudiced beliefs.

Stress and Human Psychopathology

GEORGE SERBAN

A major contribution to modern understanding of diseases resulting from stressful interaction between the environment and the individual has been made by Selye, who introduced a new concept in medicine known as "disease of adaptation."

The notion of the general "adaptation syndrome" mediated by the pituitary-adrenal-cortical system and produced by nonspecific stress, derived from Selye's experimental work, has opened new perspectives for the study of human reaction to stress and widened the understanding of psychosomatic disorders.

Although Selye has concentrated his studies more on the impact of physical stressors on the bodily processes and less on the psychological factors leading to stress, he has always inferred their presence. Mason, as a result of his research on the specificity of stress, came to the conclusion that Selye's general syndrome of adaptation is primarily mediated by psychological mechanisms. Mason reported that while some physical stressors are inducing stress responses in the organism, as evidenced by the elevation of 17 OHCS, most other physical stimuli, such as exercising and fasting, do not, unless they are accompanied by psychological stress. In addition, as Mason has shown during this symposium, hormonal responses to stress are patterned by this type of agent stressor.

Mason's findings are of particular importance for broadening our understanding of how psychological factors in stress responses contribute to various noxious environmental conditions. Supporting data for these findings came from

---

GEORGE SERBAN • Associate Clinical Professor, Department of Psychiatry, New York University Medical Center.

various sources. For instance, Symington (1955) observed that (due to the sustained bodily trauma) unconscious animals did not develop pituitary-adrenal-cortical stress responses. The same observation was made in relation to humans, who, when in the comotose state due to fatal injury or disease, did not show any general signs of adaptational syndrome demonstrable by changes in adrenal-cortical glands.

The "general adaptation syndrome," as conceived by Selye and reformulated by Mason in terms of specificity, apparently represents two different levels of adaptive process related to the quality and quantity of noxious stimuli applied to the organism. According to the requirements of the adaptive process, the body might react, either as a wholly integrated functional organism on a nonspecific level, or specifically, on the target organ of stress, due to a self-regulating response involving the affected area. Bypassing the controversy of specificity versus nonspecificity for the moment, let's clarify a more important issue from the psychological point of view, that is, the coping mechanism on which the adaptation and stress are based. To start with, it is of value to know the conditions under which the physiological concept of stress attains its psychological equivalent. Psychologically, stress is conceived as resulting from an imbalance between the perceived demands and the recognized response capability of the individual (McGrath, 1970). To further qualify this view, some researchers suggest that stress is experienced only when the environment represents a threat to the individual, who, in turn, perceives disastrous consequences, should he be unable to cope with these demands (Lazarus, 1966).

Other researchers, however, are conceptualizing psychological stress in both directions of human response, pleasant and unpleasant. What they emphasize is the magnitude of change in life produced by the stressors. Based on this concept of life changes and their impact on the individual, Holmes and Rahe (1967) conceived a measurement of stress which was found by some researchers to be statistically significant when correlated with the subsequent health changes of the subject leading to morbidity and mortality. Conversely, other investigators questioned the validity of the assumption that life stress alone leads to illness (Hinkle, 1974). At best, they found a correlation between "life stress and a tendency towards treatment-seeking behavior." The limitation of this approach is self-evident when we consider that life changes exclude an evaluation of long-term difficulties or present and future reactions produced by these events. Since a criticial report describing the serious shortcomings of the Rahe Assessment of Life Stress, as it relates to mental illness, has been fully presented by Sarason *et al.* (1975), there is no further reason to dwell on it. As long as we assume, however, that "life changes" reflect the wear and tear of the body in response to environment, the assessment of these changes might help us to evaluate the individual state depending on the number of experienced crises up to that time; but this measurement cannot tap the psychological stress as a "distress," the true indicator of one's emotional responses to the totality of events which comprise one's life.

In a study done at New York University-Bellevue Medical Center, I have evaluated the psychological distress sustained by normals and schizophrenics in the last 6 months prior to the testing, with a new approach to the measurement of stress based on an inventory of social functioning and stress of the individual (Serban, 1975). With this instrument it was possible to measure quantitatively and qualitatively the amount of stress sustained by the individual in daily life in response to his psychosocial environment. The results indicated that the maximum stress of the three groups under study (normals, and acute and chronic schizophrenics) was experienced by the chronic pseudoambulatory schizophrenics who, interestingly enough, had little significant life change, except for an inability to adjust, even on a marginal social scale of functioning, outside of the hospital.

Important to note as well was the fact that 62% of the acute and chronic schizophrenics were unable to identify any major precipitating cause leading to their hospitalization, although they were admitted in a high state of stress. These data raise a further question as to the reliability of life changes in tapping the stress experienced by an individual.

In this respect the quantitative and qualitative measurement of stress, which reflects the relationships between the two components of stress equation by the demand and response capabilities, appears to be a more exact method.

Furthermore, this comprehensive approach permits us to solve another problem in dealing with stress, that is, the fact that environmental demands are not perceived equally by all individuals and, conversely, not all individuals have the same adaptive capabilities to meet these demands. The inventory of the Social Stress and Functionability for Psychotic Disorders (SSFIPD), which can be applied to "normals" as well, was constructed to meet these criteria. In addition, the test made it possible to measure not only the general stress sustained by the individual but also the stresses experienced in various areas of his life. As a result of this study it was observed that although "normals" experience more stress than schizophrenics in specific categories of life situations, their total stress was lower than that of psychotics. Apparently it was the ability of the normals to cope with stress that differentiated them from the schizophrenics.

A different dimension in the psychological study of stress is that based on the qualification of a type of demand, defined as either excessive ("overload") or its opposite, deficient ("underload") demand, as experienced in sensory or social isolation, confinement, or impoverishment of the social milieu. This symposium dealt mainly with the excessive demands placed on the individual (with the exception of the presentations of Selye and Mason).

Consequently in these concluding remarks we should attempt to define more exactly the effects of this type of stimuli on the individual psychobiological programming, and to attempt to underline the terms under which they produce psychosomatic or emotional disease in the individual.

As we know, researchers have been able, so far, to separate only a general physiological and neuroendocrine mechanism induced by social-psychological stressors. Starting with Cannon's famous hypothesis of "flight or fight" the sympatho-adreno-medullary-system was implicated in determining the type of reaction to physical and psychologically stressful stimuli. Since then, there has appeared abundant literature describing the increase of catecholamine excretion due to various physical and emotional situations which support Dr. Frankenhaeuser's data presented in this meeting.

Another group of researchers found that parallel with the increased catecholamine activity, there is an increase in adrenal cortical activity which accompanies exposure to psychosocial demands as demonstrated by the work of Persky *et al.* (1958), as well as Rubin, Mandell, Mason, etc.

In summing up these biochemical factors one could say that two mechanisms are involved in response to psychosocial activities leading to stress: one, the fast-acting system of the hypothalamo-adrenal-medulla; and the second, the slow-acting one known as hypophyso-adrenal-cortical-system.

I will attempt to focus my discussion on what appears to require more attention: the psychological aspect of the individual's reaction to his psychobiological programming. Since the programming itself is influenced by demographic, genetic, and personality factors, which decide the level of reaction and adaptability to stress of the individual, these factors should be analyzed in more detail.

Psychologically they could be subsumed under two clusters of psychobiological traits: the first relating to the acquired knowledge and realistic learning capacity of the individual, and the second to his motivational interest and societal aspirations, as developed in the context of his socially accepted system of values. The first set of traits recognizes and appraises the environmental demands as threatening or not, while the second set attempts to cope with the stressors.

Certainly this is an oversimplification of psychobiological mechanisms responsible for adaptation, since its other components (such as personality constellation, self-confidence, self-esteem, and level of neuroticism) should also be considered. These elements, however, are indirectly implied in the abovementioned sets, as a part of the individual's acquired knowledge, which include his instrumental achievements (skills, etc.) and realistic learning capacities, and which presuppose a degree of ability for self-appraisal. The second set mentioned as responsible for the degree of adaptation to stress is strongly influenced by the social organization in terms of societal hierarchical values expressed by status and by the system of rewards and punishments, used by the system as reinforcements of these values.

In this light, the paper of Dr. Frankenhaeuser offers some stimulating observations related to the reaction of the male and female under stress, which require further qualification. If it is true that females do not show the same readiness as males in response to environmental demands, this may be indeed related to the significance given by women to the perceived situation as non-stressful. It is possible that the group of women under study might have been less competitive than men with less need for advancement, control, etc. Conversely, the men interpreting the situation as highly rewarding socially or financially respond in a much stronger manner to stress.

Frankenhaeuser's suggestion that women, due to the current change in the patterns of sexual role, will develop in a greater proportion, the same type-A driving personality as men (which will modify their adrenalin release), appears to be only partly correct. The implications of her statements require further clarification. The theoretical assumption behind this suggestion is that women will change their bioadaptive mechanism due to the new psychosocial stimuli with which they are confronted. Thus this assumption is highly hypothetical. It cannot explain why men who release "slowly decreased amounts of adrenalin" to baseline, after efforts do not change their adaptability to become fast releasers, though this will be more beneficial to their health, indicating a more adaptive response to stress. On the contrary, individual or sex difference patterns are biologically programmed, and it is questionable as to whether they can be significantly modified by social stimuli. This explains why the woman Ph.D. presented in the study by Frankenhaeuser excreted a high amount of adrenalin; she perceived this task to be vital to the organization of her future life. It might be more reasonable to assume that if she had a type of psychobiologically poor response to stress, she would have been unable to advance her career in receiving the title of Ph.D. This interpretation appears to support the findings of McClelland who believes that men and women start out in life with the same physiological reaction to stress, but due to different developmental conditions are subjected to different psychosocial forms of stress, which in return show men more reactive to stress. It should be emphasized, however, that these different developmental conditions between the sexes projected in the social milieu are based mainly on biological differences. For example, men (due to physical strength) have played a social role which implies defense of the territory, providing shelter and food for the family, etc., which in turn decides the psychosocial stresses to which they as a group are exposed.

This formulation raises a very basic question, whether men having been exposed evolutionarily and developmentally to stress should not be able to adapt more successfully to it. Yet, the medical literature indicates that man has more difficulty in adapting now as indicated by increased deaths due to hypertension and coronary heart disease (not to mention mental illness, suicide, and psychosomatic conditions). All these data are pointing toward two possible explanations. Either the individual limited capacity to adapt to ever increasing stress is overtaxed by social technological conditions, or stress is self-induced by a highly competitive social system which encourages an individual to assert his powers and needs, not necessarily beneficial to him. In Western society both factors are acting simultaneously with different emphasis on each individual at different times. This situation appears to be compounded by the social organization itself, which ironically in attempting to eliminate the struggle for survival has increased the competition for a better life, placing undue pressure on the individual to obtain maximum comforts and goods regardless of his capacity to secure them.

On an experimental level, examples of each one of these situations are presented or suggested by the presentations of Frankenhaeuser, Atkinson, and McClelland. For instance, the study of chronic stress responses of sawmill workers done by Frankenhaeuser raises just these issues, namely, whether the high state of stress is due to the strain of having reached each one's maximum capacity to adapt or conversely, due to their low capacity, to adapt to an otherwise acceptable amount of stress. If it is true that under high conditions of physical strain and restriction of social interaction these workers develop psychosomatic disease, it will be of interest to know what type of personality they have and to what other concomitent social and familial stress they are exposed, in order to find out other sets of variables than work, which may be contributing to their psychosomatic illnesses. It will be of value to know what are the psychological coping mechanisms or the behavioral strategies used by sawmill workers who did not develop psychosomatic disease: In other words, what are the contributing factors which make it possible for one group of sawmill workers to have a higher level of adaptability to stress without ill effects in high-risk stress situations than others? It would be methodologically unsound to draw even tentative conclusions from an isolated set of reactions without evaluating the full psychosocial profile of the individual working under conditions of high risk for stress.

Interestingly enough, the study of stress, per se, as leading to mental or physical illness, independent of other psychosocial variables, becomes even more controversial, if we consider the position held by some researchers, who, based on their observations, came to the conclusions that high stress is not necessarily associated with psychosomatic conditions or maladaptive behavior (Hinkle, 1974). In this respect, one such example are the experiences in the concentration and labor camps where previous psychosomatic conditions were abated. Supporting data come from another set of observations of line combat military units, whose men suffered less stress than those in rear areas. These facts not only question the widely held position that stress alone affects the physical health of the individual, but also suggest that under stress existent physical conditions might disappear. Thus the alleged positive correlation between stress and illness presupposes other intervening personality variables, which should differentiate the high-risk stress group from the low-risk one prone to noxious physical reactions.

In this context, another factor which appears to play an extremely important role in the process of adaptability is that of the motivational level of the individual in coping with the situation. Under normal conditions the psychosocial stressors will affect us either when the need for achievement is higher than existent possibilities inherent in our personality to fulfill it, or when the task appears to be insurmountable, in which case the individual feels helpless. When the task is perceived as leading to failure the individual experiences anxiety which might affect his performance in future tasks which will be perceived by him as identical. Atkinson drew the conclusion that individuals with high anxiety are poor performers in need achievement tests, due to the fear of failure which becomes a paralyzing force. This presents only one side of the picture. In different experiments Miller has indicated that fear-anxiety could be a highly mobilizing factor for increasing performance when strong motives are overriding fear itself. This then explains why Marks reached the conclusion that in phobias and obsessions successful treatment could be obtained by exposure with either relaxation or implosion methods, which presuppose a maximal anxiety. Moreover, if the individual learns to discriminate between real fear and the unjustified one as Miller's experiments suggest, he will cope better with any given situation. This line of argument makes the issue of the fear of failure in women as described by Atkinson (based on the experiment of Horner) even more controversial. According to Horner, fear of occupational success in women was explained as based on resistance by associating fear with the negative consequences of success. Obviously the fear of success as indicated by her testing cannot be explained on purely biological grounds. Horner's apparent explanation rests on psychosocial factors, either as a learned social fear on the part of women to compete with men or as precipitated by a different orientation of their competitive roles toward status as brought out by beauty, maternity, social acceptance, etc. But even if we attempt to accept this explanation as valid in justifying fear of success in general, and that of women in particular, the methodology used to reach this conclusion is far from perfect, due to the fact that the TAT test itself does not capture the achievement as a finalized act, but only its direction toward possible execution. There is quite a difference between motivational disposition expressed on the level of interest and the realization of the act in the achievement itself. In this sense fantasy might give some possible clues as to the orientation of motivation, which is subject to various modifications, depending on situational incentive changes in the direction of the goal, which, in final analysis influence the performance itself.

For example, the study of Opton and Lazarus (1967) which refers to the interaction between the stimuli and the subject in a stress situation is highly edifying. Based on this study, the researchers were able to separate two groups of subjects, one responding to the fantasy world of films with stress reactions and the other to the world of reality through electric shock experience. These distinctive types of personality experience stress situations in a different manner. When Horner's subjects were tested with TAT they might have affected the data accordingly. In the same line of thinking is the treatment approach to maladaptive behavior of Wolpe, in this country, and Marks and the School of London, who found a more successful result in treatment with "vivo exposure" than the one based on fantasy.

Another point of interest in Atkinson's presentation is related to his findings which concern an existent correlation between motivation and performance based on an inverted U-shaped curve, similar to the curve found by other researchers related to performance under stress. A particular aspect of this observation appears of significance, that is, overmotivation reduces performance which in exchange produces anxiety stress. These findings are valuable to the extent to which they indicate another dimension of personality difficulty leading to maladaptation when the motivation is misapplied. However, stress and individual functioning-performance could be negatively correlated, in the sense that the higher the stress the lower the functioning. In the Bellevue study we have found this inverse relationship between performance and stress. The chronic schizophrenic with the highest stress functioned at the lowest level, while comparable social-demographic normals experiencing low stress had the highest functioning level. The acute schizophrenic who experienced an amount of stress which fell midway in comparison to the other two groups functioned on an intermediate level in relation to the normals and chronics (Serban, 1975). The relationship between stress and misdirected motivation leading to maladaptive behavior becomes even clearer in the experimental work of McClelland, who shows the association between drive for power and cardiovascular disease (hypertension and coronary heart disease). His findings suggest that need for power produces stress by increasing epinephrine production. According to McClelland, when needs for power are associated with a low degree of inhibition, the end result could be alcohol abuse, while the needs for power associated with high inhibition can lead to cardiovascular disease. Based on these findings, McClelland reached the conclusion that a high need for power is detrimental to health, which as such represents a maladaptive functioning of the individual.

Yet, this position is highly debatable to the extent to which the drive for power apparently represents a basic innate drive, evolutionarily developed, and present from the pecking order of chickens to the complex hierarchical organization of primate societies. If power is an innate drive, then it presupposes that the individual should be capable of adapting to it. The fact that need for power as expressed in a particular act results in an increase in epinephrine levels does not say too much because any state of overstimulation or understimulation will produce the same increase in the hormone. The long-term pursuit of power might lead to maladaptive behavior when it is unsupported either by positive personality capacities or socially available possibilities. On the other hand, if the social system is overcompetitive and places high burdens on the individual to produce, or the need for power is blocked by social discrimination, then apparently the maladaptive process starts in some individuals. Let us assume that the individual tested by McClelland functions in a corporate system which pushes him forward beyond his real capacity to cope with new responsibilities. Then the fear of failure will become an important factor in leading him to either neurotic maladaptive responses, such as psychosomatic conditions, or possibly overindulgence in alcohol consumption. This position is supported by the same recent research related to stress and cardiovascular disease (Hinkle, 1972). Furthermore, when the same competitive system reinforces unrealistic concepts of equal opportunities it might create a serious problem of adaptation for some individuals with strong needs for affirmation who are taking these socially relative formulations too literally. Another group of individuals predisposed to stress is the one whose members overestimate their possibilities and who have a high motivation level in the presence of low performance and efficiency. In such cases the interaction between the psychosocial environment, personality structure (with individual perception of reality, ego defenses), behavioral strategies, and genetic background will determine presence or absence of a chronic stress response. This means that psychopathology of stress is determined by multivariate interactions of psychobiological mechanisms which cannot be explained either on a pure biological basis as attempted by some researchers, or solely on environmental factors, as assumed by others.

The necessity for a multidimensional approach becomes even clearer for psychiatrists and clinical psychologists when they try to understand and treat maladaptive behavior secondary to the stress reaction and experienced clinically as anxiety state accompanied by phobic behavior, compulsive reaction, or sexual deviations. Anxiety, as a clinical manifestation of stress, is produced by the individual's inability to solve an emotional psychosocial conflict. Evidence suggests that this conflict is based on a faulty interpretation of reality (Serban, 1974). Anxiety appears to be aroused by the individual's inability to control the outcome of an act by the anticipation of remote possibilities, which in his judgment might interfere with the execution, or realization of that act. His perception of reality, based on patterns of thinking developed most likely since childhood, leads to an incorrect appraisal of specific present events associated with anxiety-arousing situations related to his past (Serban, 1974).

In this context maladaptive patterns of behavior learned in the developmental years, or under certain conditions of stressful life situations, could be relearned by various behavioral techniques. Based on his experiments, Wolpe reached the conclusion that neurotic anxiety habits are learned, and as such, can be treated or cured by reciprocal inhibition.

Yet not all behavioral therapists agree either with his formulation of reciprocal inhibition or with his claims of a 90% cure rate. For instance, Wilkins citing accumulated research literature on the subject questions the conceptualization of Wolpe's method based on deep muscle relaxation, greater hierarchy of instructed imagination, or their combined use. According to Wilkins, only instructed organization of fear-provoking scenes is necessary (Wilkins, 1971).

Marks, in a well-documented presentation, reduces all behavioral manipulation to the exposure of the patient in need of anxiety reduction to his fear-provoking scene.

What is interesting about Marks' study are his findings concerning the ability to control or reduce anxiety either by exposure of the patient to intense stress as in the flooding techniques or to low stress as in relaxation. This raises very important questions as to whether or not other factors besides exposure itself operate in the therapeutic process responsible for the reduction of anxiety. In this context Wilkins advances an interesting hypothesis, that is, that the expectancy of the therapeutic gains, and the use by the patient of the therapist as a reinforcer of nonfear behavior, might play an important role in behavior therapy. He comes to the conclusion that the skill of the therapist acting as a social reinforcer contributes greatly to the success of this type of therapeutic intervention.

The complexity of the problem of learning nonmaladaptive patterns by behavioral means was amply developed by Hunt in discussing the use of behavioral therapy with psychotics. Summing up the state of behavioral therapy, one should be in full agreement with the conclusion of Marks who admits the paucity of complete knowledge of the treatment mechanisms involved in behavior therapy.

A better explanation of this mechanism might come through a clearer understanding of underlying neurophysiological processes of adaptation. In this sense the work of Grossman, Valenstein, and, in particular, Olds appears to be extremely helpful.

For instance, Olds' experiments with rats have located the existence of drive and reward neurons which might show high potential in clarifying intimate aspects of behavioral learning. Olds identifying the neurochemical mediators of these neurons appears to be even more important in explaining the mechanism of controlling behavioral priorities and defining their actions and their neural pathways. The association of noradrenalin and dopamine with basic drives and reward behavior and of serotonin with diminution of arousal requires further study for evaluating their exact role in the psychopathological conditions of adaptation such as schizophrenia, affective disorders, etc.

The role of dopaminergic systems in the process of learning and organization of behavior was fully discussed by Valenstein and Grossman. Grossman's suggestion that the striatum might be related to an associative process could explain some aspects of psychopathological behavior for which there has as yet been no clear neuroanatomical or neurophysiological interpretation. The implications of these findings are far-reaching for the understanding of some neurological and psychiatric syndromes. For example the experimental destruction of nigrostriatal systems leading to a lack of response to either distal or tactile stimuli reminds us of similar phenomena which sometimes appear in schizophrenia. In the same context, Valenstein pointed out that the dysfunction of dopaminergic systems could be associated with disruption of thought processes. The nigrostriatal dopamine systems dysfunction attempts to explain as well psychotic behavior (as in amphetamine model psychosis). In a sweeping generalization, it may be said that the balance between various neurotransmitters–dopamine, or epinephrine, methyl choline–plays a crucial role in the organization of normal behavior.

The problem of stress and human adaptation is an extremely complex one with multidimensional levels and facets which cannot be reduced to any one neurophysiological or psychological dimension. This symposium has attacked the whole concept elegantly and attempted to integrate it on an interdisciplinary plane in terms of mechanisms and manifestations. Obviously the intricacy of the problem is far from being resolved, yet the direction and framework for further study is clearly indicated.

Only by searching for the fundamental answers at the neurophysiological and psychological levels will we be able to understand the ultimate process of the psychobiology of human adaptation and thus hopefully to control psychopathology.

REFERENCES

Hinkle, Jr., E. L. (1972). An estimate of the effects of stress on the incidence and prevalence of coronary heart disease in a large industrial population in the U.S. In *Thrombosis: Risk Factors and Diagnostic Approaches* (M. K. Brinkhouse *et al.*, Eds.). Stuttgart, New York: Schattauer Verlag, 16–65.

Hinkle, Jr., E. L. (1974). The concept of stress in the biological and social sciences. *Int. J. Psy. Med.* *5*(4), 335–337.

Horner, M. S. (1974). The measurement and behavioral implications of fear of success in women. In *Motivation and Achievement* (J. W. Atkinson and J. O. Raynor) Washington, D.C.: V. H. Winston and Sons, pp. 91–120.

Lazarus, R. S. (1966). *Psychological Stress and the Coping Process.* New York: McGraw-Hill.

Levi, L. (1972). Psychosocial stimuli, psychophysiological reactions and disease. In *Stress and Distress in Response to Psychosocial Stimuli* (L. Levi, Ed.). *Acta Med. Scand.* [Suppl. *528*], pp. 11–27.

Mason, J. W. (1968). A review of psychoendocrine research on the pituitary adrenal cortical system. *Psychosomatics 30*, 576–631.

McGrath, E. J. (1970). *A Conceptual Formulation for Research on Stress in Social and Psychological Factors in Stress.*

Opton, E. M., Jr., and Lazarus, R. S. (1967). Personality determinants of psychophysiological responses to stress: A theoretical analysis and an experiment. *J. Personal. Soc. Psychol. 6*, 291–303.

Persky, H., Hamburg, D., Basowitz, H., *et al.* (1958). Relation of emotional responses and changes in plasma hydrocortisone level after a stressful interview. *Arch. Neurol. 79*, 434–439.

Rahe, R. H., McKean, J. D., and Arthur, R. J. (1967). A longitudinal study of life changes and illness patterns. *J. Psychosomat. Res. 10*, 355–366.

Sarason, G. I., de Mondraux, C., and Hunt, T. (1975). Methodological issues in the assessment of life stress. In *Emotions, Their Parameters and Measurement* (L. Levi, Ed.). New York: Rowen, pp. 499, 509.

Serban, G. (1974). The process of neurotic thinking. *Am. J. Psychother. 28*(3), 418–429.

Serban, G. (1975). Stress in schizophrenics and normals. *Br. J. Psy. 126*, 397–407.

Serban, G. (1975). Relationship of mental status, functioning and stress to readmission of schizophrenics. *Br. J. Soc. Clin. Psychol. 14,* 291–300.

Symington, T., Currie, A. R., Curran, R. S., and Davidson, J. N. (1955). *The Reaction of the Adrenal Cortex in Conditions of Stress. Ciba Foundations Colloquia on Endocrinology Vol. VIII. The Human Adrenal Cortex.* Boston: Little, Brown, pp. 70–91.

Wilkins, W. (1971). Desensitization: Social and cognitive factors underlying the effectiveness of Wolp's procedure. *Psychol. Bull. 76*(5), 311–317.

Index of Names

Subject Index